Construction Machines

Dinko Mikulić · Draško Mück

Construction Machines

Springer

Dinko Mikulić
University of Applied Sciences Velika Gorica
Zagreb, Croatia

Draško Mück
University of Applied Sciences Velika Gorica
Zagreb, Croatia

ISBN 978-3-031-67847-9 ISBN 978-3-031-67845-5 (eBook)
https://doi.org/10.1007/978-3-031-67845-5

Reviewers: Vladimir Koroman, Ph.D., B.Eng., Croatian Academy of Engineering, Zagreb; Andrija Šaban, Senior Lecturer, mag.ing.mech., University of Applied Sciences Velika Gorica
Proofreader: Marina Manucci, Professor, University of Applied Sciences Velika Gorica

This Springer imprint is published by the registered company Springer Nature Switzerland AG
The registered company address is: Gewerbestrasse 11, 6330 Cham, Switzerland

Foreword

The book *Construction Machines* shows the *theory and design* of these machines. This is a professional book that provides knowledge on how to model, calculate, and design the machinery for construction operations. The key technical solutions of mobile construction machines are presented.

Carrying out the construction works is unthinkable without construction machines. Therefore, knowing the role and properties of construction machines is very important for their construction, production, application, and maintenance. The fundamental properties of the basic construction machines, bulldozers, loaders, excavators, vibratory rollers, cranes, telehandlers, and construction trucks are defined.

The authors of the book used an expert methodical approach that connects theory, construction and practice of using machines, in a gradual, expert and illustrative way, with the help of pictures, schemes and examples.

The content of the book covers the requirements of professional literature that enables an efficient approach to the application of technical knowledge in the field of mechanical engineering. The book can also be used in vocational educational institutions and businesses.

We hereby thank the authors for creating this publication. This should help the development of construction machines in the world and their use in construction operations.

Zagreb, Croatia

Vladimir Koroman
Andrija Šaban
Reviewers

The original version of the book has been revised. A correction to this book can be found at https://doi.org/10.1007/978-3-031-67845-5_13

Preface

The book *Construction Machines* describes the properties of modern construction machines and their design. There are three important factors that influenced its writing; the lack of professional literature for the education of engineers for the development, application and maintenance of these machines.

The technical solutions of the machines are processed step by step in an expert and illustrative manner, with the help of pictures, diagrams, tables, and examples. The properties of representative machines for earthworks, bulldozers, loaders, excavators, vibratory rollers, as well as other machinery, truck cranes, telehandlers, construction trucks, and their maintenance systems are explained.

The knowledge about the construction machines is very important for their efficient use and maintenance. With partial or complete automation of work, the construction works are carried out precisely, quickly, and safely. For work operatives and entrepreneurs, the machines are, along with manpower, a decisive factor in the execution of contracted jobs.

The book is primarily intended for students of professional studies, and it can be used in other educational institutions and businesses. For some of the images used, we sincerely thank their authors, all for the purpose of educating experts. We would like to thank the reviewers of the book who willingly accepted the review and offered useful advice.

Special thanks to cooperating institutions: *University of Applied Sciences Velika Gorica, DOK-ING Company, Teknoxgroup Hrvatska Croatia.* Also, we thank the collaborators for their support in writing the book, Andrija Šaban, Josip Rauker, Goran Košir, and Emil Hnatko.

We believe that the book will help in the education of engineers and technicians in solving the problems of construction machinery.

Zagreb, Croatia
February 2024

Dinko Mikulić
Draško Mück

Contents

Abbreviations

CG	Center of Gravity
CI	Soil Cone Index
E	Electric transmission
EBS	Electronic Braking System
FOPS	Falling-Object Protective Structure
GP	Global Positioning System
H	Hybrid transmission
HD	Hydrodynamic transmission
HS	Hydrostatic transmission
M	Medium active time of preventive and corrective maintenance
MDT	Mean downtime time due to maintenance
MMP	Mean maximum ground pressure
MTBF	Mean time between failures
MTBM	Mean time between maintenance
MTTR	Mean active repair time
PTO	Power Take-Off
ROPS	Roll-Over Protective Structure

Chapter 1
Introduction

1.1 Role and Importance of Construction Machines

Construction works are carried out on various buildings, civil engineering and high-rise construction, and other facilities. Civil engineering works are construction works on the objects of transport and water management projects. The main feature of these works is the extent of earthworks over a larger area. Construction works are works on residential, commercial and social facilities. The main feature of these works is that they are performed in relatively small spaces. Other construction works on objects are communal works, works on bridges, viaducts, tunnels, and the like.

The wide scope of construction works increases the use of construction machinery. Advanced construction technology requires the construction of new construction machines. Only by using construction machines can demanding work be done in a short time. Therefore, the role of construction machines in the construction of buildings is very important.

The construction operative is equipped with a variety of machines. The efficiency of the construction machines is their basic indicator. The efficiency of the machine is the amount of material that can be processed in a unit of time, expressed by volume, mass, per piece (m^3/h, t/h, pc/h, etc.). On the example of efficiency of several construction machines for earthworks, their importance for the execution of construction works is recognizable. The efficiency of a medium-class crawler bulldozer, for digging and pushing the earth, in a radius of 15–30 m, is up to 100 m^3/h. The efficiency of a wheel loader when excavating the earth is approximately 100 m^3/h, and when loading loose material by truck it is 150 m^3/h. The efficiency of a rotary backhoe on the construction of canals in waterworks with a depth of up to 1.5 m is 200–300 m/h, etc.

D. Mikulić and D. Mück, *Construction Machines*,
https://doi.org/10.1007/978-3-031-67845-5_1

Therefore, in construction development and the humanization of human work, it is not possible to achieve social progress without construction machines. That is why the development and production of construction machinery is developing intensively. New machines are appearing adapted to new construction technologies, with high performance and quality of work performed. By partial or complete automation of the work process, the construction works are performed precisely, quickly, and safely. For construction work organizations and entrepreneurs, the machines are often a decisive factor in the performance of contracted work. Therefore, it is almost unthinkable to talk about construction and not think of the construction machinery at the same time.

1.2 Division of Construction Machines

Considering the required variety of construction works, construction companies use a large number of dedicated machines. The division of construction machines according to purpose, classifies the machines into four main groups:

- Earthmoving machines;
- Machines for working in stone and concrete;
- Machines for lifting and moving cargo;
- Transport machines / cargo vehicles.

Machines can be divided into self-propelled, attached, light, medium, and heavy machines, as well as machines with cyclical and continuous work.

Earthmoving machines

Bulldozers
Graders
Scrapers
Loaders
Excavators
Vibrating rollers and plates

Machines for working in stone and concrete

Compressor machines
Drilling rigs
Crushers
Milling machines

Machines for lifting and moving cargo

Mobile cranes
Telehandlers
Tower cranes

Transport machines/freight vehicles

Conveyor belts and the like
Construction trucks and trailers.

1.3 Construction Machines Structure

A construction machine consists of the basic machine (undercarriage) and one or several working devices for performing works, Fig. 1.1. The necessary power for the working devices is provided by the drive engine, which can be located in the undercarriage or on the superstructure, or sometimes the power is provided by the engine of the attached means.

The chassis of the machine is tracked or wheeled. Tracked machines are more mobile off-road than wheeled machines. However, wheeled machines intended for off-road movement also provide good mobility on soft ground, because the surface pressure on the ground approaches that of tracked machines. Considering the trafficability of the soil, modern 6 × 6 and 8 × 8 wheeled vehicles can be compared to

Fig. 1.1 Construction machine structure

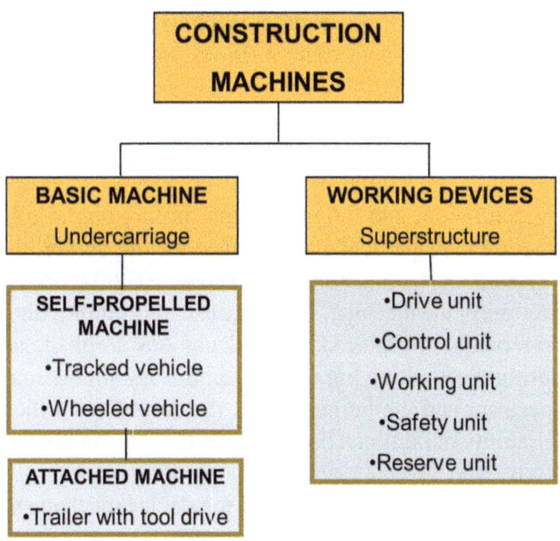

tracked vehicles. Therefore, tracked and wheeled machines represent the basis of the development of construction machines.

The working device is directly related to the purpose of the construction machine. The executive body of the working device directly performs the technological process. The working body is controlled by hydraulic and electronic system components. The required working pressure is created by a hydraulic pump driven by a drive motor. The safety of the machine operation depends directly on the installed measuring equipment, which prevents the consequences of disturbances due to poor handling of the machine.

Basic machine drive (undercarriage)

- Diesel engine and mechanical transmission
- Diesel engine and hydraulic transmission
- Diesel engine and hydro-mechanical transmission
- Diesel engine and electric transmission
- Hybrid transmission.

Working device drive (superstructure)

- Diesel engine and hydrostatic transmission

1.4 Lifetime of Construction Machines

An important property of construction machinery is its working life or durability. The working life of a construction machine is based on the resistance of key assemblies and parts of its construction. In accordance with the purpose of the machine, the computer calculation of the structure is carried out for static and dynamic load, wear resistance, and in the case of special machines, for resistance to mine destruction.

When the machine is overloaded, parts can fracture in the mechanical construction, which can be plastic, rigid, and fatigue. Rigid fracture occurs at places of stress concentration, and fatigue fracture occurs due to highly variable stresses. Fatigue fracture is of particular importance for the durability of parts. The working life of key parts of the machine is defined as 10,000 working hours, which determines the lifetime of the machine. These are usually complex and expensive parts, which cannot be repaired or replaced. The calculations are made according to the load spectrum during the working life of the machine. By simulating the load according to the load spectrum and the fatigue curve, the durability of the parts can be determined. The reliability of parts can be determined based on the probability of working and critical loads.

For the key parts of the working device mechanism, the materials with increased tensile strength, increased strength, high toughness at low temperatures, and resistance to cracking are used.

The cause of a higher load on the machine may be due to the blunting of the blade of the working tool. That is why it is important to evaluate the durability of parts according to the rate of wear. The durability indicator is the number of machine hours at which wear reaches the limit value. This increases the efficiency of the machine, i.e. the ability to perform tasks at minimal costs.

1.5 Development of Construction Machines

Based on the requirements of construction technology, various machine designs have been developed. From the point of view of development or procurement of the machine and its economy, three directions of development can be distinguished:

A—development of a special machine
The development of a special machine is economical for large construction organizations, when contracting large jobs, where the machine can be used (large and complex jobs—they can accept an expensive machine).

B—development of a standard machine
The development of a standard machine is economical when performing medium- and large-scale intensive works (e.g. bulldozers, excavators, etc.), where the degree of machine utilization is high.

C—development of a multipurpose machine
The development of a multi-purpose machine is economical for performing communal work, for smaller work organizations, and entrepreneurs (various jobs, small-scale works, combiners).

The development of the machines is closely related to the automation of the work process. Computer systems are installed in the machines to increase quality and work performance. Information reception systems, management programs, remote execution of works, control of work safety and efficiency, and others, are installed. A higher level of automation increases the price of the machine and requires higher level of training for operators and mechanics.

Technical requirements

Basic technical requirements that can be set before the development or purchase of a construction machine include:

- machine weight;
- machine dimensions;
- level of automation;
- machine efficiency/performance;
- working life of the machine / durability;
- universality of use;
- soil trafficability;
- fuel consumption and working fluids;

- environmental acceptability of the machine;
- machine performance/mobility;
- typification and unification of the machinery park;
- reliability of the machine (low intensity of failure);
- availability of the machine (short repair time);
- transportation of the machine;
- maintenance equipment;
- spare parts;
- qualification of machine operators.

The operation of construction mechanization requires, first of all, the reliability of the machines in various conditions (terrain, temperature, moisture, energy, etc.), the safety and durability of the main assemblies and parts during the depreciation of the machine, and a high degree of utilization of the machine in a group of several different machines.

Artificial Intelligence

Environmental directives and incentives for the use of clean vehicles and machines accelerate the process of automation. It is believed that smart machines will contribute to higher quality, efficiency, work safety, and environmental cleanliness. Therefore, construction technology is looking for a new vision of using machines that will provide greater utilization. The basis for this is provided by artificial intelligence applied to construction machines.

The Global Positioning System (GPS) uses multiple satellites in the Earth orbit. A technique such as kinematic positioning in real time (*real-time kinematics*/RTK) provides precise coordinates of the object on the ground. The coordinates are determined based on the distance between the satellite and the receiver on the object.

Well-known platforms on which machine operation management is performed are *AccuGrade* and *Trimble*, which use GPS technology for simpler and more complex surface shaping works. Automatic knife management reduces the burden on the machine operator and ensures precision and higher productivity. Artificial intelligence makes it possible to perceive the environment. Based on the correction signal, the hydraulic control system raises or lowers the knife to maintain the elevation of the cutting edge of the tool.

Driving systems

Battery machines are primarily recognized for the fact that they do not pollute the environment by emitting harmful exhaust gases. The electric drive provides favorable performance in the most difficult working conditions. The direct current from the batteries (*lithium-ion batteries*) is converted into alternating current to run the three-phase electric motors that drive the wheels or the tracks, Fig. 1.2.

Fig. 1.2 Battery mini bulldozer (*DOK-ING, Zagreb*)

Hybrid electric machines use a combination of electricity and fossil fuel. The combination of batteries and internal combustion engines contributes to lower consumption of fossil fuels and, therefore, lower harmful emissions of exhaust gases.

Chapter 2
Soil Properties

2.1 Division of Soil

Soil is the surface layer of the Earth crust. It is formed by the decay and erosion of rocks under the influence of atmospheric phenomena, water and the work of organisms. The action of organisms creates loose layers on the surface of which the organic world develops. Soils differ in appearance and properties.

Complex physical, chemical, and biochemical processes are continuously taking place in the soil, causing constant changes, and altering soil quality and forms. Soil science, or pedology, has determined that the dependence is based on the interaction of the parent substrate, climate, plant and animal life, the duration of soil formation, and relief.

Soil consists of solid and liquid substances. Solid components are÷ mineral substances (45%) and organic substances (5%), and liquid substances are÷ water (20–30%) and air (20–30%). Their constant mixing and changes alter the appearance, composition and structure of the soil. Soil can be divided into several layers: biologically active soil layers (in which organisms live), followed by biologically inactive layers.

The structure of the soil, its potential states, and its physical and mechanical properties are determined by soil mechanics. One of the fundamental laws of soil mechanics is the law of soil shear strength, as proposed by *Charles-Augustin de Coulomb*. The choice and dimensions of construction facilities depend on the load that the soil can bear.

The old builders solved the tasks very successfully, as evidenced from many historical legacies. There have been failures, such as the *Leaning Tower of Pisa* (fourteenth century), which were the result of incorrect assessment of the properties of the soil.

Classical soil mechanics (nineteenth century) is based on two idealizations of actual soil. The first assumes that the soil is a rigid plastic body when considering the limit load problem, and the second assumes that the soil is an elastic body when

D. Mikulić and D. Mück, *Construction Machines*,
https://doi.org/10.1007/978-3-031-67845-5_2

Table 2.1 General division of soil [1]

Soil	Type of soil	Grain size (mm)
Monolithic (Rock)	Crumbs, rocks	70–300
Incoherent	Gravel	2.0–70
(Loose)	Sands	0.06–2.0
	Dust	0.002–0.06
Coherent	Organic soils	0.002–0.05
(Bonded)	Clay	0.0002–0.002

considering the problem of deformations. The actual soil does not correspond to any of the models, but under certain typical conditions, an idealization of the actual soil is appropriate. By applying modern methods of soil mechanics, the solution of soil loading tasks is more comprehensive.

The properties of natural soil differ from those of a homogeneous substance. The force that loads one contour of the granular soil mass is transmitted within that space by individual forces in contact between adjacent particles which crush and move the particles. The number of contacts per grain depends on the grain size, compactness, and granulation of the sample. The stress in the granular mass is not distributed homogeneously over the cross-sectional area, but it is the resultant of individual forces. The increase in stress causes deformations, not only from forces in contact with the grains, but also from the sliding and rotation of the neighboring grains, as well as crushing and changing the arrangement of the grains, thus reducing the volume of pores between the particles.

By reducing the load on the soil, the volume is only partially restored, ~~because~~ since the sliding, rotation and crushing of the grains cause the process to be irreversible. The behavior of such materials when the stress changes determines the type of medium in the pores: air or water, i.e. air and water. Soil saturated with water is less compressible than a granular structure with air in its pores. Volume changes are possible if part of the water is removed from the pores. The process of transferring stress from pore water to the soil skeleton is referred to as consolidation. For coarse-grained soil, the duration of consolidation is shorter than that of fine-grained soil.

The general division of soil is shown in Table 2.1. All types of rocks whose particles are firmly connected by a cementing substance form a special type of hard monolithic soil. When the soil is destroyed under load or blasting, it cannot be restored to its original state. Smaller forms are formed by crushed stone ranging from 70 to 300 mm. Incoherent soil is loose, permeable, and changes shape under load. Gravel and sand can be classified into several subtypes individually.

Coherent soil has solid, interconnected particles, so it is usually impermeable. Fine-grained cohesive soil can be classified into several groups. Organic soils contain large quantities of organic and plant matter, which is why they bind large amounts of water and are highly unstable (silt, humus).

By sieving the mixture through a sieve, the grain diameter is determined and displayed with a granulometric curve, which gives a picture of soil properties such as permeability and capillarity of sandy and dusty materials.

Atterberg limits determine more precisely the properties of fine-grained soil, which can be found in four stable states according to water content; the soil is solid when dry, semi-solid when mixed with water, and can change into a plastic or liquid state. According to the water content, the soil is divided into dry (up to 5% water), moist (10–25% water), and wet (25% water and more).

According to the difficulty of carrying out construction works and the density of the soil, the soil can be divided into a number of categories. The first four categories of soil can be processed mechanically using machines, Table 2.2.

Table 2.2 Soil categories [1]

Soil category	Soil type and characteristics	Density (kg/m^3)
Loose soil I.	– Hummus without roots	1200
	– Sand of natural humidity with an additive gravel or crushed stone up to 20%, sandy and clayey light soils	1600
	– Loose slag	750
II. Ordinary soil	– Hummus with roots	1200
	– Clay of natural humidity soil with addition of gravel	1800
	– Clay soil, loam	1750
	– Greasy, soft or loose clay,	1700
	– Settled with addition of gravel, pebbles	1800
III. Solid soil	– Hard settled clay	1800
	– Heavy and slate clay with addition gravel, pebbles up to 10%	1950
	– Construction mortar	1850
IV. Hard soil	– Hard clay, heavy	2000
	– Soft chalk	1550
	– Soft marl	1900

2.2 Physical and Mechanical Properties of Soil

In the field of application of construction machines in the changing climatic conditions, the physical and mechanical properties of soil are important, since they affect the design and performance of the machines. The basic physical and mechanical properties of the soil [2] are:

- Soil density;
- Clay content;
- Looseness of soil;
- Soil humidity;
- Soil strength and bearing capacity;
- Shear strength of soil;
- Soil cohesion;
- Plasticity and abrasiveness of the soil;
- Stickiness of the soil;
- Natural ground angle.

Soil density

Soil density in its natural state:

$$\rho = m/V[\text{kg/m}^3]$$

γ	Volume weight of soil, natural state: $\gamma = G/V$ [N/m^3],
γ_{pr}	Volume weight of soil, loose state: $\gamma_{pr} = G/V_{pr}$ [N/m^3].
m	Mass of soil,
G	Weight of soil,
V	Volume of soil in natural state,
Vpr	Volume of soil in temporary loose state.

Content of clay particles

The granulometric composition determines the ratio of clay particles to the mass of the sample:

Up to 3%	Sands
3–12%	Clayey sand
12–25%	Sandy clay
> 25%	Clay
50%	Loam.

Looseness of the soil

Looseness of the soil is determined by the factor of temporary (k_{pr}) and permanent volume increase (k_{tr}) factors. The average values of temporary and permanent looseness factors according to soil category are:

$$I.cat: \quad k_{pr}/k_{tr} = 1.15/1.02$$
$$II.cat: \quad k_{pr}/k_{tr} = 1.20/1.04$$
$$III.cat: \quad k_{pr}/k_{tr} = 1.25/1.05$$
$$IV.cat: \quad k_{pr}/k_{tr} = 1.30/1.08$$

Soil moisture

Soil moisture w is determined by the ratio of the mass of water in the soil to the mass of dried soil, i.e. its solid components. Relative humidity w_r is the ratio of the mass of water to the mass of wet soil. Normal soil moisture is 10–25%. The soil humidity corresponding to the highest density of the soil during compaction is called the optimum moisture content w_o.

$$w = m_v/m_0;\ w_r = m_v/m$$

m_v Mass of water in the sample, $m_v = m - m_0$.
m Mass of wet soil sample.
m_0 Sample mass after drying during 24 h at temperature 105–110 °C.

Soil strength and soil bearing capacity

Soil strength is its resistance to external forces that want to break it due to pressure, tension, bending, and shearing. Hardness is the resistance that the soil offers to the penetration of another body, during which plastic or short-term destruction of the surface layer occurs.

The strength of the soil determines its bearing capacity, i.e. the amount of load it can support. The highest soil load that causes its deformation is called the ultimate bearing capacity or compressive strength of the soil. The bearing capacity of the soil depends on the ~~soil~~ humidity content, the amount of clay, and the compaction of the soil. By compacting the soil, its carrying capacity increases, which is essential for the construction of buildings and the mobility of machines.

The bearing capacity of the soil decreases with increasing soil humidity. The ultimate bearing capacity depends on the humidity, the amount of clay, and the compaction of the soil. Soil with a clay content of about 10% has low strength. Compacting the soil increases its strength. When determining the permeability of the soil, the compressive strength of the soil is important, as it deforms and crushes at a certain depth. The force on the surface of 1 cm^2 at which deformation of the soil at 1 cm depth occurs is called the unit crushing resistance σ_0. Under the wheels of construction machines, the permissible soil crushing depth of 6–12 cm corresponds to the permissible soil load, i.e. the limit bearing capacity of the soil (σ).

Shear strength of soil

The shear resistance of the soil depends on the cohesion force of the particles and the force of internal friction between the particles [2]. The soil shear stress τ has a linear dependence (*Ch. A. Coulomb*):

$$\tau = c + \sigma_n tg\varphi[\mathrm{Pa}]$$
$$t = c + \sigma_n \mu[\mathrm{Pa}]$$

c Cohesion force (20–30 kPa), loose incoherent soil $c = 0$,
σ_n Normal stress on the shear surface, normal pressure ($p = G/A$),
μ Coefficient of internal friction.

Soil cohesion

Soil cohesion is the degree of internal connection between particles. The forces of cohesion oppose external forces that tend to disrupt their bond. The cutting resistance of the soil depends on the cohesion force. Coherent soils with higher cohesion are clayey soils. By increasing the surface pressure in non-coherent soils, for example sand, its ability to bond increases.

Soil plasticity and soil abrasiveness

Plasticity of the soil is its ability to change shape under the load of external forces and to retain it after the cessation of the action of those forces. Plasticity depends on soil humidity, which reduces strength and shear resistance. Such is the case with clay soils. With a small amount of sand, wet plastic soils become sticky and difficult to shake out of the shovel of the earthmoving machine.

The abrasiveness of soil is a property that causes rapid wear of tools (drill bits, excavator shovel teeth). There are weakly abrasive soils, significantly abrasive soils, and extremely abrasive soils. For example, significantly abrasive soils include quartz sandstones and granites.

The stickiness of the soil

The stickiness of the soil is its property of adhering to another material. Shear resistance of such soil on an adhesive metal surface:

$$F = \tau_1 A\ [\mathrm{N}]$$

A Surface to which the soil adheres.
τ_1 Shear stress of bonded soil on metal /sandy clay $\tau = 0.5$–$0.7\ \mathrm{N/cm^2}$; fat clay, $\tau = 0.7$–$1.0\ \mathrm{N/cm^2}$

Natural ground angle

When pouring from a certain height, the loose material is arranged in an approximately conical pile. The natural angle of the ground refers to the internal friction angle of the material during its motion, which is the shear angle of the loaded soil. This angle corresponds approximately to the pouring angle, $tg\varphi \approx \mu$, where μ is the material-specific friction coefficient. For example, the natural angle of sandy clay is: φ_p ~ wet, moist, dry/30°, 35°, 45°.

2.3 Soil Trafficability

The mobility of the machines depends on the soil trafficability. Soft soils can create a problem for the mobility of machines, as they can impede their movement. Machines typically move on wheels or tracks. Construction machinery and construction trucks require off-road mobility. The assessment of the machine ability to work is based on the load capacity of the soil and the maximum pressure exerted on the soil. The greater the bearing capacity of the soil and the lower the pressure of the machine on the soil, the greater the soil trafficability. The trafficability of the soil for machines and vehicles can be assessed based on the measurement of the bearing capacity with a penetrometer (*Cone Index,* CI) and using a model for estimating the maximum pressure of the vehicle on the ground (MMP) [3].

Evaluation of soil trafficability using *CI* index

CI—soil cone index, an indicator of the bearing capacity of the soil, is measured using a penetrometer. The penetration resistance of the penetrometer cone into a specific soil is measured, as shown in Fig. 2.1. *The standardized value of measuring resistance to cone penetration at a depth of 15* cm *(ASAE EP542) is known as the soil cone index (CI).* The bearing capacity of the soil and asset mobility is shown in Table 2.3.

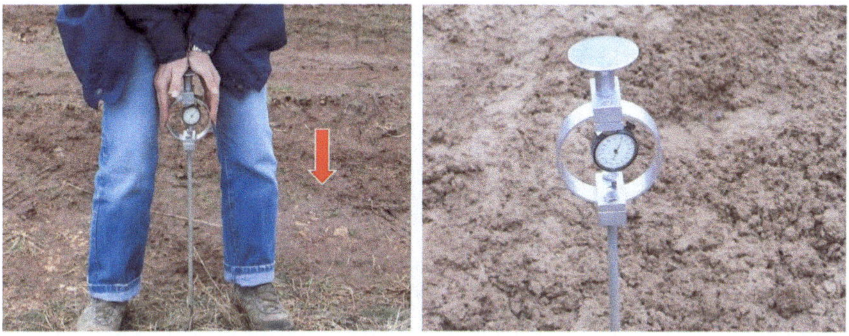

Fig. 2.1 Penetrometer—measurement of the bearing capacity of soil (*Cone Penetrometer*, kPa/psi; *Soil trafficability*)

Table 2.3 Scale of soil bearing capacity and vehicle mobility [4]

CI (kPa)	Description
0–21	No practical movement value
40–62	A man has difficulty walking on soil without sinking
103–165	A special tracked vehicle can make about 50 passes
186–228	A light tracked vehicle can make about 50 passes
276–352	A medium tracked vehicle can make about 50 passes
372–497	A Jeep-type vehicle can make about 50 passes
517–662	Heavy vehicles
683–935	Passenger vehicles
1000	All vehicles pass

Soil cone index:

- For very soft soils: $CI < 300$ kPa (3 bar).
- For medium hard soils: $CI = 300$–500 kPa (3–5 bar).
- For solid soils: $CI > 500$ kPa (5 bar).

Evaluation of soil trafficability based on soil pressure

With wheeled construction machines, there is a problem of high wheel pressure on soft ground, which causes decreased ground trafficability. With tracked construction machines, this problem of pressure on the ground is generally much less severe.

An estimate of the maximum ground pressure of a wheeled or tracked machine can be shown using the MMP model, which stands for *Mean Maximum Pressure*—the mean maximum pressure of the driving force on the ground.

Wheeled machine

The MMP model of the maximum pressure of a wheeled machine [3], depends on the following parameters:

$$MMP = k\, W_T / n\, b^{0.85} d^{1.15} (\delta / d)^{0.5} [\mathrm{Pa}]$$

W_T Machine weight (kN),
k Factor of the number of drive axles,
n Number of vehicle wheels,
b Tire width, unloaded tire (m),
d Tire diameter, unloaded tire (m),
δ Tire deflection due to load (m).

The MMP model allows for comparison of the mobility of machines with different construction solutions. The lower the MMP, the greater the trafficability, thus making the machine more mobile. A greater number of passes on the same track significantly affects the trafficability of the ground because the depth of the ruts increases, which reduces the height of the undercarriage from the ground.

Tracked machine

Normal pressure of the tracked machine on the ground:

$$NGP = W_T/2L\,b(\mathrm{Pa})$$

W_T Machine weight (kN),
L Track length (m),
b Track width (m).

Mean maximum ground pressure [3]:

$$MMP = 1.26W_T/2Ln\,b\sqrt{D}t[\mathrm{kPa}]$$

W_T Machine weight (kN),
L Track length (m),
b track width (m),
n Number of support wheels of one track.
D Wheel diameter (m),
t Track pitch (m).

Limiting Cone Index, CI_L

$$CI_L = 0.827\,MMP[\mathrm{kPa}]$$

The limiting cone index determines the minimum bearing capacity of the soil at which a machine with a certain MMP index can be mobile. In other words, the bearing capacity of the soil should be at least approximately 83% of the calculated MMP for a particular machine in order for it to successfully complete the pass.

Wheel machine rut depth

To estimate the depth of soil deformation (z), the wheel index is defined:

$$N_k = CI/p$$

p Tire inflation pressure

Wheel rut depth for one pass of the machine:

$$z_1 = d\,(0.224/N_K^{1.25})$$

Wheel rut depth for multiple passes:

$$z_n = z_1 n^{1/a}$$

d wheel diameter,
a coefficient of multiple passes (2–3 for low soil load capacity; 3–4 for medium soil load capacity; 4–5 for high load capacity).

Some examples of soil trafficability [5]: MMP/z_1

$$\begin{aligned} \text{Off - road vehicle } 4 \times 4(3\ \text{t}):&\quad 340\,\text{kPa}/5.3\,\text{cm} \\ \text{Construction truck } 6 \times 6(11\ \text{t}):&\quad 320\,\text{kPa}/8.0\,\text{cm} \\ \text{Construction truck } 8 \times 8(22\ \text{t}):&\quad 300\,\text{kPa}/8.8\,\text{cm} \end{aligned}$$

References

1. Mikulic D (1998) Construction machines. Design, calculation and use (in Croatian), itG, Zagreb
2. Alekseeva TV et al (1985) Machines for earthmoving work. Theory and Calculation, Third Revised and Enlarged Edition, Amerind Publishing Co
3. Nell S (1998) Experimental evaluation of mean maximum pressure (MMP) in clay. Wheels and Tracks Symposium Cranfield University
4. Sarrilahti M (2002) Soil interaction model. Project deriverable D2, Development of a Protocol for ECOWOD, University of Helsinky
5. Mikulic D, Marusic Z, Stojkovic V (2006) Evaluation of Terrain vehicles mobility. Strojarstvo 48, 5–6:309–319, Zagreb

Chapter 3
Soil Digging Resistances

3.1 Introduction to Soil Digging Theory

Digging the soil is based on a sharp wedge. The wedge is the basic element for achieving the thrust force of the machine for the purpose of digging the soil. The starting dimensions of the wedge are: the angle of the blade β, and the width of the wedge b, Fig. 3.1. The position of the wedge in space is determined by the angles: the back angle, i.e. the angle of inclination towards the ground γ, the engagement angle towards the direction of cutting φ, and the angle of cutting α in relation to the horizontal position.

The development of the digging blade resulted in its various forms. The high-quality wedge blade was achieved by adding a special cutting knife, which can be replaced in case of wear or damage. The knife is a direct part of the working tool of the construction machine for digging the soil. The knife can be designed in different ways: straight, curved, disk-shaped, or as a plow knife with or without teeth. Sometimes the entire working mechanism for digging the soil is referred to as a knife. By adding additional teeth on the knife, the cutting through hard layers of soil is made easier.

The basic forms of working tool of earth-moving machines are shown in Fig. 3.2, as well as the ideal shapes of blades in Fig. 3.3.

Phases of soil digging

The process of excavation consists of cutting the layer of soil (***cutting phase***) and displacement of soil along the executive body (***displacement phase***). The soil cutting phase is the first and most important phase in the excavation process. The separation of the soil layer is performed with a knife blade under the action of force, and depends on the physical properties of the soil and the shape of the knife and the state of its blade, Fig. 3.4. In the phase of cutting the soil, compaction of the soil pushed in the direction of movement occurs, followed by shearing of a part of the soil layer in the

D. Mikulić and D. Mück, *Construction Machines*,
https://doi.org/10.1007/978-3-031-67845-5_3

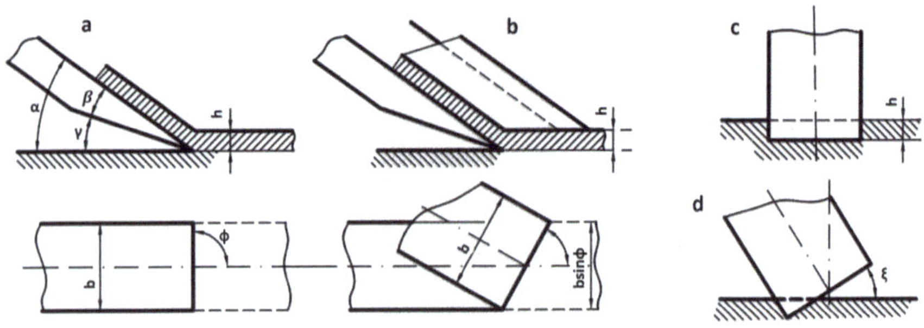

Fig. 3.1 Wedge parameters. α—cutting angle, β—wedge blade angle, γ—back angle, φ—angle of engagement according to the direction of soil cutting, ξ—tilting angle

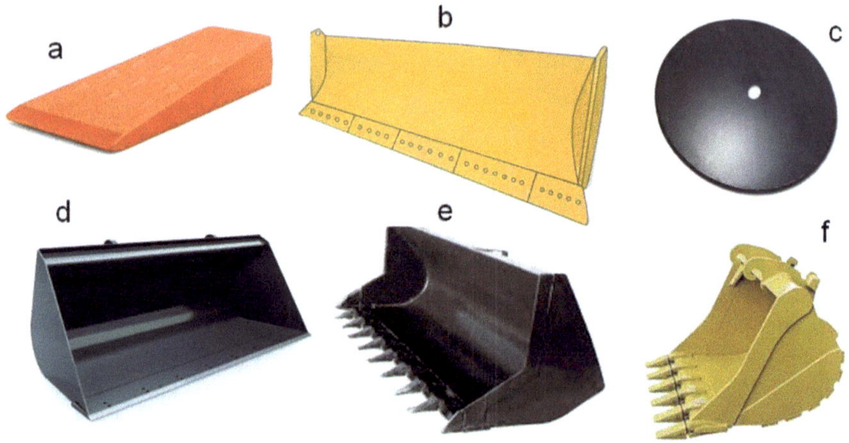

Fig. 3.2 Shapes of working tools. **a**—wedge, **b**—flat dozer knife, **c**—disc-shaped knife, **d**—shovel with knife without teeth, **e**—shovel with teeth, **f**—excavator shovel with teeth

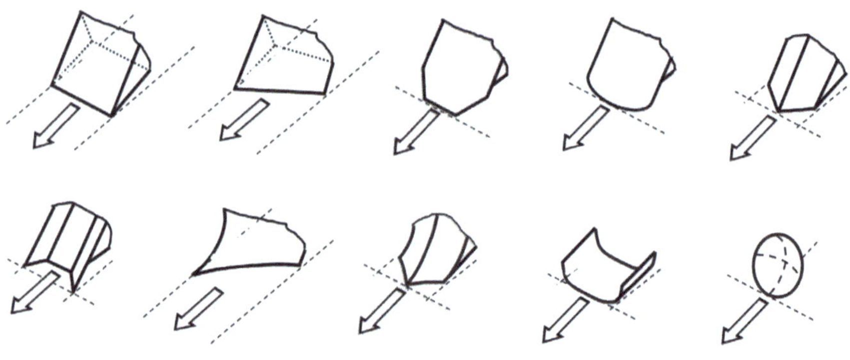

Fig. 3.3 Ideal knife shapes for cutting soil [1]

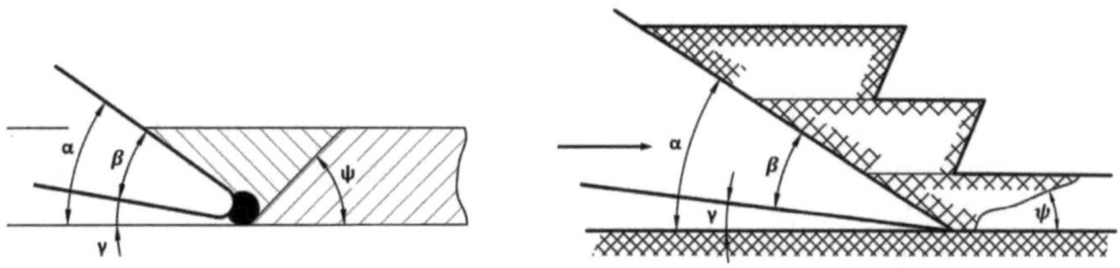

Fig. 3.4 Cutting the soil layer and removing the layer by shearing [2]

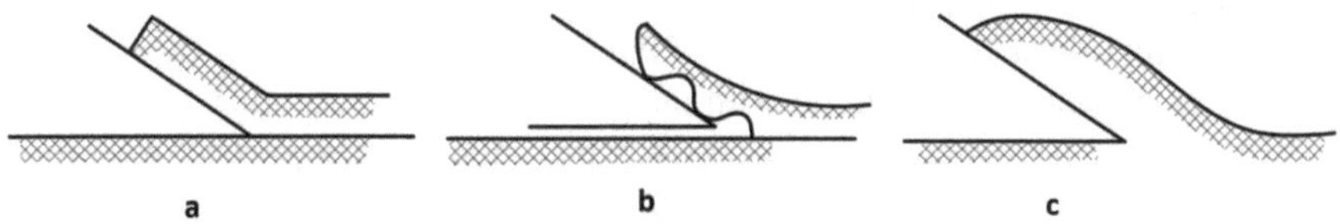

Fig. 3.5 Forms of soil digging layer. **a**—bound wet soil, **b**—bound dry soil, **c**—unbound soil

plane of the highest stresses at shear angle ψ ($\approx tg\varphi$). This is followed by alternating compaction and shearing of the soil.

During soil displacement, the material is moved in front of and behind the knife in the shovel. Depending on the type and properties of the soil, the separated layer of soil takes on different shapes. This displacement of soil makes determining the soil resistance complex. However, the resistance can be distinguished between the cutting phase and the displacement phase. In the case of coherent wet soil, the plastic layer of material is moved along the blade in the form of a continuous strip, Fig. 3.5a. In the case of coherent dry soil, the layer breaks into pieces (b), while a non-coherent soil is scattered and begins to accumulate in front of the knife (c).

Soil digging resistance R_{kop} is equal to the sum of the resistance of the cutting phase and the soil displacement phase. Soil cutting resistance R_{rez} with earthmoving machines mostly depends on the properties of the soil and the shape of the blade. The resistance of soil displacement ($R_{i,\ i=1\div n}$) is of variable value and depends on the working mechanism for soil excavation and working conditions.

The cutting resistance of the soil has the highest value. For example, during the operation of a bulldozer, the resistance to cutting the soil is 60–80% of the total digging resistance, and the rest is made up of the resistance of moving the soil prism in front of the blade. In excavators with a front shovel, cutting resistance in category IV of soil accounts for 40–70% of the total digging resistance, while the shovel loading resistance is 6–10%, and the earth prism displacement resistance is 3–5%. Therefore, soil cutting resistance is the key factor in digging resistance analysis.

3.2 Resistance to Cutting Soil with a Knife Blade

Figure 3.6 shows the resistance forces of the soil, on the blade of the knife during digging. The cutting resistance of the soil can be calculated or measured (by gradually loading the tool).

The soil cutting resistance R_{rez} is the result of the following resistance forces:

- Resultants of soil compaction resistance at the tip of the knife R_z (component R_T in the direction of movement of the tip of the knife and normal component R_N);
- The weight of the soil at the tip of the knife Q;
- Friction resistance on the upper side of the knife, $R'_t = \mu_1 Q \cos\alpha$
- Friction resistance of the knife with the ground on the lower side of the knife, $R''_t = \mu_1 F_2$, μ_1—friction factor between the knife and the ground, F_1 , F_2—tangential and normal force, R_1—soil cutting resistance, reactive tangential force on the knife blade.

The force of cutting the soil on the knife blade F_{rez} required to overcome$F$ the cutting resistance R_{rez} is the resultant force acting on the knife. The cutting force components F_1 and $_2$ correspond to resistances R_1 and R_2 of the opposite direction. Force F_2 is the result of the weight of the working device and the normal force pushing the knife into the ground, and force F_1 is the tangential cutting force at the tip of the knife that overcomes the tangential cutting resistance R_1. This cutting resistance mostly depends on the category and type of soil being cut (k_1), as well as the width (L) and depth (h) of the cutting layer.

Soil cutting resistance:

$$R_1 = k_1 L h [\mathrm{N}]$$

k_1 Specific soil cutting resistance (N/m^2),
L Soil cutting length/B—knife blade width (m),
h Depth of the soil layer to be cut (m).

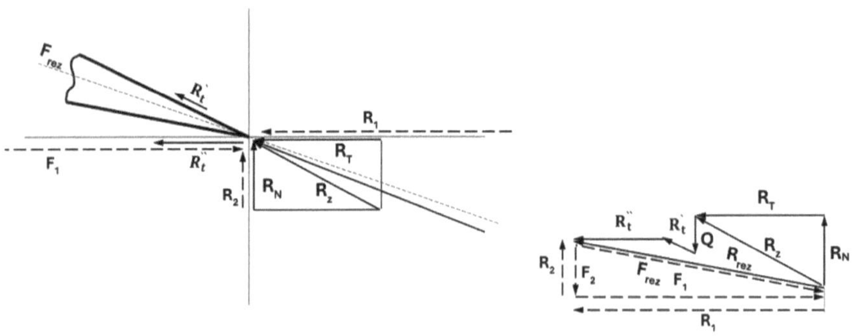

Fig. 3.6 Graphic scheme of determining soil cutting resistance force [1, 2]

The specific cutting resistance k_1 of a certain category of soil can take a wider range of values (Table 3.1). Based on the working conditions, shape, and condition of the knife, the starting value of the specific resistance can be taken.

The cutting length of the soil is understood as the width of the blade of the knife, which is normal to the direction of movement of the machine. The cutting length in the normal cutting mode is equal to the width of the knife blade, $L = B$. In case of an oblique position of the knife at an angle φ, the cutting length is: $L = B\sin\varphi$, Fig. 3.7.

Total soil digging resistance for specific tools

For specific tools (excavator front shovel, trailing shovel, and scraper shovel), the calculation of the total digging resistance can be set in the form of an equation [3]:

$$R_{\text{kop}} = k_{\text{k}} L h \,[\text{N}]$$

Table 3.1 Specific cutting resistance k_1

Soil category	Soil type	k_1 (kN/m^2)
I	Sand, light medium humidity, and loose sand	10–30
II	Sandy clay, fine and medium gravel, light moist and loose clay	27–60
III	Sandy compacted clay, medium clay, light moist clay or loose clay	55–130
IV	Sandy clay heavy with gravel or pebbles, heavy clay, conglomerate weakly cemented	130–250
V	Medium shale, heavy dry clay, compact loess hard, chalk and gypsum, soft marl	230–320
VI	Soft porous limestone, chalk, shale, marl and gypsum of medium hardness, solid coal, cemented conglomerate	300–550
VII	Slate, marl, chalk and gypsum solid, limestone of medium hardness, frozen soil	600–2000

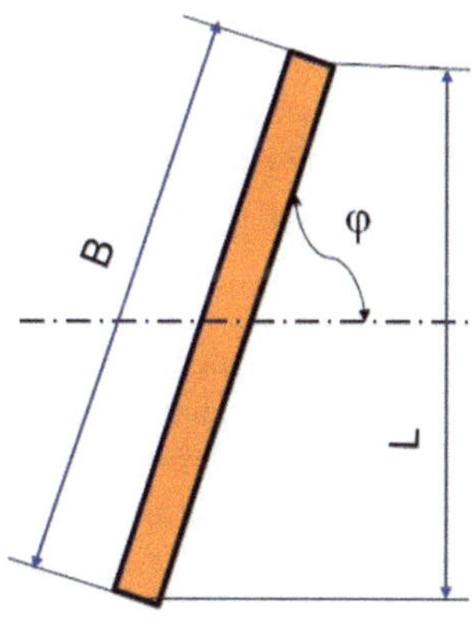

Fig. 3.7 The width of the knife blade and the length of cutting the soil at an angular position of the knife, B—width of the knife blade, L—length of soil cutting

Table 3.2 Specific soil digging resistance k_k, (kN/m^2)

Soil category	Soil type	Front loading shovel	Trailed loading shovel	Scraper loading shovel
I	Sand, clayey sand	16–70	28–120	25–105
II	Sandy clay, gravel	60–130	100–190	95–180
III	Sandy compacted clay	115–195	160–260	175–286
IV	Heavy clay	200–300	260–400	320–495

k_k Specific soil digging resistance (N/m^2),
L Soil cutting length/B—knife blade width (m),
h depth of the soil to be cut (m).

The specific soil digging resistance k_k of a certain category of soil, depending on the shape of the tool and the state of the blade, can have a wider range of values (Table 3.2). Due to the mode of cutting the soil, an increase in the cutting depth compared to the cutting width of the knife blade has a significantly greater impact on the cutting resistance.

3.2.1 Soil Cutting Resistance Factors

Knife geometry

With an increase in the cutting angle of the knife, the cutting resistance of the soil gradually increases $\alpha = 10$–$30°$ (35°), Fig. 3.8. By increasing the cutting angle of the knife from 30 to 50°, the resistance curve gets a greater increase [1, 4]. When cutting different types of soil, the favorable cutting angle of the knife is about 25–30°. For example, with bonded materials, the resistance increases by up to 20%. The smallest value of the angle of the knife blade is $\beta = 5°$, for wet plastic soil, $\beta = 20$–$25°$. In order to avoid contact of the back side of the knife with the soil, and thus friction resistance as well, the back angle is usually $\gamma = 5$–$10°$. Considering the above, and the necessary mechanical strength of the knife for working in soil category IV, the optimal value of the knife cutting angle is $\alpha = 25$–$35°$.

The shape of the front blade of the knife

Cutting resistance depends on the shape of the front knife blade: rectangular, pointed, or semicircular shape, with or without teeth. The best results are obtained by a knife with a curved semicircular shape and a protruding part at an angle $\gamma = 12$–$15°$, where the blade is arched, Fig. 3.9. The cutting resistance of the knife blade is 20–25% lower than that of a rectangular flat knife. The greatest pressure of the knife is

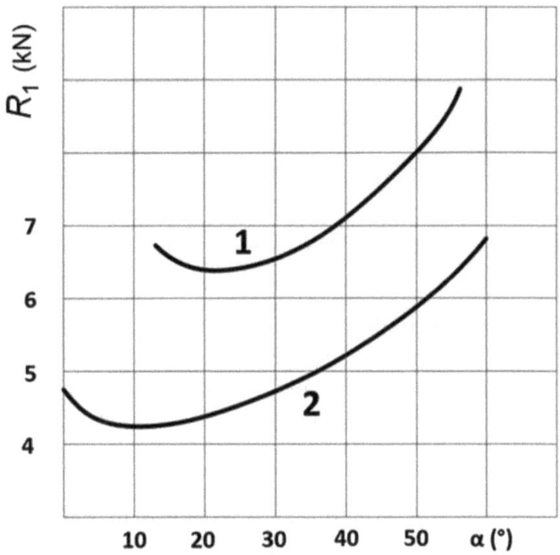

Fig. 3.8 Cutting resistance R_1 as function of knife cutting angle (α), when tested in two soil hardnesses (*1, 2*) of a certain section (*L*, *h*), adapted [1]

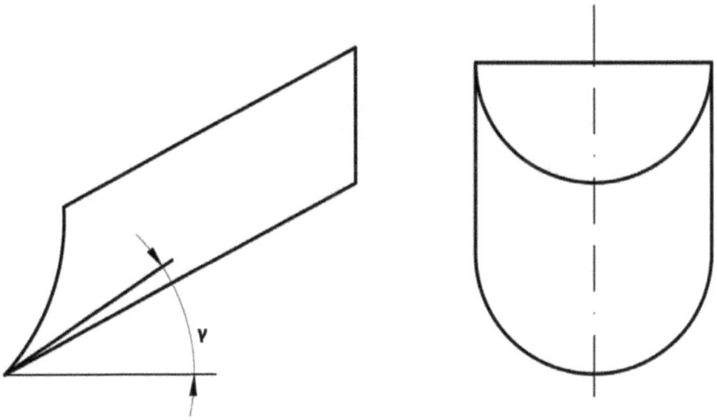

Fig. 3.9 Semicircular knife

transmitted through the middle part of the blade. The knife quickly penetrates the soft soil, the pressure spreads slightly to the sides of the knife, and a soil prism forms in front of the knife. Raised front knife blade is reinforced with teeth for cutting hard ground. For these reasons, in case of backhoes the shovels are made with a protruding semicircular blade without teeth or with a trapezoidal blade with two teeth.

Knife blade wear

The specific cutting resistance depends on the wear of the knife blade. A worn knife blade causes a significant increase in cutting resistance. Due to the blunting of the blade in backhoes, the resistance increases up to 25% (30%). At the same time, there

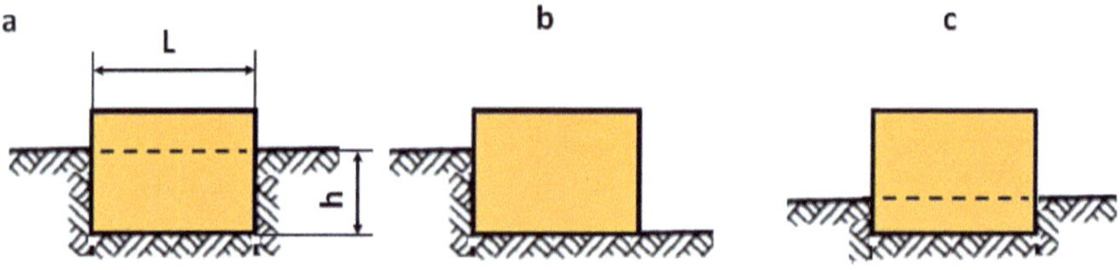

Fig. 3.10 Method of cutting the soil. **a**—blocked, **b**—semi-blocked, **c**—unblocked/normal

is an increase in the transverse force that tends to eject the knife blade. This can be prevented by increasing the vertical pressure of the machine on the knife, but this causes increase in the component of friction resistance on the underside of the knife and the wear of the blade.

Mode of soil cutting and cutting depth

The method of soil cutting and the depth of the cutting layer have the greatest influence on the specific cutting resistance. There are three methods of soil cutting: *blocked, semi-blocked, and unblocked,* Fig. 3.10a–c. The cutting resistance is the highest in blocked, followed by the semi-blocked cutting mode. The unblocked, i.e. normal cutting mode provides the least soil cutting resistance.

The cutting resistance of the soil depends more on the depth of the cut and less on its length [1, 4], Fig. 3.11. For example, the specific cutting resistance of the backhoe is lower when the cutting length $L < 0.8$ m, for a larger length $L > 0.8$ m the change is insignificant. By increasing the ratio of section to cutting length A/L, the cutting resistance of the soil is reduced, which is explained by the "relatively smaller shredding of the layer", which means less work spent on loosening the soil. This ratio can be increased by the dimensions of the loading tool—loading shovel. With an increase in the volume of the excavator loading shovel by 0.15–15 m^3, the soil cutting resistance decreases by 25–30%, but it does not decrease in the case of increased influence of the blocked cutting mode.

Influence of cutting speed

The influence of the cutting speed on the change in specific cutting resistance is significant in continuous-duty machines, such as rotary backhoes. At normal cutting speeds (0.5–2 m/s) there are no significant changes in resistance. However, when increasing the cutting speed 3–5 times, the cutting resistance increases by 30–40%. By placing the teeth on the knife blade, the soil cutting resistance (R_1) in hard soil is reduced by about 20%.

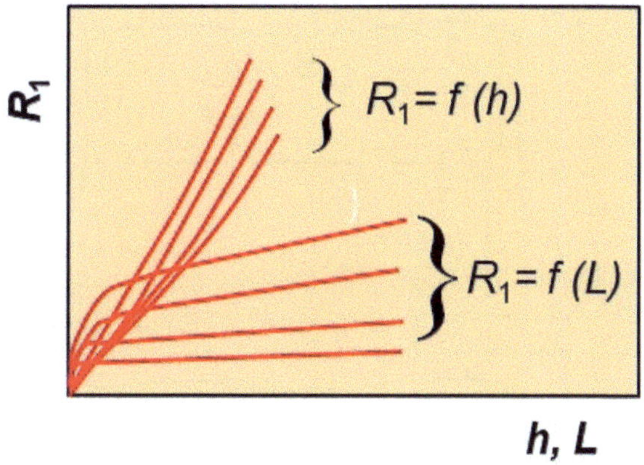

Fig. 3.11 Cutting resistance R_1 as a function of soil cutting length (L) and depth of soil cutting (h), adapted [1]

3.3 Soil Cutting Resistance with Teeth

When cutting hard soil, the knife blade is strengthened with sharp teeth. The teeth reduce the cutting resistance by 10–20% (30%) and the digging resistance by 5–15%. With soft soil, the impact of the teeth is negative because they increase the frictional resistance and the digging resistance. Wet soil sticks both to the teeth and to the loading shovel, which hinders its shaking. The importance of placing the teeth on the knife blade is to protect the base of the knife blade from wear, because it is the teeth that receive the initial load first. The teeth are replaceable by design. If the worn teeth are not replaced on time, the cutting resistance increases, and the wear is transferred to between the teeth, i.e. the knife blade.

Profile and arrangement of teeth

The profile of the teeth and their arrangement along the knife blade are shown in Fig. 3.12. The arrangement of the teeth on the knife blade is very important. The teeth are also placed at the ends in order to expand the loosening of the soil layer and prevent friction of the sides against the side edges of the excavation. The depth of the trench formed by the tooth during cutting is $f = l \sin(\beta/2 + \gamma)$, where l is the length of the tooth. Assuming that the shear surfaces of the layer between the teeth will be at an angle $\psi = 45°$, the distance between the teeth that should completely protect the knife blade from wear is $b_0 = 2f$. In the case of a larger distance between the teeth, the blade of the knife will wear more. In the case of a smaller distance between the teeth, the cutting resistance increases.

From the conditions of the highest load on one tooth, the tooth width is determined ($b = 60$–80 mm), Fig. 3.13. The ratio between teeth and tooth width (bo/b) is set in the range 2.0–2.5, and the trench depth $f = 0.7b$. In backhoes, the teeth are more closely spaced: $bo/b = 1.25$–1.50.

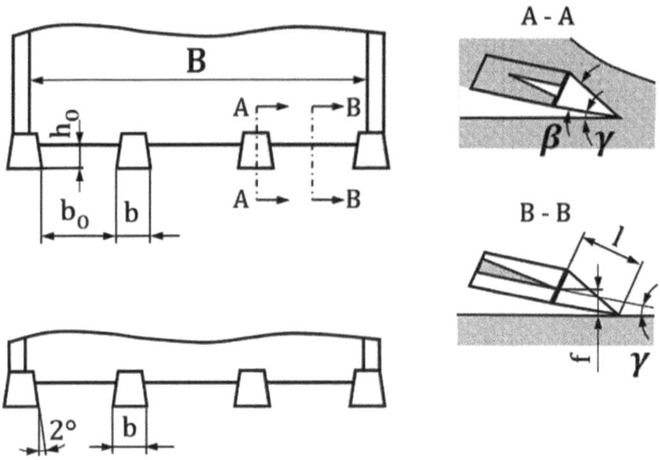

Fig. 3.12 Profile and arrangement of teeth along the knife blade [5]. $b = 0.11V^{1/3}$, $l = 0.26\ V^{1/3}$, $f = l\ \sin\ (\beta\ /2 + \gamma)$, $b_0 = 2f$

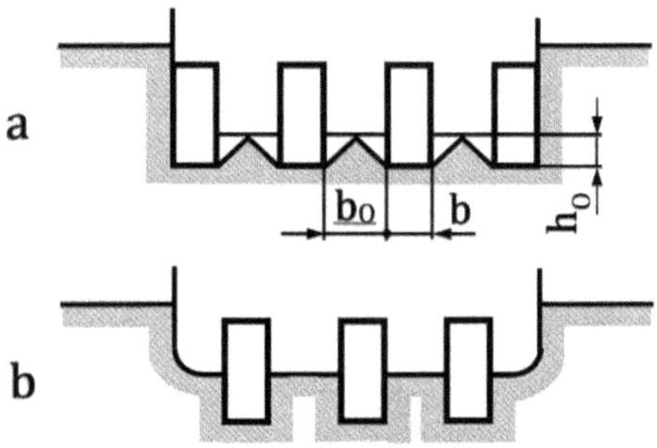

Fig. 3.13 Alternating arrangement of teeth on consecutive excavator shovels [5], loosening the soil over the entire width, **a**—first shovel, **b**—second shovel

Tooth blade

While digging the soil, the tooth blade wears out and changes shape, Fig. 3.14. With the new tooth blade, the upper and lower sides of the tooth close the radius $r = 1.5$–3.0 mm. After greater wear of the tooth blade, the cutting resistance can increase 1.5–2 times. The transition from the upper to the lower side of the tooth is irregularly rounded. At the same time, wear of the knife in between the teeth may also occur. It is considered permissible to wear the front surface up to $z = 5$–7 mm, after which they are replaced or repaired. With a variable cutting angle, the knife blade wears out faster.

Soil cutting resistance with teeth [6]

$$R_1 = k_1 h(B + h)(0.55 + 0.015\alpha) + \sigma(z\,n\,b + \mu y\,n\,b)\ [\text{daN}]$$

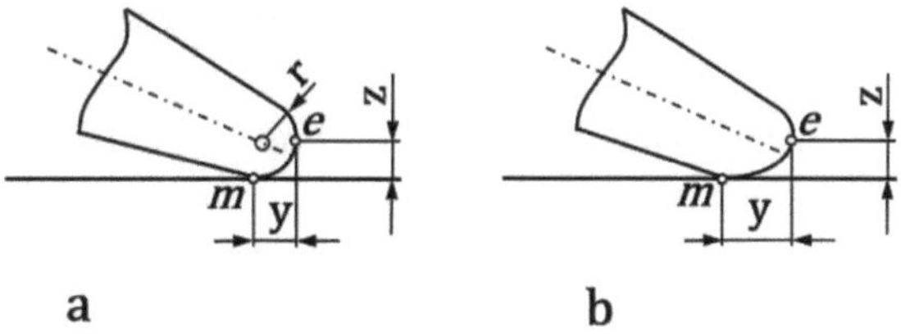

Fig. 3.14 Tooth blade (**a**—new, **b**—worn out) [5]

B Width of the knife blade (cm).
h Cutting depth / h = (1/5 − 1/7)B, (cm).
α Soil cutting angle (rad).
k_1 Specific cutting resistance (dN/cm^2).
σ Limit bearing capacity of soil (daN/cm^2).
n Number of teeth, b—tooth width (cm).
y, z Tooth wear (cm).
μ_1 Steel-soil friction coefficient.

For $n = 4$, $y = 5\,z$, $\mu = 0.5$: example

$$R_1 = k_1 h\,(B + h)(0.55 + 0.015\alpha) + 14\sigma z\,b[\text{daN}]$$

If the teeth are removed from the knife blade, the expression of the cutting resistance of soil with teeth takes on an expression equal to the resistance of cutting soil without teeth.

Short conclusion

The calculation of soil cutting resistance depends on several factors, with a knife without teeth and with teeth, so it is necessary to determine good starting values. The width of the knife (*B or L*) and the depth of the soil cutting layer (*h*) are the values that can be safely counted on. The cutting resistance of the soil (R_1) changes the most with the change in the cutting depth, and thus the working performance of the machine.

In tabular data of specific cutting resistance (k_1) and specific digging resistance (k_k), there is a wide range of resistance values. The aim is to adopt a value as close as possible to the specific cutting resistance, and tests will confirm or somewhat modify the data. Actual values of cutting resistance and digging resistance can only be obtained by tests in real conditions.

3.4 Other Soil Digging Resistances

Depending on the shape of the digging tool, the machines are loaded with additional soil digging resistances:

R_3 Shovel charging resistance,
R_4 Resistance to moving the soil prism in the shovel,
R_5 Frictional resistance between the towed shovel and the ground,
R_i Other resistances.

Shovel charging resistance

The knife blade is inclined towards the path of movement at the cutting angle α, Fig. 3.15. During the movement, the first layer of soil gradually moves forward. The new cutting layer climbs on top of the previous layer and so on. The material is gradually piled up in the shovel. The accumulation of soil in the shovel and in front of the shovel depend on the shape of the loading shovel, the type of soil and its humidity.

In case of coherent ground, the layer does not disintegrate until the back wall of the shovel, when the layer breaks and the formation of a new layer occurs. With weakly bound soil, the layer disintegration and accumulation start earlier, more in the front part of the shovel and in front of the shovel in the form of a prism. The greatest resistance occurs when the filling of the shovel is being completed. Shovel filling resistance consists of the weight of the pile of material being lifted and the frictional resistance when pushing such a pile upwards [1, 6]:

$$R_3 = BhH\gamma + \mu_2' BH^2\gamma\,[\mathrm{N}]$$

B Inner width of the shovel (m),
H Shovel height (m),
h Depth of the cutting layer (m),
γ Volume weight of soil (N/m^3),
μ'_2 Coefficient of internal friction, $\mu'_2 \approx 0{,}5\ \sin 2\rho$, clay $\rho = 15$–$20°$, $\mu'_2 = 0{,}25$–$0{,}30$, loam $\rho = 25$–$30°$, $\mu'_2 = 0.35$–0.45, sand $\rho = 35$–$40°$, $\mu'_2 = 0.45$–0.50

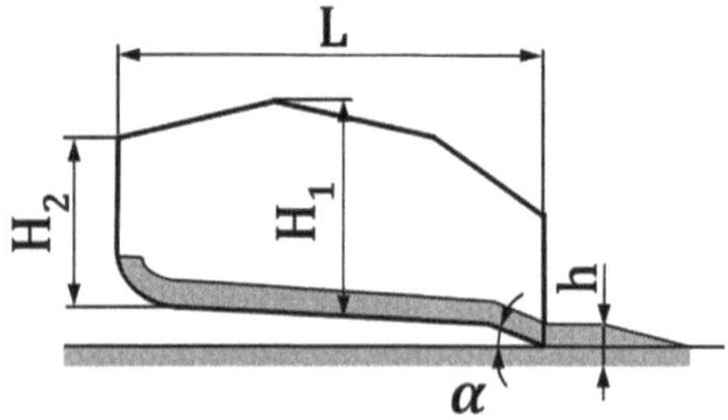

Fig. 3.15 Movement of soil layer in the shovel. α—soil cutting angle, h—cutting depth, $H_{1,2}$, L—shovel dimensions

Shovel loading resistances are highly variable, such as the frictional resistances of the towed shovel, the resistances of moving the soil prism, and the resistances when filling the shovel. Moving the earth makes the process of analyzing drag forces complex. The lowest resistances are at the beginning of cutting and loading the shovel, and the highest when the shovel is full. In order to balance the available machine force with the digging resistances, the machine operator or control unit regulates the thickness of the soil cutting layer.

Resistance to moving the soil prism inside the shovel

$$R_4 = y_1 B H^2 \gamma (\mu_2 \pm i)[\mathrm{N}]$$

y_1 0.5–0.6 volume factor of the prism, which depends on the construction of the shovel and knife
μ_2 0.3–0.5 friction coefficient soil—soil
i tgα, factor of resistance to the movement of the soil prism on the sloping ground
α Ground slope (+ uphill, – downhill).

Frictional resistance of the towed shovel against the ground

The towed shovel of the machine is pulled along the ground by a steel rope and in the process it excavates the soil. The highest resistance of the towed shovel is:

$$R_5 = (G_\mathrm{k} + V\gamma)\cos\alpha(\mu_1 \pm i)[\mathrm{N}]$$

G_k Weight of the towing shovel (N)
V Shovel volume (m^3)
γ Volume weight of soil (N/m^3)
μ_1 Metal-earth friction factor.

3.5 Soil Digging Resistance with Continuous Tool Operation

With a rotary backhoe with continuous operation, two movements are required: the movement of the basic machine as a shear movement and the rotation of the rotor, which digs the soil with the cutting blades. The result of combining the two motions on the tool blade is described by the epicycloid equation, Fig. 3.16:

$$x = R_\mathrm{r}\sin\varphi + k\varphi$$
$$y = R_\mathrm{r}(1 - \cos\varphi)$$

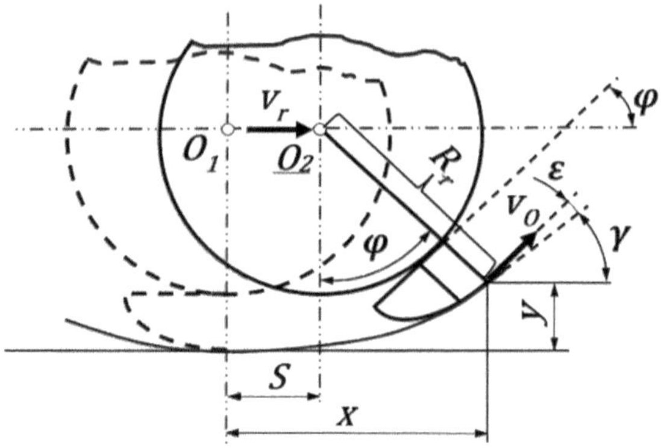

Fig. 3.16 Path of the rotor cutting blade. R_r—radius of the blade (tooth), φ—turning angle of the shovel ("geometric cutting angle of the shovel"), $k = R_r\, v_r\, /\, v_0$—constant, v_r—working speed of the rotary backhoe, v_0—peripheral speed of the tip of the blade, $\varepsilon = \varphi - \gamma$, angle difference, circle—epicycloid, $S = k\,\varphi$, rotor shear

Cutting angle of the blade

From the entry of the blade until its exit from the grip, the slope of the tangent to the movement path changes (γ). This angle is the kinematic cutting angle at a given point of the path. This angle is always smaller than the geometric cutting angle φ, which is formed by the tangent of the rotor. The angle difference (ε) is 4–6°, which is important when determining the back angle, in order to avoid touching the back side of the tool in the excavation.

Cutting layer depth and rotor speed

During cutting, the depth of the cutting layer changes and takes a sickle-like shape, from $S = 0$ to the highest value of S, Figure 3.17. The peripheral speed of the rotor is approximately equal to the speed of the machine (2–3 m/s).

Current depth of the cutting layer

$$S_t = S \sin \varphi [\mathrm{m}]$$

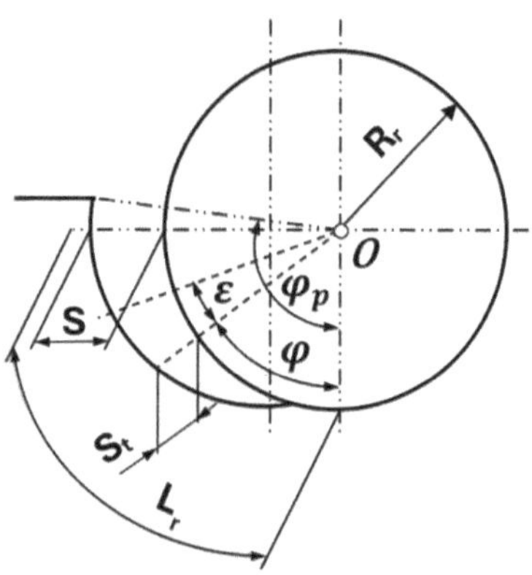

Fig. 3.17 The shape of the cutting layer. S—maximum depth of the cutting layer, S_t—current depth of the layer

S Maximum depth of the cutting layer, i.e. the rotor shear corresponding to the movement of the rotor in the time between the passage of two consecutively placed blades.

$$S = v_r / z_l n_r [\mathrm{m}]$$

z_l Number of blades on the rotor
n_r Number of revolutions of the rotor (min^{-1})

Cutting path

$$L_r = R_r \varphi_p / \cos\varepsilon [\mathrm{m}]$$

φ_p Full angle of rotation of the rotor (*rad*) for which the blade is in the grip of the trench, $\cos\varepsilon = 1$, $L_r = R_r\,\varphi_p$

Cutting resistance

When digging a trench, a certain number of rotor cutting blades grip the soil, and the actual number of blades depends on the digging depth, Fig. 3.18. The different number of rotor blades in the trench grip causes uneven soil cutting resistance.

The resistance forces acting on the rotor blades can be set, in *counter* or *direct* digging mode. Actually, in the counter-directional mode of operation, the digging resistance can be broken down into the cutting resistance, i.e. the tangential component ($R_t = R_1$) and the normal component ($R_n = R_2$). The highest resistance acts on the rotor blade that comes out of the grip of the soil.

Tangential cutting resistance on one blade [6]:

$$R_1 = k_1 B\, S_t [\mathrm{N}]$$

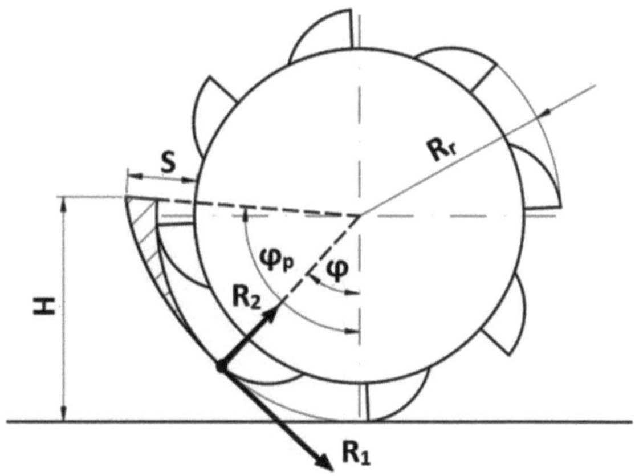

Fig. 3.18 Resistance forces on the rotor blades in the counter mode of soil digging. H—trench depth, R_t—rotor radius, R_1 and R_2—cutting resistance components

k_1 specific cutting resistance
B width of blade/trench
S_t current layer thickness ($S\ sin\varphi$)

Normal resistance component: $R_2 = (0.2–0.5)R_1$ [N]

Soil cutting resistance with multiple rotor blades

$$R_{1u} = \sum_{1}^{m} k_k B S_{ti} [\text{N}]$$

$$R_{1u} = BSk_k[\sin \varphi_p + \sin(\varphi_p - \Delta\varphi) + \sin(\varphi_p - 2\Delta\varphi) + \cdots + \sin(\varphi_p - m\Delta\varphi)$$

m The number of rotor blades in the soil cutting process
$\Delta\varphi$ Angle of grip of one blade, $\Delta\varphi = 360°/z_l$.
φ_p Angle of grip of the rotor $\cos\varphi_p = (R_r - H)/R_r$.
k_k Specific cutting resistance (for a blade with teeth, at digging speeds of 2 m/s and cutting angle α =25–30°): Category I, 300–600 kN/m^2; Cat. II, 600–1000 kN/m^2; Cat III, 1200–1800 kN/m^2;. Cat. IV, 2000–3000 kN/m^2.

Cutting the soil with discs

Discs for soil cutting are of specific design [7], in the shape of a circular cutter (mass about 1 kg), attached to a chain of radius (about 0.5 m), Fig. 3.19. The cutting blade of the disks, of a certain number of discs (z_n, about 40), are arranged on an about 2 m long shaft in a spiral arrangement (*helix system*). The rotation speed of the impact discs (500–900 rpm) ensures continuous cutting of the soil layer. The tool blades cut and discard the soil (or crush it).

The shape of the disk is a circular blade (diameter, d = 50–100 mm). The rotor can rotate in both directions, in the counter-direction or in the same direction. Counter-rotation is used to clear the road and terrain or discard a certain layer thickness, and

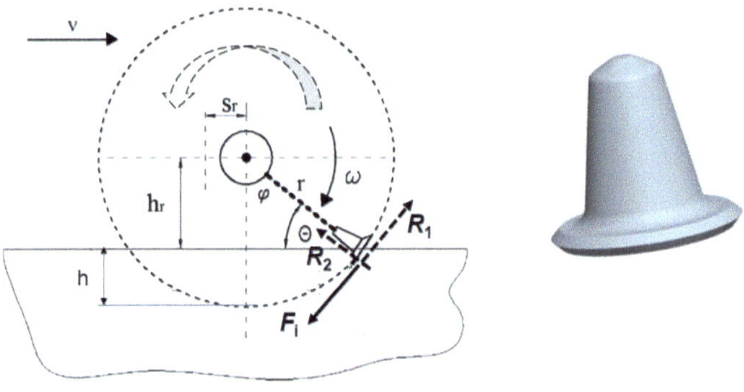

Fig. 3.19 Rotary disc for soil digging [7]. h—cutting depth, F_i—force impulse, R_1—tangential cutting resistance

direct rotation is used for digging the soil or breaking the buried objects, as a *flail*. At the same time, the speed of soil processing is $v = 0.5$–2 km/h, the thickness of the cutting layer $S_t = 20$–30 mm, the depth of cutting $h = 10$–20 cm.

Impacts of disc against the soil are collisions of two bodies. According to the law of energy conservation, the force impulse is:

$$F_i = m\,v(1-k)/\Delta t$$

m	Disc mass
v	Peripheral speed of the disc before the impact
k	Restitution factor, $k = u/v$ (ratio of speed after and before impact)
Δt	Time interval of the disk in the digging process
Excavation criteria	Impulse force greater than cutting resistance
$F_i \geq R_{1u}$	For bound soil/soil cutting ($R_{1u} = z_n\, k_1\, d\, \Sigma S_{ti}$)
$F_i \geq R_\sigma$	For loose soil/soil breaking ($R_\sigma = \sigma A$)

References

1. Markovic V (1975) Gradevinske masine za zemljane radove. Naucna knjiga, Beograd
2. Mikulic D (1998) Gradevinski strojevi. Konstrukcija, proračun i uporaba. itG Zagreb
3. Plavsic M (1984) Gradevinske masine. Fakultet tehnickih nauka, Novi Sad
4. Alekseeva TV et al (1985) Machines for earthmoving work. Theory and calculation, Third Revised and Enlarged Edition, Amerind Publishing Co
5. Kelic VN (1990) Bageri rovokopaci. Naucna knjiga, Beograd
6. Mikulic D (2024) Gradevinski strojevi. Teorija i konstrukcija. Veleuciliste Velika Gorica, Velika Gorica
7. Mikulic D (2013) Design of demining machines. Springer, London

Chapter 4
Traction Properties of Construction Machines

Abbreviations

F_v [N]	Traction force
M_m [Nm]	Engine torque
i_{tr} [–]	Transmission ratio
r_d [m]	Dynamic wheel radius
v_i [km/h]	Machine speed
R_1 [N]	Soil cutting resistance
R_u [N]	Climb resistance
R_i [N]	Inertia resistance
R_k [N]	Rolling resistance
G_{ad} [N]	Adhesion weight of the machine
F_v[N]	Machine adhesion force
f_k [–]	Rolling resistance coefficient
M_{hm} [Nm]	Hydro-motor torque
V_{hm} [cm^3/rev]	Hydro-pump volume
η_u [–]	Degree of usefulness of transmission
P_{hm} [W]	Hydro-motor power
P_p [W]	Hydro-pump power
M_2 [Nm]	Turbine torque
M_1 [Nm]	Pump torque
M_S [Nm]	Stator torque
i_{21} [–]	Hydraulic transmission ratio
k [–]	Torque increase factor
η_{HDM} [–]	Degree of converter efficiency
P_K [W]	Total braking power
P_M [W]	Braking power of the engine
P_{HK} [W]	Braking power of the retarder

D. Mikulić and D. Mück, *Construction Machines*,
https://doi.org/10.1007/978-3-031-67845-5_4

4.1 Traction Characteristic of the Construction Machine

The properties of construction machines are based on their traction characteristics. Both engineers and machine users use traction indicators for the analysis and selection of the machines. The traction properties of the machines are the result of the joint work of the engine, transmission, and final drives of the machine. By connecting the operation of the engine and power transmission with the resistances of movement and operation of the machine, one can analyze the influence of the structure on the traction properties of the machine and on the performance of the machine. The machine users use the traction diagram (*Rimpull-Speed*) to evaluate the efficiency of the machine.

The requirements for power transmission are placed first on the continuity of the power flow from the drive motor to the tracks or wheels and on the controllability of the machines. Complex requirements can be ensured by the following power transmissions:

1. Hydrostatic transmission (HS)
2. Hydrodynamic transmission (HD)
3. Electric transmission (E)
4. Hybrid transmission (H)

Considering the application in modern construction machines, hydrostatic and hydrodynamic power transmissions are equally represented, while electric and hybrid power transmissions are experiencing strong development. These power transmissions can be compared and conclusions can be drawn about the choice and purchase of the machines. It is especially important for the operation of construction machinery to increase power in time and to have uninterrupted transmission from the engine to the wheels. For example, when the machine cuts and moves the soil with a knife, any power interruption can cause a stronger inertial load and high stress on the main structure components, as well as change in the quality of the processed soil surface.

Typical representatives of the construction machines for which traction characteristics are extremely important are earthmoving machines. These are slow-moving construction machines on wheels or tracks. In their construction, the power transmission from the engine to the final drives provides the available traction force (F_v) to overcome the movement resistance, Fig. 4.1.

Traction characteristic of the machine

The traction characteristic of a construction vehicle represents a diagrammatic representation of the traction forces on the drive wheels and the resistance to motion in relation to the speed of movement (F–v), as shown in Fig. 4.2. Practically, the working range of the machine can be assessed for all working conditions and all gear stages in the transmission.

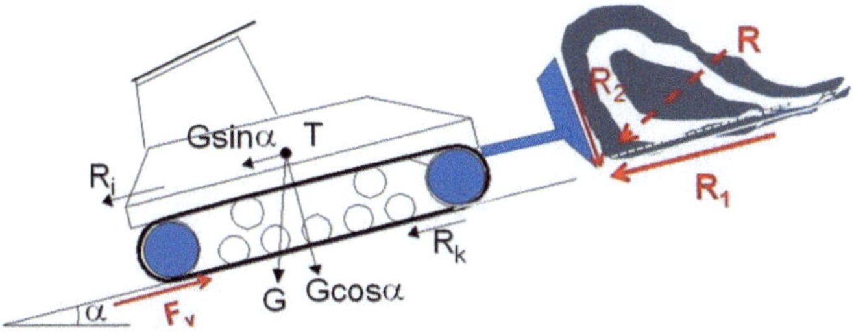

Fig. 4.1 Work resistance and traction force of the earthmoving machine. F_v—traction force, R_1—soil cutting resistance, $Gsin\alpha$—climbing resistance, R_k—rolling resistance, R_i—inertial resistance

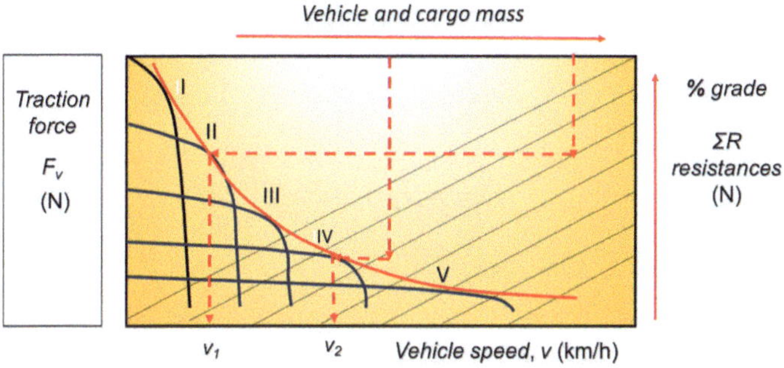

Fig. 4.2 Scheme of using the traction characteristic of the machine

The actual traction characteristic of a medium-class bulldozer (F–v) with hydrostatic (HS) and hydrodynamic (HD) power transmission are shown in Fig. 4.3. A difference in the continuity of the traction force can be observed for both transmissions. For a certain soil category, a favorable range of force can be determined for overcoming the soil cutting resistance (from penetration, soil cutting, dosing, to the speed of soil displacement).

Indicators of tractive characteristics of construction machinery [1]

- traction force on the drive wheels

$$F_v = M_{pk}/r_d \text{ [N]}$$

$M_{pk} = M_m\, i_{tr}\, \eta_u$, torque on the drive wheels
M_m—engine torque
r_d—dynamic wheel radius
i_{tr}—transmission ratios in each gear
η_u—efficiency ratio of the transmission in a particular gear
$F_v \le F_\mu$—machine adhesion force.

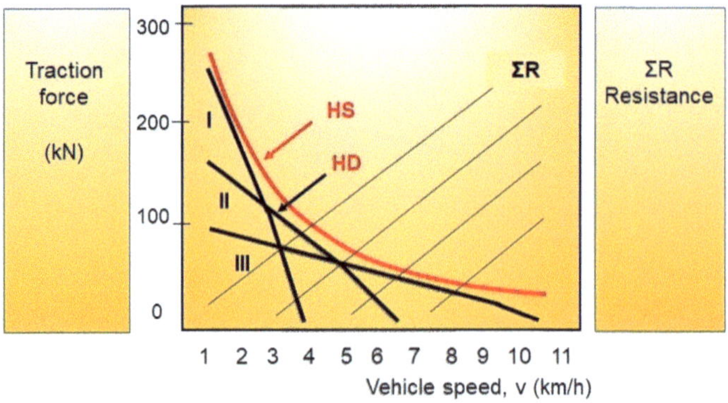

Fig. 4.3 Traction characteristic of bulldozers with HS and HD power transmissions

- machine speed

$$v_i = n_m r_d / i_{tr} \text{ [km/h]}$$

- machine movement resistances

$\Sigma R = R_1 + R_k \pm R_u + R_i$ [N] (at the machine)
$\Sigma R = R_k + R_z \pm R_u + R_i$ [N] (at the vehicle)
R_1—soil cutting resistance, $R_1 = f_k\, L\, h$
R_u—climb resistance, $R_u = G\, sin\alpha$
R_z—air resistance, R_i—inertia resistance
R_k—rolling resistance, $R_k = f_k\, G_{ad}$
G_{ad}—adhesion weight of the machine
f_k—rolling resistance coefficient.

4.2 Hydrostatic Power Transmission

Hydrostatic drive of mobile construction machines is a closed hydraulic circuit. The basic elements of such a transmission are: variable flow hydraulic pump, piping, control elements, constant flow hydro-motor, and hydraulic oil. The concept of installing a hydrostatic power transmission between the drive diesel engine and the axle transmission in wheeled vehicles is shown in Fig. 4.4.

The concept of hydrostatic transmission of a tracked machine is shown in Fig. 4.5. One hydraulic control circuit for one track and another hydraulic circuit for the other track.

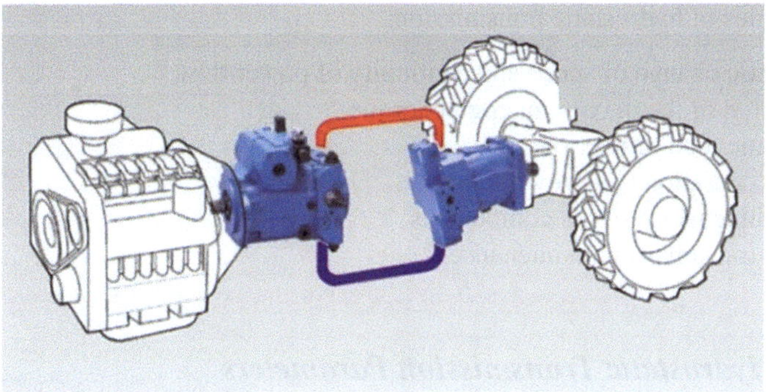

Fig. 4.4 Model of hydrostatic transmission of wheeled machines [2] (hydro-pump variable volume and constant volume hydro-motor)

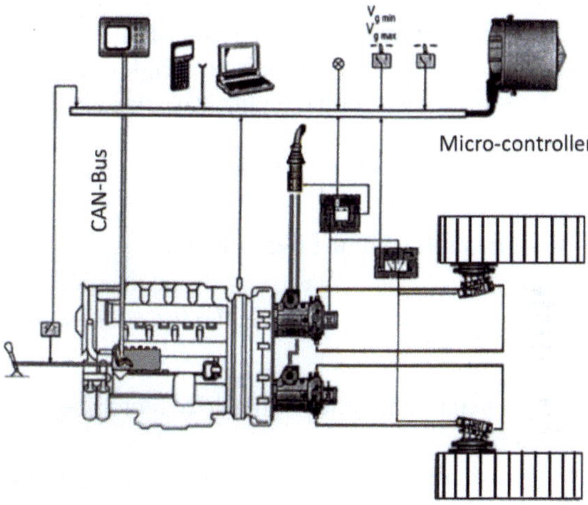

Fig. 4.5 Hydrostatic transmission of tracked machines (adapted [5])

The control regulation is performed by changing the working volume of the hydro-pump and/or hydro-motor (variable pump/motor). This is done electronically by controlling the flow with a *joystick* and a microcomputer, based on the shift of the *servo-piston* which changes the angle of the rotary plate. This enables easy changes in speed and direction of the machine. Usually, there is a difference between the working mode of movement (low speed) and the transport mode of driving (higher speed).

The components of the power transmission occupy relatively small space in the machine, and provide simplicity of control through pressure and flow. The hydrostatic transmission is a practical way of energy transmission in construction machinery.

Properties of hydrostatic transmission.

- automatic change of speed and continuity of power flow,
- utilization of the maximum engine torque,
- automatic load protection,
- simple operation (*joystick*),
- small dimensions of the components,
- simple transmission maintenance.

4.2.1 Hydrostatic Transmission Parameters

(a) Flow of hydro-pump and hydro-motor

F_μ—maximum traction force on the drive wheels from the adhesion conditions.
$F_v \leq F_\mu$—adhesion force/μ—adhesion.
$F_\mu = \mu\, G_{ad}$ [N], G_{ad}—adhesion weight on the drive wheels.
M_{pk}—maximum torque on the drive wheels [Nm]

$$M_{pk} = F_\mu r_d \eta_m = M_{hm} i_r \eta_v \text{ [Nm]}$$

Required torque of the hydro-motor

$$M_{hm} = M_{pk}/i_r \eta_m \text{ili} M_{hm} = M_m i_h \eta_v$$

M_m—maximum torque of the drive motor [Nm], n—number of revolutions of the motor
i_r—transmission ratio of the shaft reducer, i_h—transmission ratio: $i_h = n_p/n_{hm}$
n_p -number of revolutions of the pump, n_{hm}—number of revolutions of the hydro-motor
Selection of hydraulic components/*manual-brochure*
Hydro-motor volume: V_{hm} [cm^3/rev], flow: $Q_{hm} = V_{hm}\, n_{hm}/\, \eta_v$ [l/min],
Maximum pressure (bar), maximum speed (rpm)
Hydro-pump volume: $V_p = V_{hm}/\eta_v$ [cm^3/rev], flow: $Q_p = V_p\, n_p/\, \eta_v$ [l/min].

Degree of usefulness of transmission

$$\eta_u = \eta_v \eta_h \eta_m$$

η_v—volumetric losses, clearances (e.g. $\eta_v = 0.95$)
η_h—hydraulic losses, piston resistances, pressure drop
η_m—mechanical losses, parts that undergo friction.

Total degree of transmission efficiency: $\eta_{u} = 0.7\text{–}0.9$, $\eta_{u} = P_{hm} / P_{p}$

Hydro motor and pump power

- hydro-motor power: $P_{hm} = M_{hm} \cdot \omega_{hm}$ [W]
- hydro-pump power: $P_{p} = M_{p} \cdot \omega_{p}$ [W]

(b) Working oil pressure

The working pressure of oil determines the required force (F) and torque (M).

High oil pressures significantly affect the service life of components, so the pressure is limited (100–450 bar).

The working pressure (p) is set, and the pump flow rate (Q) is determined: $\boldsymbol{p_1 Q_1 = p_2 Q_2}$.

Oil flow (min/rev and max/rev) determines the working speed (v) of the machine.

The oil speed in the suction / pressure pipeline is, $v_u = 1.5\text{–}2$ m/s; $v_p = 3\text{–}6$ m/s.

Hydraulic oils have the necessary physical and chemical characteristics: density, viscosity, compressibility, lubricity, solubility of gases, flammability, and toxicity. The most important property is viscosity, which is defined as the ability of layers of liquid to resist shearing and sliding. Viscosity is evaluated by the coefficient of kinematic or dynamic viscosity. In hydraulic systems for working temperatures higher than 80 °C, synthetic fluids are used.

The application of hydraulic oil of nominal viscosity: ISO VG 10–100 (VG 22, VG 32, VG 46, VG 68). For example, *Hidraol HDS VG 46* mineral oil has a high viscosity index, and additives protect its stability at high and low temperatures. *Shell Tellus S2 VX 68* mineral oil can be mixed with other mineral oils. *Shell Tellus* oils offer improved wear protection and longer oil life. The oil viscosity characteristics are shown in the diagram, Fig. 4.6.

(c) Control function parameters

ϕ_p—angle of the hydro-pump turn plate (0–1)
ϕ_m—angle of the hydraulic motor turn plate (0–1).

The regulation of the hydro-pump and hydro-motor shows the volume change diagram, $\phi_p = 0\text{–}1$, Fig. 4.7. The power increases linearly by increasing the angle of the turn plate ϕ_p, while the torque remains constant. When regulating the hydraulic motor, the torque decreases linearly, while the power is constant. The hydrostatic transmission achieves linear regulation of the speed of rotation of the output shaft, which can achieve the highest torque.

By changing the volume of the pump, the power and speed of the engine are affected, while the torque is kept constant. At the maximum working volume of the pump, the phase of reducing the working volume of the motor begins, during which higher rotation speeds are achieved, the torque decreases, and the power remains constant.

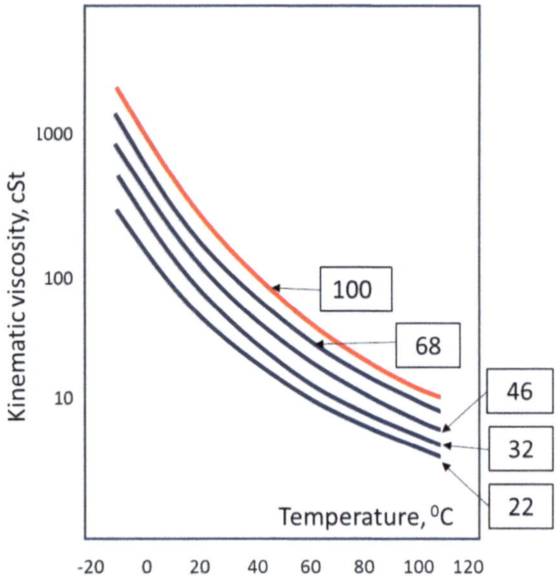

Fig. 4.6 Viscosity—temperature diagram for *Shell Tellus S2 VX* [3]

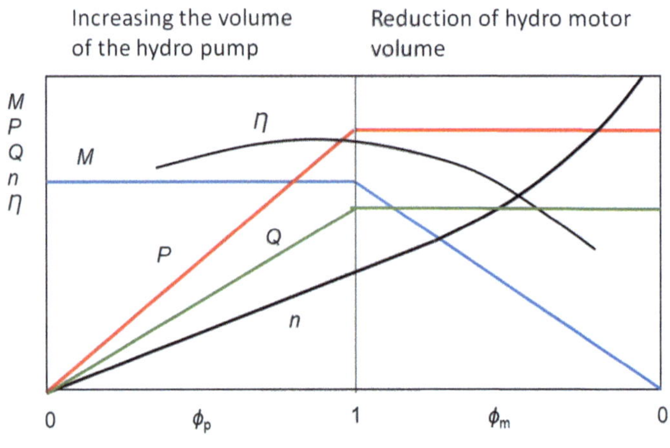

Fig. 4.7 Parameters of the hydrostatic transmission in the regulation function [2, 4]

4.2.2 Traction Characteristic of the Machine with Hydrostatic Transmission

In the transmission of engine power to the wheels or tracks, a closed hydraulic circuit with variable flow enables power regulation (multiplier of pressure and flow, *p-Q curve*), Fig. 4.8a. For example, this results in adequate traction force and movement

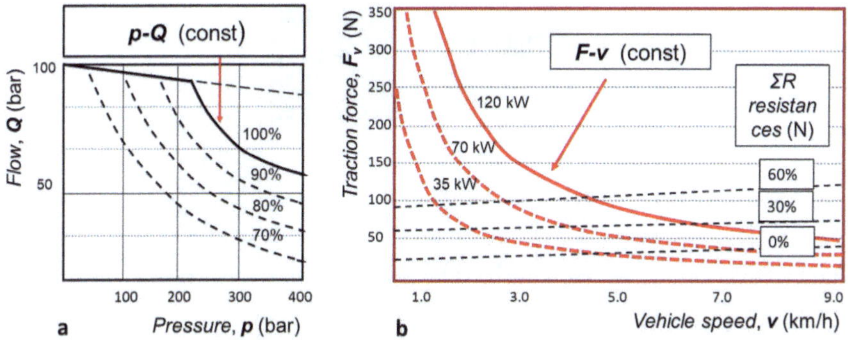

Fig. 4.8 *p-Q* power regulation (**a**) and *F-v* traction characteristic of the machine (**b**)

speed of the construction machine (*b*), which corresponds to the ideal hyperbola of using the maximum constant power of the diesel engine ($F_v\ v = const.$).

Figure 4.9 shows the control scheme of the hydrostatic drive of the machine using a microcomputer. It is possible to synchronize the movement, rotation of the machine and change the direction of movement, i.e. automatically adjust the parameters of the hydraulic motor to overcome external movement resistance. The operation is controlled by the power regulation system (*Load Sensing*), which detects the change in pressure, with which the pump automatically regulates the flow in order to maintain a constant pressure difference. For example, the cutting resistance on the blade of the tool that is transferred to the machine is very variable. Any change in load is balanced by a hydraulic pump.

The hydrostatic transmission of the machine satisfies the set requirements of the working and transport mode of movement. By engaging the working device, multiple functions are automatically executed sequentially (movement of the working tool, positioning of various tools, monitoring, etc.). Considering the difference between the minimum and maximum working volume of the hydraulic motor, it is necessary to achieve transport speeds of moving the machine with less input power. By increasing the speed of rotation of the hydraulic motor, the speed of the machine increases, the torque, or traction force, decreases. Manufacturers of hydrostatic drive offer optimal characteristics of the working device, for example for digging the ground, as well as for moving the machine over long distances with minimal fuel consumption.

4.2.3 Properties of Hydraulic Pump and Hydraulic Motor

The design of variable hydraulic pumps and hydraulic motors is very similar. Axial-piston pumps and hydraulic motors are intended for systems with very high pressures. They are structurally more demanding than hydro pumps and hydro motors of

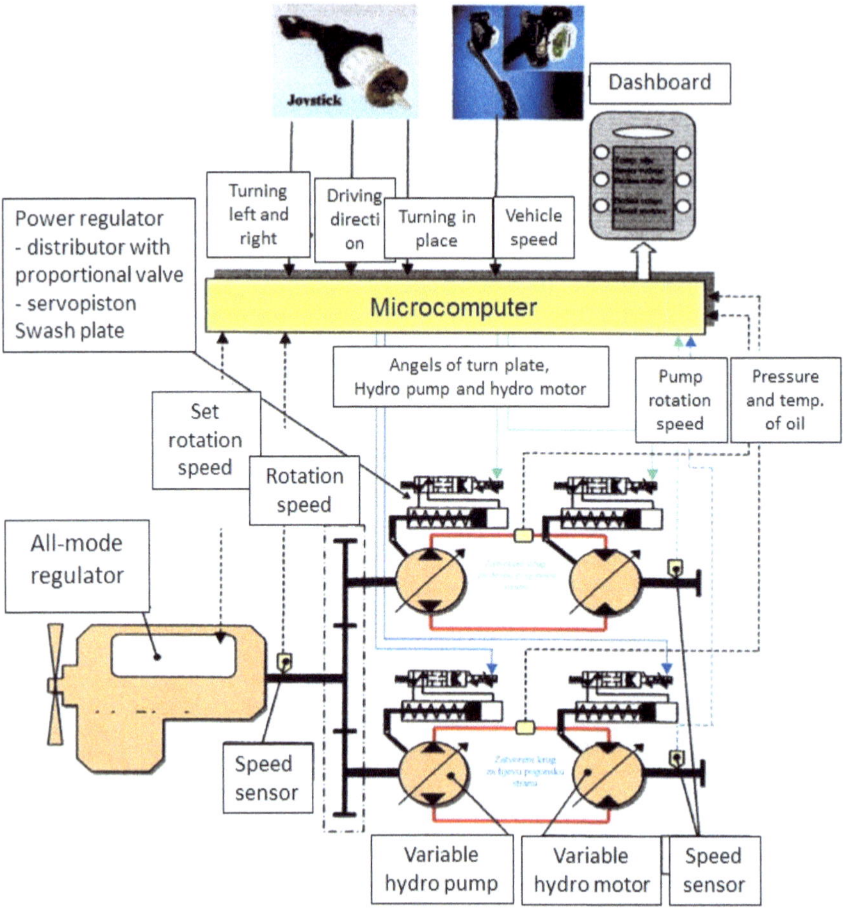

Fig. 4.9 Scheme of controlling the hydrostatic drive of the machine [5]

constant working volume. There are several versions with a swivel plate (the plate is rotated relative to the shaft) and a swivel drum (the drum with pistons is tilted relative to the shaft). An illustration of the operation of an axial-piston pump with a turn plate is shown in Fig. 4.10.

In the pump housing (1), the pistons (2) move parallel to the axis of rotation of the shaft (3). The pistons are located in the drum (4), and they are connected to the turn plate via sliding plates (5). The plate is tilted at a certain angle that is adjusted electronically, whereby the servo piston acts on the plate and changes the tilt angle. Due to the inclination of the turn plate, the pistons located in the drum move back and forth, thus sucking in and pressurizing the oil. The distribution plate (6) (section A–A), serves to separate high (P) and low pressure (T) oil.

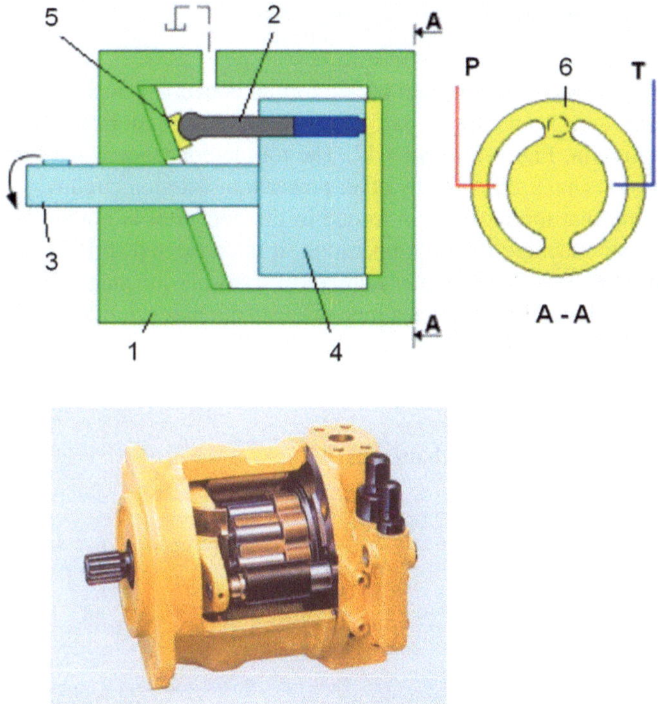

Fig. 4.10 Axial-piston pump with turn plate (*Bosch-Rexroth*) [4]. 1—pump housing, 2—pistons, 3—shaft, 4—drum, 5—turn plate, 6—distribution plate

A greater tilt of the turn plate results in a greater flow at the same number of revolutions of the diesel engine, whereby more favorable operating characteristics of the drive can be obtained. These components require a higher level of oil purity than constant displacement components.

Some other advantages and disadvantages of comparing hydraulic pumps and hydraulic motors: hydropumps with a turn plate are generally cheaper, and their advantage is the possibility of arranging several pumps on one drive shaft; they are easier to adjust, so they are a more common choice for servo pumps in hydrostatic systems of machines. Rotary drum assemblies generally have a higher degree of useful action, better starting characteristics for hydro motors and a higher rotational speed.

4.3 Hydrodynamic Power Transmission

Hydrodynamic power transmission uses a torque converter. Such a hydrodynamic power transmission is an integral part of a modern transmission, as part of an automatic transmission, Figs. 4.11 and 4.12. The torque converter is the first circuit in the transmission chain, followed by other power transmission circuits.

Considering that the increase in torque in the converter is not sufficient for all the needs of the machine range of operation, it is connected serially with a planetary gearbox that provides a greater number of transmission stages. The automatic transmission of the construction machine consists of subassemblies:

- hydrodynamic torque converter,
- planetary gearbox,
- hydrodynamic brakes (retarder),
- servo-hydraulic device (mechatronics).

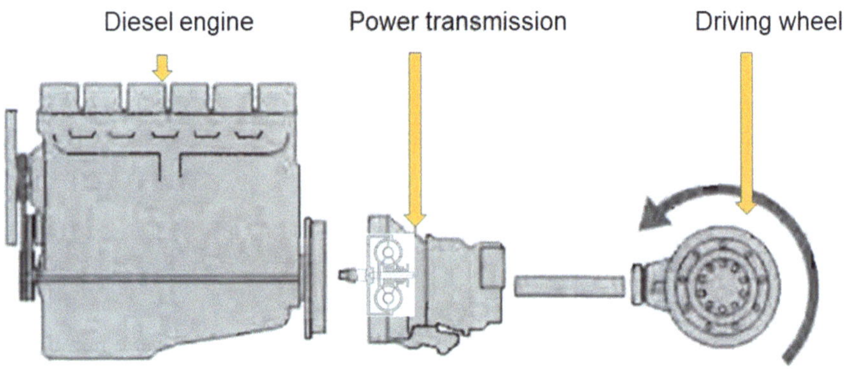

Fig. 4.11 Transmission with torque converter for wheeled machines

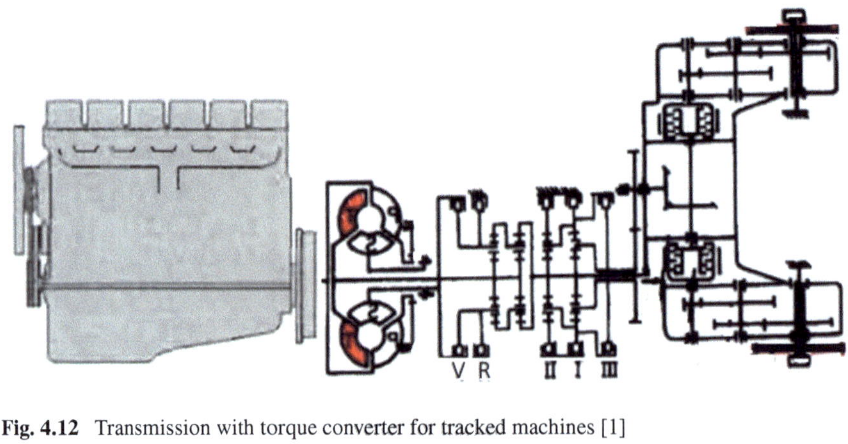

Fig. 4.12 Transmission with torque converter for tracked machines [1]

4.3.1 *Traction Characteristic of the Machine with Torque Converter*

Typical traction characteristics of a tracked and wheeled machine with a converter are shown in Fig. 4.13. In a bulldozer-type ***tracked machine***, the machine uses three gears. In the display of the available power *F–v* (*traction force-speed, drawbar pull-speed, rimpull-speed*), the bulldozer achieves in the first gear the highest traction force for overcoming the soil cutting, with the working speeds not lower than 5 km/h. In the second and third gears, the speeds are higher for soil displacement and for backward movement.

In the case of a bulldozer-type ***wheeled*** machine, the traction forces are significantly lower, but the speeds are higher, so the use of the machine is foreseen for lighter works and moving earth material. Gear changes follow well the hyperbole of utilizing constant maximum engine power. The machine user is given the possibility of applying actual movement resistances and a realistic assessment of movement speed and performance.

4.3.2 *Torque Converter Properties*

The torque converter is the key component of a hydrodynamic power transmission, which enables the transmission and transformation of the engine torque using oil that circulates through the working space of the elements, Fig. 4.14. The converter consists of three parts: pump, stator, and turbine. The impeller pump (P) is connected to the engine crankshaft. The stator/reactor (S) is connected to the housing by means

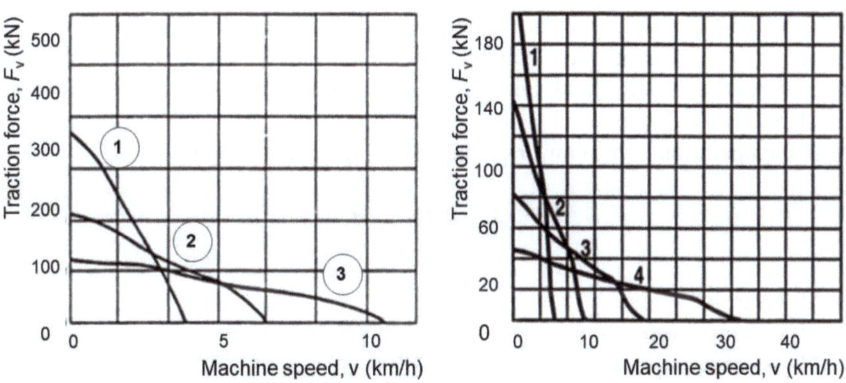

Fig. 4.13 Traction characteristics of a medium-class tracked and wheeled bulldozer [6]

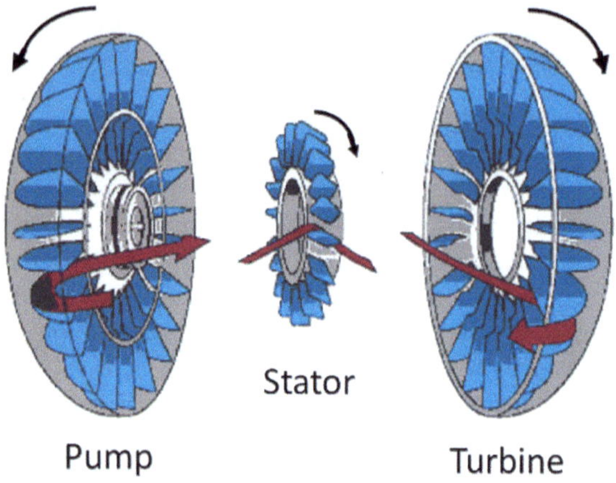

Fig. 4.14 Hydrodynamic torque converter [7]

of a freewheel clutch, which allows it to rotate in the direction of the pump (P) and turbine (T).

When the torque on the turbine is equal to the torque on the pump, then the torque on the reactor equals zero. However, when the speed of the turbine is higher, that is, when the torque of the turbine is lower than the torque on the impeller pump, the reactive torque of the stator changes direction. At the same time, the reactor starts to rotate in the direction of the pump and at that moment the torque converter turns into a hydraulic clutch. This is necessary for a higher degree of efficiency of the converter at higher machine speeds. The degree of useful action can be further increased by mechanically connecting the pump and the turbine (*lock-up clutch*).

The change in torque and angular speed is automatic, depending on the load on the output shaft. At the beginning of the movement of the machine, the greatest increase in torque occurs, which then decreases with the increase in movement speed, until the input and output torques are equal. Then the torque converter switches to the operation of the hydrodynamic clutch, which increases the degree of useful action. The reactor then rotates freely via the freewheel coupling, so that it does not create resistance to the flow of oil.

In the operation of the converter, the following equality applies:

$$M_2 = M_1 \pm M_s$$

M_2—turbine torque
M_1—pump torque
M_S—stator torque/ reactor.

Advantages of using a converter compared to a mechanical transmission:

- quieter start and more even acceleration of the machine,
- self-regulation of power transmission transformation,
- continuous gear change under load,
- transmission reliability, no dynamic shocks and torsional vibrations,
- greater trafficability of the terrain (without sharp jerks and damage to the tread surface).

Some disadvantages:

- complexity of production and maintenance,
- degree of efficiency,
- special hydraulic fluid ATF,
- higher fuel consumption.

Higher fuel consumption is usually not estimated well enough. In this case, the increase in fuel consumption is largely offset by savings in spare parts, working life and greater machine performance. Because of all the favorable properties, torque converters have found application in the transmissions of construction machines and trucks.

4.3.3 Parameters of Hydrodynamic Power Transmission

On the pump and turbine torque diagram, Fig. 4.15, the number of revolutions of the turbine n_2 is plotted on the abscissa, and the torque of pump M_1 and turbine M_2 is plotted on the ordinate. With a change in the output speed of turbine n_2, the pump torque changes slightly, while the turbine torque changes intensively. If the output shaft of the transmission is loaded with torque M, then the torque converter will establish stationary operation at point I, in which the turbine torque M_2 will be equal to the load torque M. At the same time, the number of revolutions of the output shaft will be n_2' and the pump torque will be M_1'. When the external load operating conditions change and the load torque increases by ΔM, the torque converter will establish a new steady state at point II, in which the turbine torque will increase by ΔM_2, in order to equalize the resistance torque, while the turbine speed will fall by Δn_2. During this time, the pump torque will increase in that interval from M_1' by ΔM_1. Therefore, depending on the external load, the torque converter automatically adjusts the output torque and the number of revolutions, keeping the power parameters within the limits that allow the engine to operate at the optimal mode.

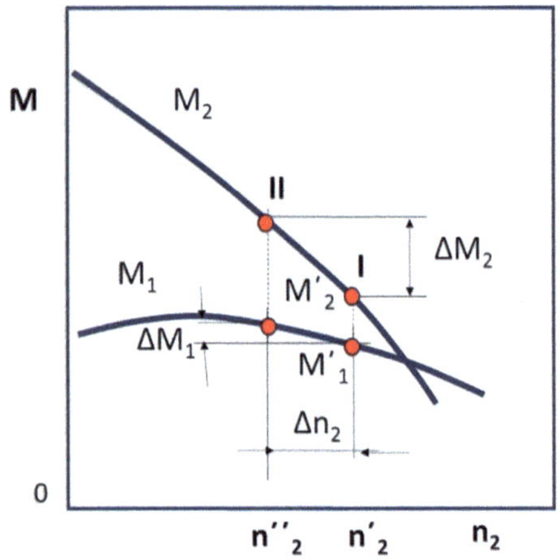

Fig. 4.15 Torque converter pump and turbine torque change [1]

Torque converter parameters

When analyzing and choosing a torque converter, the following parameters are defined:

- hydraulic transmission ratio:

$$i_{21} = \frac{n2}{n1}$$

n_2—number of revolutions of the turbine (n_t)
n_1—number of revolutions of the pump (n_p).

- torque increase factor:

$$k = \frac{M2}{M1}$$

M_2—torque on the turbine.
M_1—torque on the pump.

From the theory of hydromachines:

$$M_1 = \lambda_1 \rho n_1^2 D_c^5;\, M_2 = \lambda_2 \rho n_1^2 D_c^5;\, k = \lambda_2/\lambda_1$$

λ_1; λ_2—torque proportionality factors

- degree of converter efficiency:

$$\eta_{\mathrm{HDM}} = \frac{P2}{P1}$$

P_2—power on the turbine.
P_1—power on the pump.

The characteristics of the torque converter are shown in Fig. 4.16a. At the start, $n_2 \cong 0$, the torque increase factor is the largest $k = k_{\mathrm{maks}}$, the efficiency ratio η is at its lowest. At higher values of n_2, the torque increase factor tends to zero ($k < 1$), the turbine accelerates ($i = 1$). Complex torque converters are made in order to work in the field of high loads in the η_{HDM} converter mode, and in the field of low loads in the η_{HS}, clutch mode. The transition from one operating mode to another is automatic with a change in the load on the turbine shaft ($i = i_{\mathrm{HS}}$). This ensures a higher degree of efficiency of the converter. The middle field of action of the converter is significant, ranging from $\eta_{\min} = 0.75$ to $\eta_{\mathrm{maks}} = 0.85/0.9$ (without oil overheating). By testing the converter on the test bench, its characteristics are determined. When the number of revolutions of the turbine (n_2) is about 70% of the number of revolutions of the pump (n_1), the torque on the turbine is equal to the torque on the pump ($k = 1$), so the reactive torque on the stator equals zero, then the torque converter (HDM) switches to clutch operation (HS).

Automatic transmission fluid
ATF fluids (*Automatic Transmission Fluids*) are special mixtures of oils, lubricants and additives, developed for optimal operation of torque converters and lubrication of planetary gearbox parts. Figure 4.17 shows the change in viscosity and density of ATF with temperature. ATF fluids ensure reliable operation, performance and protection

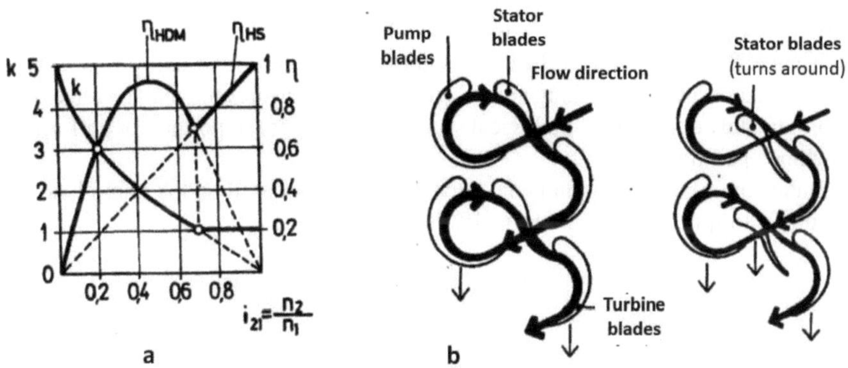

Fig. 4.16 Characteristics of the converter (**a**), operation at lower and higher revolutions (**b**)

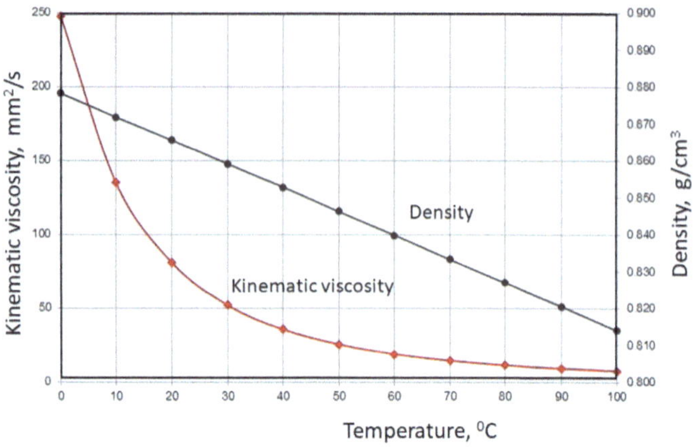

Fig. 4.17 Viscosity and density of ATF III fluid as a function of temperature [8]

of the automatic transmission. Their basic properties are: stability by maintaining fluid viscosity and density in a wide range of temperatures, resistance to wear and durability of gears and maintenance of interaction between frictional and smooth gear changes, and reduction of sediment and sludge.

4.3.4 Engine and Converter Matching

To balance the joint operation of the engine and the torque converter, it is necessary to determine their joint working area. The control characteristic of the diesel engine is shown in Fig. 4.18. Balancing with the operating resistances on the regulatory characteristic is provided by the all-mode regulator by automatically injecting a certain amount of fuel into the engine cylinders according to the current operating resistance. The real characteristic has significantly bent torque and power curves, which ensures a wider working area. Such regulation ensures the maintenance of an approximately constant technological speed, which is very important for performing machine operations.

Balancing of the motor and converter is carried out based on the pump load and the motor torque curve. The torque that loads the pump:

$$M_1 = \lambda \rho n_1^2 D_c^5 \text{ [Nm]}$$

λ—torque coefficient, N /kg m $(\text{min}^{-1})^2$
ρ—fluid density at operating temperature, kg/m^3
n_1—number of pump revolutions, $n_{\min} - n_{\text{maks}}$, min^{-1}
D_c—largest circulation diameter of the converter, m

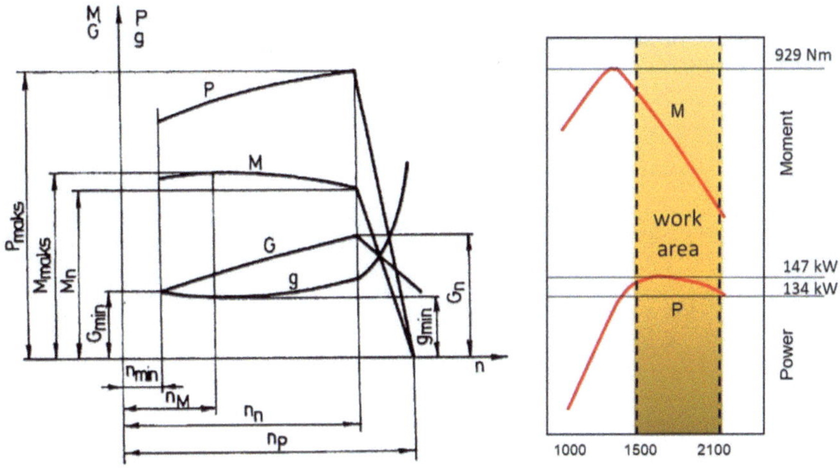

Fig. 4.18 Regulatory characteristics of diesel engines (adapted [6]) P_{maks}—nominal motor power, M_{maks}—nominal motor torque, $M_{\mathrm{n}} = M_{\mathrm{Pmaks}}$—motor torque, $n_{\mathrm{n}} = n_{\mathrm{Pmaks}}$—engine speed at maximum power, n_{M}—engine speed at maximum torque, n_{ph}—number of revolutions of the engine at idle speed of regulatory characteristics, n_{min}—minimum number of revolutions of the engine, G_{n}—nominal hourly fuel consumption, G_{min}—minimum hourly fuel consumption, g_{min}—minimum unit fuel consumption

M_2—turbine torque.
$M_2 = M_1 + M_s$
M_s—stator load moment

Torque on the drive wheels

$$M_{2\mathrm{pk}} = M_2 i_{\mathrm{tr}}$$

$$M_{2\mathrm{pk}} = M_1 k\, i_{\mathrm{tr}}$$

The load moment curves of the pump (M_1) are graphically applied to the regulatory characteristic of the motor, Fig. 4.19. The points of intersection of the load curves with the motor torque curve determine the area of joint operation of the motor and converter (the highest power in the interval of the number of revolutions at which the machine works continuously). The machine works most of the time on the regulatory characteristic, while the remaining part of the power remains in reserve. Transmission manufacturers usually provide converter load torque curves, and by plotting the speed and control characteristics of the motor, the user can quickly estimate the area of joint operation.

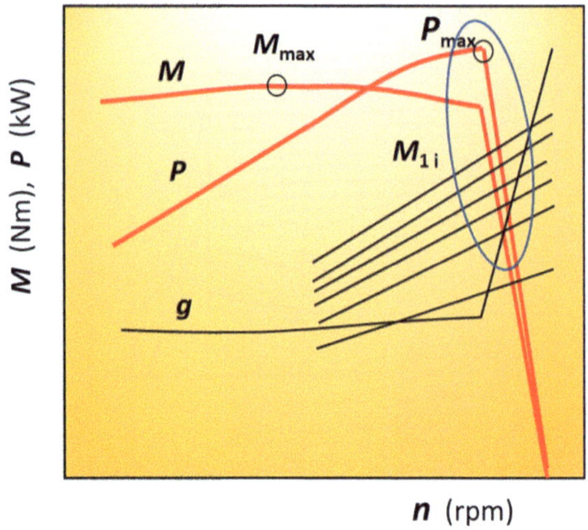

Fig. 4.19 Area of joint operation of diesel engine and torque converter

The area of optimal operation of the engine and torque converter can be evaluated according to the following criteria:

- the highest degree of utilization at maximum engine power (the gearbox that gives the most power on the output shaft in the interval of the number of revolutions at which the machine will work permanently works best);
- the highest torque (M_2, k, i_{21}) in the engine torque stability regime and the highest machine load (start, ascent, descent, soil digging, loading, …).

Planetary gearboxes

The planetary gearbox consists of a certain number of planetary rows of gears, with two friction lamellar clutches (A, B) and four brakes (C, D, E, F), Fig. 4.20. Two clutches are used to change the direction of movement (A—forward or B—backward), and four brakes are used to change gears. By mutual engagement of the planetary rows, with the activation of the brakes, the necessary transmission ratios of the gearbox are achieved. The transmission is done by friction between the lamellae, which creates oil pressure. Mechatronics combines the joint operation of the converter and the planetary gearbox. enables maintaining the working pressure of the oil in the converter (e.g. 1.5–3.0 bar), switching on multi-plate clutches for continuous gear changes, gear lubrication, and oil cooling.

The mechatronic system for controlling the gearbox and gear change is performed first by engaging and disengaging clutches and brakes. The next gear is engaged first, and then the previous gear is disengaged. As there is a very short overlapping time of both clutches, this does not have any significant consequences on the service life of the clutches. The construction requires an optimal covering time of two friction clutches, in addition to the criteria of continuity of power transmission and equality

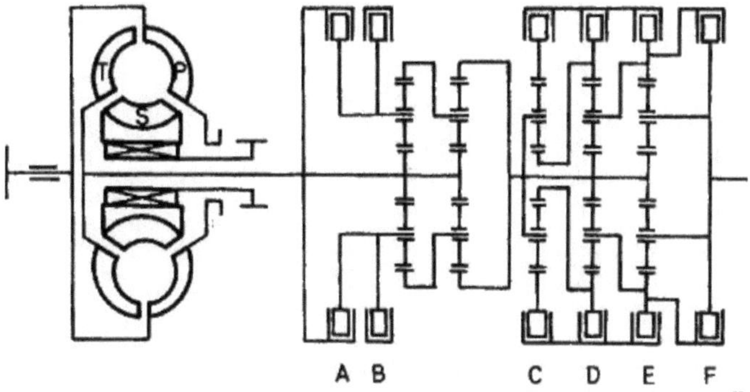

Fig. 4.20 Scheme of torque converter and planetary gearbox

of moments, which require a greater number of sensors and actuators. If the torque of the friction clutch is M_{f} and the torque of the clutch following the transmission stage is $M_{\mathrm{f+1}}$, then the criteria are met because:

$$M_{\mathrm{f}} i_{\mathrm{k}} + M_{\mathrm{kf+1}} i_{\mathrm{k+1}} > M_{\mathrm{otpora}}$$

Microprocessor management ensures:

- continuous gear and direction changes, automatically or manually,
- maintenance of working oil pressure in the torque converter,
- turning on the retarder / hydrodynamic brake,
- lubrication and cooling of gears,
- monitoring of all gearbox functions.

4.3.5 Hydrodynamic Brake - Retarder

Installation of retarders for long-term deceleration on heavy vehicles is legally prescribed. Such a brake must maintain constant speed of the loaded vehicle of 30 km/h, in the appropriate gear on a 7% descent and a distance of 6 km. The retarder converts part of the vehicle kinetic energy into heat.

Within the automatic transmission, a retarder (3) is built in between the converter and the planetary transmission, as shown in Fig. 4.21. The torque converter has an overrun clutch (1) with which it is automatically engaged and disengaged under certain conditions, the so-called *lock-up clutch* connected to the planetary gearbox automatically performs changes in higher gears (e.g. for the 3rd and the 4th). The one-way clutch (5) serves to establish a solid connection between the engine and the transmission during braking. The clutch automatically enables engine braking at the moment when the engine speed is lower than the output speed of the torque converter. The retarder (3) works as a one-way brake. It consists of a rotor and a stator part.

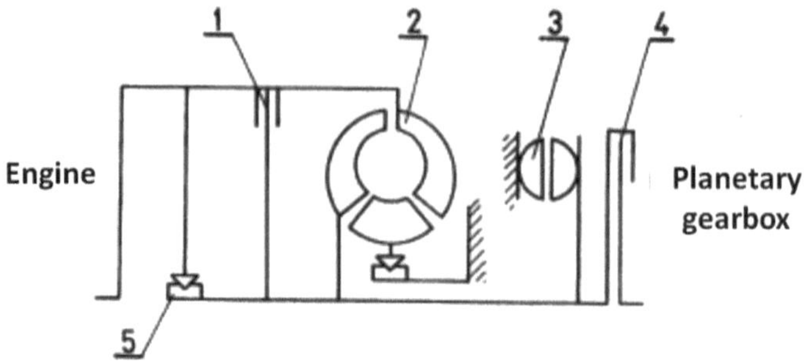

Fig. 4.21 Place of installation of long-term brake—retarder. 1—clutch for bridging (*lock-up*), 2—torque converter, 3—hydrodynamic brake (retarder), 4—gearbox clutch, 5—one-way clutch

The rotor wheel is connected to the drive shaft; the stator wheel is stationary because it is connected to the torque converter housing. When braking, one circular flow is used, which is controlled by the permanent brake valve.

The braking process is activated by a manual or foot valve, or automatically thus releasing the oil flow circulation and the intensity of the braking torque. At the same time, fuel injection into the engine is reduced. The rotor wheel is driven by the drive wheels, which accelerates the oil filling. As a result of the deflection of the oil flow in the stator blade space and the reactive moment, the rotor is braked, and thus the vehicle is braked. The continuous change in the braking power depends on the engaged gear, i.e. the number of revolutions of the rotor (for different values of the air pressure used for steering). The consequence of current friction is the heating of the oil, which must be cooled. The oil is cooled in a heat exchanger which is an integral part of the cooling system of the torque converter.

Required braking power.

Required braking power and vehicle speed reduction:

$$P_K = (R_a + R_i - R_k - R_z)v \text{ [W]}$$

R_α—downhill resistance, $R_\alpha = (\alpha/100)\, m\, g$ [N].
α (%)—downhill, m—total mass of the machine, g—acceleration of the earth gravity.
R_i—inertia resistance, R_i = (a/100)mg [N]; $a = x/g$, %; x—deceleration, m/s^2
R_k—rolling resistance, $R_k = f_k\, m\, g \cos\alpha$ [N], e.g. $f_k \approx 0.01$ - asphalt / 0.02 - other
R_z—air resistance, Rz $\approx R_z \approx 0{,}323\, v^2$ [N], v—vehicle speed

The braking power necessary to maintain constant movement speed:

$$P_K = (R_a - R_k - R_z)v \text{ [W]}$$

Since the one-way clutch is installed in the converter, braking by the vehicle engine is enabled, as long as the gearbox is engaged in a specific gear. Partial braking power from the engine and retarder depends on the number of revolutions. Therefore, engine braking power does not include vehicle braking by closing the engine exhaust pipe. Consequently, for braking of the working machine or vehicle, the total braking power of the engine (P_M) and retarder (P_{HK}) is available:

$$P_K = P_M + P_{HK} \text{ [W]}$$

a. engine braking

$$P_M = p_{mk} V_h \omega_m / \iota\pi \text{ [W]}$$

p_{mk}—mean pressure of engine braking resistance
V_h—working volume of engine cylinders, m^3
ω_m—angular engine velocity, rad/s
ι—engine tact (4)

b. retarder braking

The diagram of the braking torque $\boldsymbol{M_{HK}}$ and the number of revolutions of the retarder $\boldsymbol{n_{HK}}$ for different control pressures is shown in Fig. 4.22. At a pressure of 6 bar, the braking torque is the largest, which increases linearly in the area of lower revolutions (500–1000 min^{-1}), before stabilizing. At a pressure of 6 bar, at a speed of 1000 min^{-1}, 10% of the maximum braking torque is reached in 0.2 s, while 90% of the maximum braking torque is reached in 0.7 s. As the number of revolutions increases, the time shortens, so for a number of revolutions of 2000 min^{-1}, the times amount to about 0.2 and 0.45 s, respectively. The shutdown time is about 2.5 s, and it does not depend on the number of revolutions. In the off state, the retarder has a so-called residual braking torque that depends on the number of revolutions and represents the so-called ventilation losses. These power losses are not significant and at the maximum number of revolutions of 2500 min^{-1} they can be up to 4 kW.

Braking characteristic

The braking characteristic of the retarder provides an estimation of the vehicle deceleration capabilities on long downhill stretches. The braking characteristics of a 37-ton dump truck are shown in Fig. 4.23, in the diagram of the total braking power of the engine and retarder, depending on the vehicle speed. At lower gear levels, it is possible to drive on steeper downhill stretches at lower speeds. To maintain constant dumper speed, for example 30 km/h on a 7% downhill slope, the required braking power is 180 kW. Depending on the available braking power, this can be achieved in gear IV, with a considerable power reserve of 60 kW.

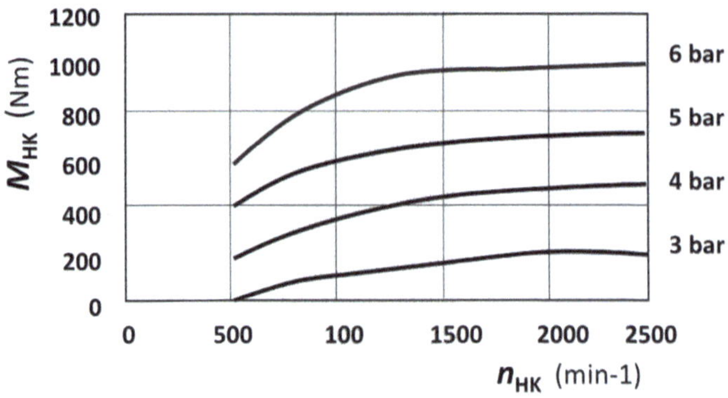

Fig. 4.22 Diagram of braking torque and speed of the retarder for different pressure values [1]

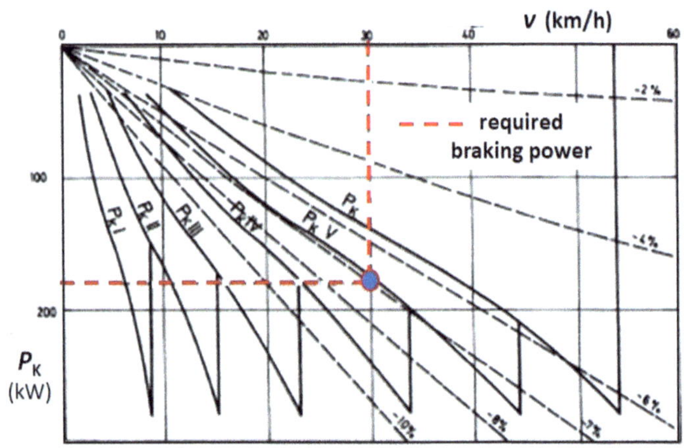

Fig. 4.23 Braking characteristics of a dumper [1]

4.4 Comparison of Hydrostatic and Hydrodynamic Power Transmission

The principal difference between hydrostatic (HS) and hydrodynamic power transmission (HD) is shown in the diagram, Fig. 4.24. The linear increase of the hydromotor torque with the number of revolutions ensures the realization of the highest torque of the drive engine, but not the highest power. The exponential increase of the torque converter of the torque ensures the realization of the maximum power of the drive engine, which means higher performance of the machine movement and efficiency.

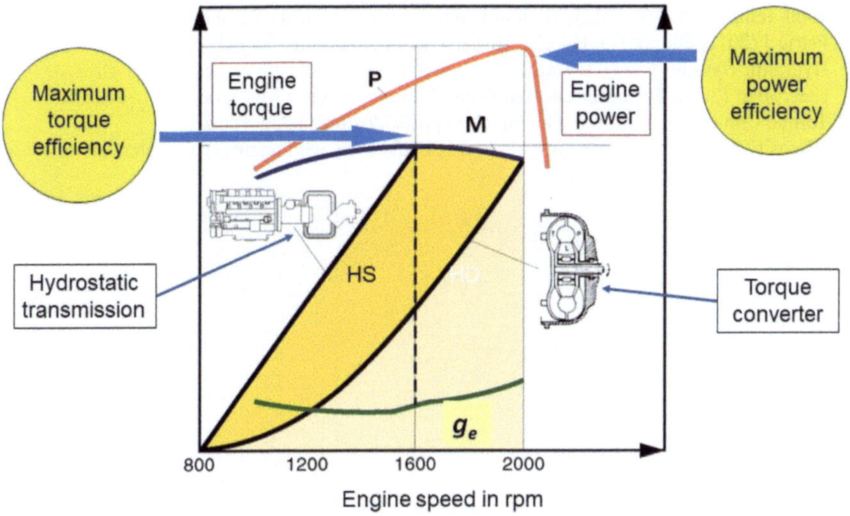

Fig. 4.24 Comparison of hydrostatic (HS) and hydrodynamic power transmission (HD)

Short conclusions about the traction capabilities of the machines:

- HS transmission ensures the use of maximum engine torque, continuous power change, automatic engine overload protection, and simple control (*joystick*).
- HD transmission ensures the use of maximum engine power, continuous power change, automatic gear change, engine overload protection, and easy management.
- The disadvantage of both transmissions is the degree of efficiency, which is relatively higher in HD transmissions. A number of manufacturers of construction machines prefer hydrostatic transmission for lighter machines, while other manufacturers prefer hydrodynamic power transmission for heavy machines.

The key component of the HD power transmission is the torque converter. Due to its favorable features, it is widely used in the transmission of construction machinery and vehicles. In addition to enabling automatic adjustment of power, torque, and number of revolutions, the torque converter allows for increased vehicle mobility on difficult terrain. Braking properties of the heavy machine on downhill slopes are provided by the hydrodynamic brake, i.e. the retarder in the HD power transmission.

References

1. Mikulic D (2024) Gradevinski strojevi. Teorija i konstrukcija. Veleuciliste Velika Gorica, Velika Gorica
2. Cipek M (2023) Auditorna vježba, stupanj korisnog djelovanja. Fakultet strojarstva i brodogradnje Zagreb. https://titan.fsb.hr/~mcipek/base/novosti/m1m2.pdf

3. Shell Tellus S2 VX 68 (2023) Hydraulic fluids. https://www.shell.co.uk
4. Petric J (2012) Hidraulika. Fakultet strojarstva i brodogradnje, Zagreb
5. Hydraulic Systems (2000) www.liebherr.com/en/int/products/components
6. Caterpillar (2020) Performance. Handbook, Edition 29. www.wheeler.com
7. Torque Converter (2023) www.learnpick.in/prime/documents/ppts/details
8. Viscosity of Automatic Transmission Fluid (2023) www.wiki.anton-paar.com/en/automatic-transmission-fluid-atf

Chapter 5
Bulldozers

Abbreviations

U [m^3/h]	Working efficiency
V_d [m^3]	Volume of material pushed by the machine
T_c [min]	Working cycle time
H [m]	Height of the dozer knife
L [m]	Length of the dozer knife
R_0 [m]	Radius of curvature of the blade
α [^{0}]	Angle of soil cutting
γ [^{0}]	Back angle
β [^{0}]	Blade angle
N [N]	Normal soil reaction
R_1 [N]	Soil cutting resistance
R_2 [N]	Blade pushing resistance
R_3 [N]	Friction resistance of knife blade
R_4 [N]	Resistance to movement of the soil in front of the blade
R_5 [N]	Resistance to moving the soil by blade
R [N]	Resultant resistance
R_u [N]	Total soil digging resistance
v_r [km/h]	Machine operation speed
P_e [W]	Machine power
G_{du} [N]	Weight of the dozer device
F_h [N]	Hydraulic cylinder force
M_a [Nm]	Adhesion moment of resistance to turning
R_z [N]	Tooth penetration resistance
R_x [N]	Soil ripping resistance
R_u [N]	Total resistance
P_r [W]	Ripping power

D. Mikulić and D. Mück, *Construction Machines*,
https://doi.org/10.1007/978-3-031-67845-5_5

5.1 Purpose and Division of Bulldozers

Bulldozers are machines for excavation and relocation of soil by pushing with dozer tools. The bulldozer consists of a base machine on tracks or wheels and a dozer blade with a knife, Fig. 5.1. The back of the machine can be connected to a ripper for breaking hard soil or a winch for pulling and self-extraction.

The working cycle of the bulldozer consists of digging the soil, pushing and unloading into a landfill, and returning to the starting point. In the working cycle, the blade is lowered to the required cutting depth. When the earth prism reaches the upper edge of the blade, the device is raised to the height of the terrain and continues to move the earth to the unloading site at the landfill. In the return cycle, the machine returns backwards to the excavation with the blade raised. The dozer method of earth moving is economical up to 100 m for tracked bulldozers, and up to 150 m for wheeled bulldozers. The speeds of the machine are usually: 2–4 km/h (cutting) and 6–12 km/h (pushing), optimal transport distances: 30–60 m (max. 100 m).

Work efficiency of a bulldozer

$$U = V_d k_r k_t / T_c \left[\mathrm{m}^3/\mathrm{h}\right]$$

V_d—volume of material pushed by the machine,
k_r—looseness factor,
k_t—time utilization factor,
T_c—working cycle time (excavation, pushing, return).

The working performance of a construction machine is their fundamental indicator. The heavier and stronger the machine, generally its working performance is higher. Taking as example two mid-range crawler excavators, as shown in Fig. 5.2, their working performance is visible (U):

- 350–500 m$^{3/}$h with a machine power of 160 kW, over a length of 25–50 m
- 250–300 m$^{3/}$h with a machine power of 100 kW, over a length of 25–50 m.

Fig. 5.1 Tracked bulldozer [1]

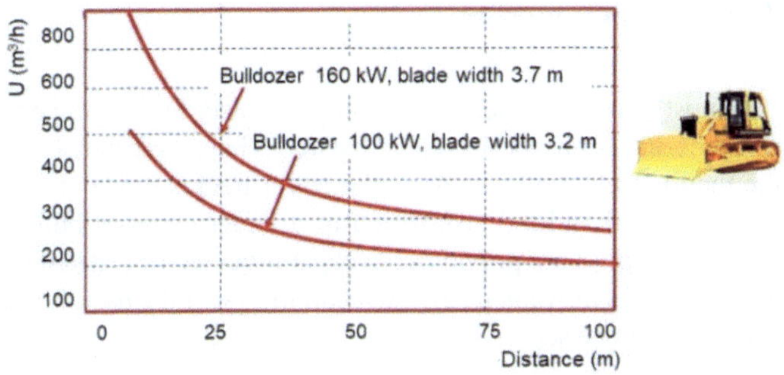

Fig. 5.2 Working efficiency of tracked bulldozers [2]

Depending on the position of the dozer blade, the machine can work in the bulldozer, *angldozer* and *tiltdozer* variants, as shown in Fig. 5.3. The bulldozer (basic variant) is used for surface stripping and cleaning of the soil, excavation of the soil, leveling and pushing-dosing into the landfill. The *angldozer* is used for throwing the soil aside, making cuts, filling trenches, clearing debris, and other tasks. The *tiltdozer* variant is used for profiling the roadbed, making channels and similar tasks.

Division of bulldozers

Construction organizations have bulldozers of different masses and engine powers, and therefore different working performance, in their composition. They can be classified into classes of tracked bulldozers: light, weight up to 10 t, power up to 75 kW; medium, 10–25 t and up to 150 kW; heavy, up to 40 t and 250 kW, very heavy, etc. In practice, the majority of bulldozers are medium class (40–50%), followed by heavy-class bulldozers (25–30%).

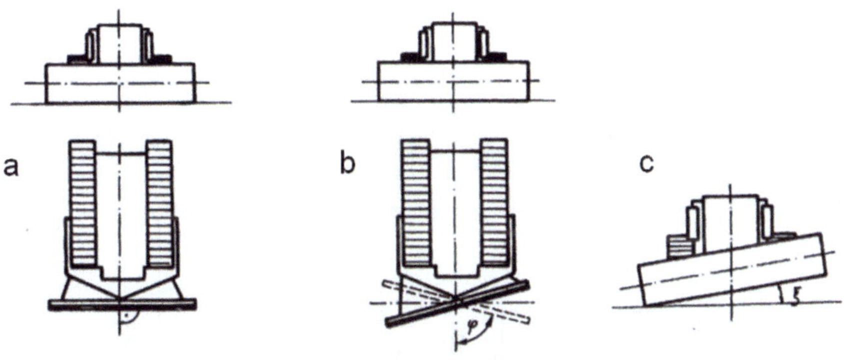

Fig. 5.3 Variants of the dozer depending on the position of the blade of the working tool. **a** *bulldozer*—blade normal to the longitudinal axis of the machine, **b** *angldozer*—blade placed at an angle, left or right, **c** *tiltdozer*—one end of the blade raised in relation to the other

5.2 Bulldozer Construction

Bulldozers are mostly tracked machines and less wheeled. Depending on the executive mover and the contact surface, it is clear that tracks have a greater ability to realize traction force than wheels, Fig. 5.4. Also, the larger track surface provides greater permeability of soft soil and the installation of a larger capacity bulldozer blade. The wheeled machine does not provide such capabilities as the tracked machine but provides higher speeds of movement. Tracked bulldozers are the basic representatives of the machine for earthworks.

Undercarriage and drive system

The main units of the machine are: diesel engine, hydrostatic or hydrodynamic transmission, frame-housing, tracked mechanism, and cab. As the engine, hydrostatic and hydrodynamic transmission have already been described in the previous text, the specifics of their design and the track system are presented here. The hydrostatic transmission (HS) of a medium-class bulldozer is shown in Fig. 5.5. Closed circuit hydrostatic transmission, the same for both left and right tracks. Lateral reducers increase the torque of the hydraulic motor. Any change in the load on the tracks is balanced with variable hydraulic pumps and hydraulic motors. Control via a stick (*joystick*) provides a quick and easy change of the flow, i.e. adjustment of the required power of the machine.

The hydrodynamic transmission (HD) of a medium-class bulldozer is shown in Fig. 5.6. The key part of the transmission is the powershift gearbox with a torque converter (2). Side couplings with brakes are used to control the machine (5). The two-stage reducers (6) provide an increase in the torque on the tracks.

The transmission of the heavy-class bulldozers is shown in Fig. 5.7. The transmission assemblies have an elevated disposition. The torque converter (1) is connected by a shaft to the remote automatic transmission (2). The distributor (3) ensures the distribution of torque to the side couplings (4) which, together with the brakes, serve

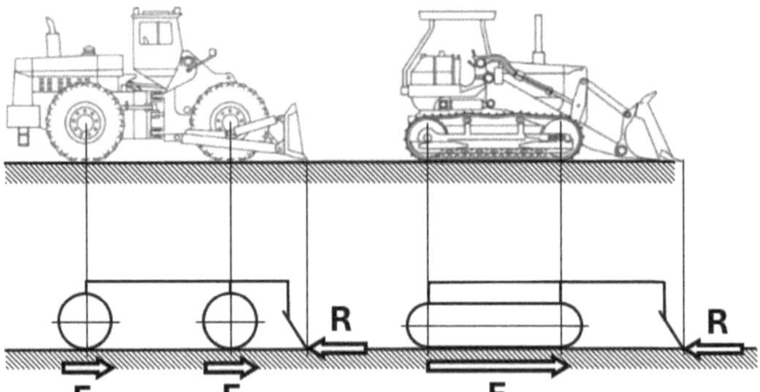

Fig. 5.4 Comparison of the contact surface of wheeled and tracked bulldozers (adapted [3])

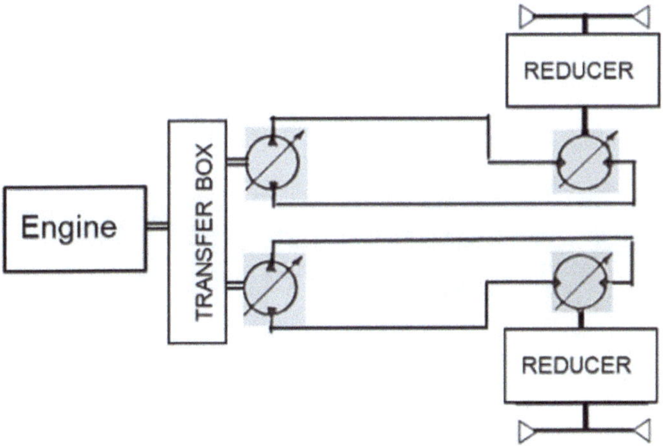

Fig. 5.5 Scheme of the HS transmission of a tracked bulldozer

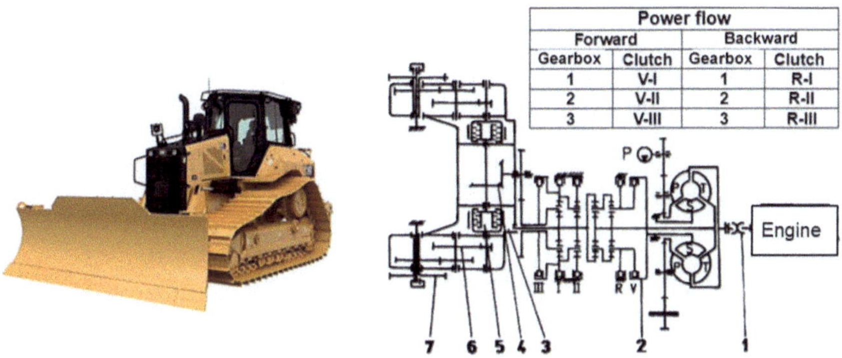

Fig. 5.6 Scheme of the HD transmission of a tracked bulldozer [2]. 1—cardan transmission, 2—torque converter and planetary gearbox, 3—housing, 4—power distributor/reducer, 5—side clutches, 6—two-stage reducer, 7—sprocket

to control the machine (4). The planetary reducer (6) provides an increase in torque to the delta tracks.

Figure 5.8 shows the traction characteristic of a mid-range bulldozer with hydrostatic (HS) and hydrodynamic (HD) transmission. The continuity of power is achieved in both transmissions. In addition to the traction forces shown in real speeds, when the ground is plowed (1) and the ground prism is pushed (2). The realization of traction force to overcome the digging resistance, from the first to the fourth soil category, amounts to 50–250 kN. HD transmission prevails in heavier bulldozers.

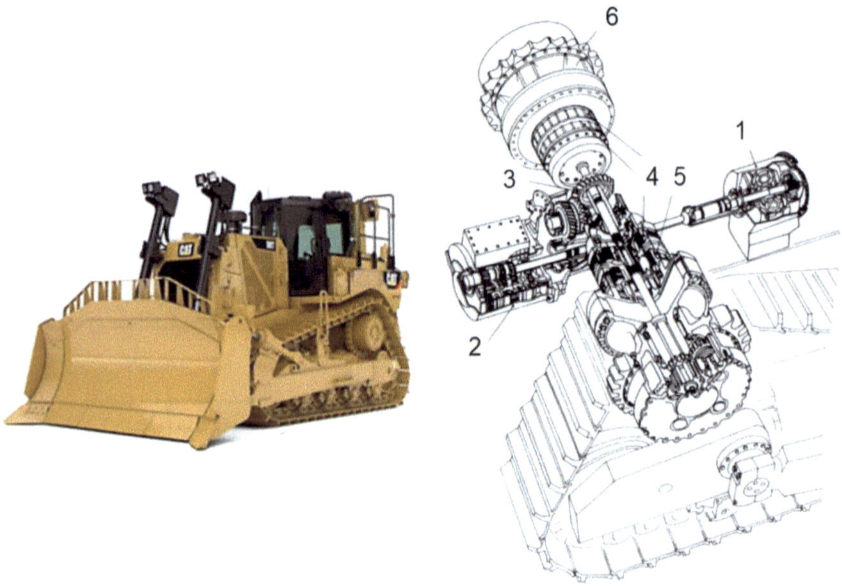

Fig. 5.7 Scheme of HD transmission of a heavy tracked bulldozer [2]. 1—converter, 2—planetary gearbox, 3—power distributor/reducer, 4—side clutches, 5—brakes, 6—two-stage reducer and sprocket

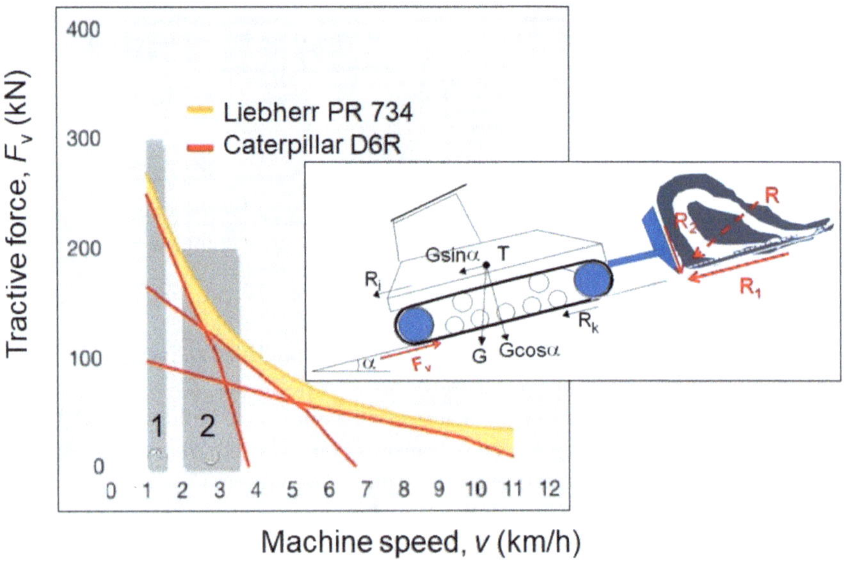

Fig. 5.8 Traction characteristics of mid-range bulldozers with HS and HD transmission (adapted [1, 4])

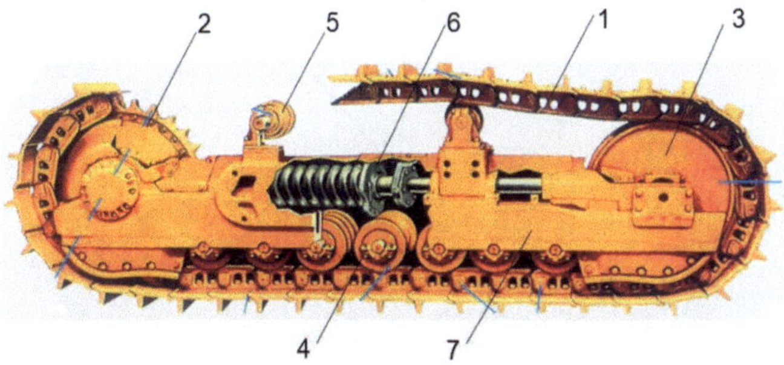

Fig. 5.9 Standard bulldozer tracked mechanism. 1—track chain, 2—sprocket, 3—guided wheel (idler), 4—carrying rollers, 5—track guide rollers, 6—spring tensioning mechanism, 7—mechanism frame

Tracked mechanism

Two types of the running part of the machine can be recognized: the *standard tracked mechanism* and the *delta tracked mechanism.* ***A standard tracked mechanism*** is shown in Fig. 5.9. The track is a track chain that encompasses the entire structure of the mechanism. The tracked mechanism consists of a drive wheel—a sprocket, a tracked chain, a guided wheel, idler, carrying rollers, guide rollers and a frame with a spring mechanism. The spring mechanism makes it possible to tighten the tracked chain and mitigate the impact of the driven wheel on an obstacle. Carrying rollers are mediators of the load between the frame and the track chain and ensure the guidance of the track. The connection between the carrying rollers and the frame can be *rigid* or *Bogie* variant. Their design is extremely important, because in addition to the load, they prevent the tracks from slipping. The carrying rollers require special maintenance attention, as does the drive sprocket, because the drive of the sprocket and track is exposed to high loads, muddy, and winter conditions.

The delta tracked mechanism defines the shape of the tracked chain (*Greek alphabet,* Δ), Fig. 5.10. Due to the raised sprocket, the structure of the mechanism forms a delta shape. In practice, such a sprocket position eliminates exposure to heavy loads in difficult mud and ice conditions, thereby avoiding jamming, which provides greater ground permeability on soft terrain. In addition, due to the large dimensions of the installed transmission assemblies, the sprocket must be raised to a greater height. Also, the elevated position of the sprocket removes the harmful effect of the normal reaction of the base (N) on the sprocket. *Bogie* suspension of carrying rollers (4) with the frame of the mechanism (7), provides good contact of the entire surface of the track with the ground and mitigation of shocks on uneven ground, which ultimately ensures an even surface pressure for the realization of the traction force. The disadvantage can be seen in the reduction of the included angle of the sprocket and the track chain.

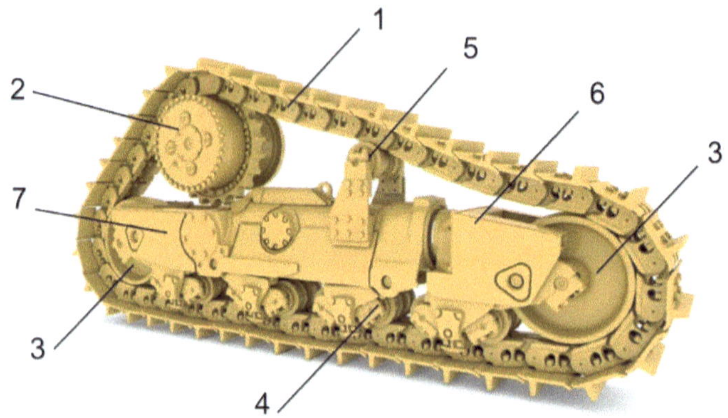

Fig. 5.10 Delta tracked bulldozer mechanism. 1—track chain, 2—sprocket, 3—guided wheel (idler), 4—carrying rollers *Bogie*, 5—track guide rollers, 6—track tensioning mechanism, 7—track mechanism frame

Track chain

The track chain consists of chain links and shoes, Fig. 5.11. The links of the chain are connected to each other by pins and bushings, through which the drive force from the sprocket is transmitted. The shoes are tightly bolted to the links, and have protruding ribs to realize the traction force. To reduce dry friction of the track and the rapid wear of the axle-sleeves, some variants lubricate the axle-sleeves. This reduces resistances in the track chain and creaking, and extends its working life.

The principle of movement of a tracked machine is created in the following way. Drive wheel-sprocket—pressing against the stationary track, forces the body of the

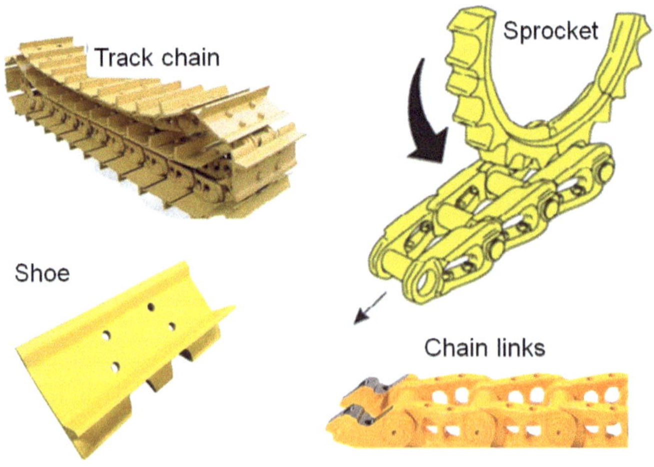

Fig. 5.11 Track chain design—track forms

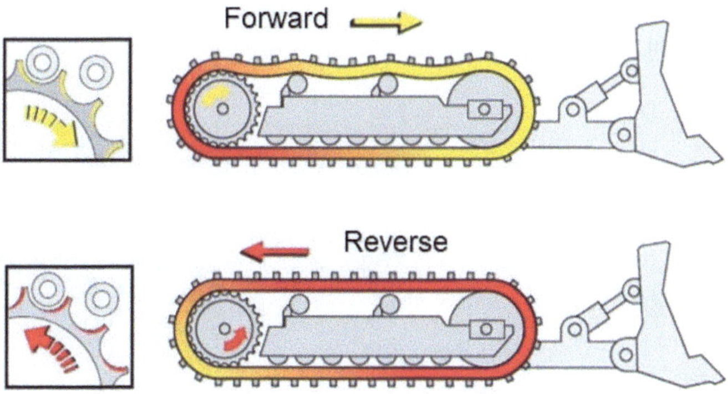

Fig. 5.12 Changing the load on the track chain [5]

machine to roll along the inner part of the track, like an endless belt. Thus, the winding of the track and the movement of the tracked machine occur. The longer and wider track provides less pressure on the ground. The lower specific pressure of the track on the ground provides greater ground trafficability. Therefore, it is possible to change the length and width of the track on the machines. The pressure on the ground with the standard track is 0.6–0.9 bar, and with the delta track 0.4–0.7 bar.

The track chain is loaded to stretch, Fig. 5.12. The highest load on the track and sprocket occurs when moving forward. In the opposite direction, the track around the guided wheel (idler) is the most loaded. The comprehensive angle of the track with the sprocket (180°, 6–7 teeth) and the guide wheel ensure less wear of the sprocket. On the other hand, with the delta track, the smaller angle of coverage on the drive wheel (90°, 3–4 teeth) causes greater wear of the track.

The trafficability of the soil also depends on the height of the undercarriage, i.e. the height between the ground and the floor of the machine, Fig. 5.13. In order for the machine to safely overcome various protruding obstacles during operation, the height of the undercarriage is 400–500 mm. The mobility of the machine and its stability may be limited in conditions of movement of ledges and trenches. In such conditions, the crawler mechanism is very loaded, which should be avoided.

Bulldozer cab

The cabin provides the operator of the machine (machinist) with an overview of the use of the machine, Fig. 5.14. From the air-conditioned and comfortable position of the operator on the ergonomic seat, easy control of the dozer blade and rake is provided using a stick, i.e. a *joystick*, with electronic monitoring of the parameters. The control console offers an intuitive interface that allows the operator to make changes to the machine settings. The cabin protects the operator of the machine from noise, vibrations, falling objects, and tipping, in accordance with the SAE safety standards for ROPS (*Roll Over Protection Structure,* ISO 3471) and FOPS (*Falling*

Fig. 5.13 Bulldozer undercarriage height

Fig. 5.14 Bulldozer cab

Object Protective Structure, ISO 3449). For preventive and corrective maintenance of the engine and transmission, the cab can be raised at one end, which provides access to maintenance parts. This fulfills the requirement of easier maintenance in off-road conditions.

5.3 Bulldozer Working Device

The working device of the bulldozer is a dozer tool, which consists of a dozer mechanism with a blade and hydraulic installations. The control of the dozer blade is performed by activating hydraulic cylinders, for lifting, turning, and tilting, using a stick (*joystick*) or automatically using GPS system. The modes of raising and lowering the dozer blade with hydraulic cylinders are shown in Fig. 5.15. The basic types of dozer mechanism with a blade are shown in Fig. 5.16.

Dozer mechanism

A-type mechanism, straight blade (S): variants, *bulldozer, angldozer, tiltdozer*. The design of the mechanism consists of: a *C-frame* that is connected to the frame of

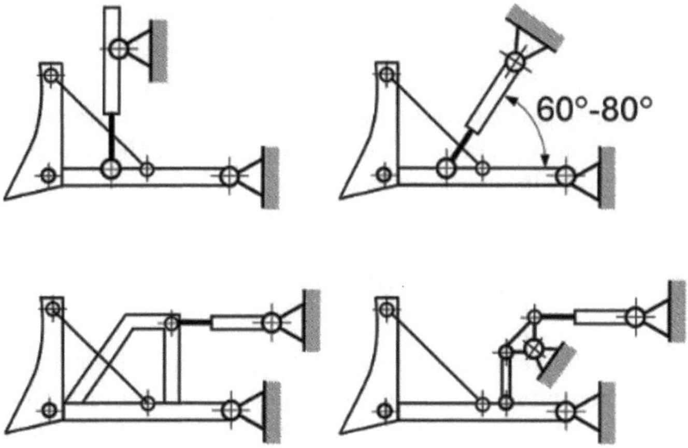

Fig. 5.15 Modes of raising the dozer mechanism with hydraulic cylinders

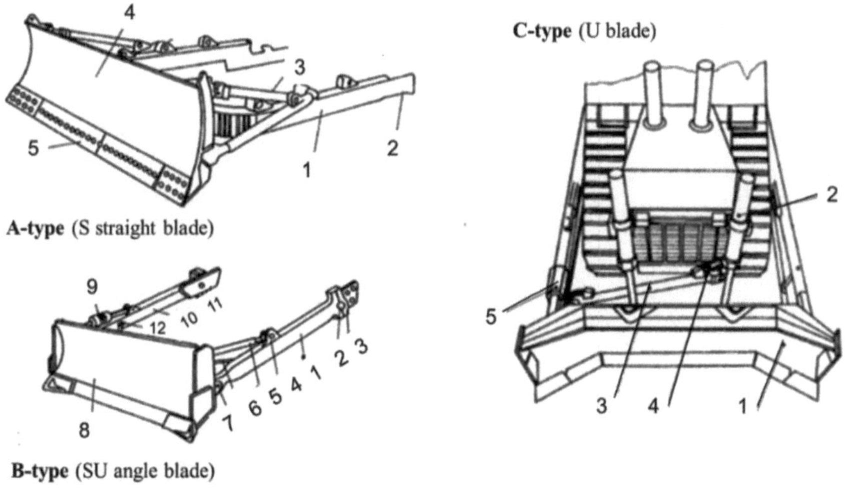

Fig. 5.16 Mechanisms of the bulldozer working device. A-type: 1,2—*C-frame*, 3—struts, upper and lower, 4,5—dozer blade; B-type: 1,10—carriers, 2,3—bearing, 4,10—link, 5,6—struts, 8—dozer blade, 9—tilt cylinder, 11—connecting plate; C-type: 1—blade, 2—cylinders for raising and lowering the blade, 3—diagonal clamp, 4—hydraulic pipes, 5—cylinder for changing the cutting angle

the tracked mechanism, connected by the *C-frame* sphere to the dozer blade, and to the side struts. Raising and lowering is usually done with two cylinders. The blade cutting and tilting angle is changed mechanically or hydraulically on the struts.

B-type mechanism, angle blade (SU): variants, *bulldozer, tiltdozer*. The design of the mechanism consists of a simpler frame with supports. It is raised and lowered by hydraulic cylinders that are connected to the blade with eyelets. The cutting and tilting angle of the blade is changed by the side cylinder.

C-type mechanism, U-blade: variant *bulldozer*. The design of the mechanism consists of a simple frame with struts and diagonal bracing. It is raised and lowered by cylinders that are connected to the blade with eyelets. The cutting angle of the blade is changed by the side cylinders.

5.3.1 Hydraulic Installation of the Bulldozer

The hydraulic installation is an open circuit. The scheme of hydraulic installation simpler construction are shown in Fig. 5.17a, b. Installation shows the main components: constant flow hydraulic pump (*1*), piston distributors for working with the blade and working with the ripper (*4*), blade cylinders (*7*), ripper cylinder (*8*), safety valve (*12*) and filters (*2,9*) and oil tank (*3*).

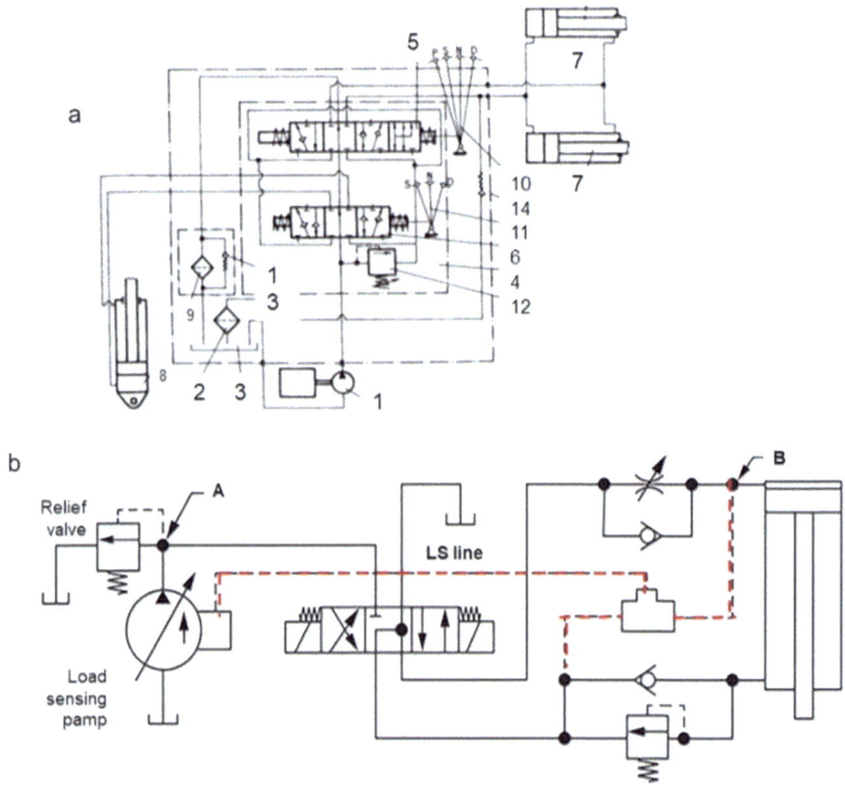

Fig. 5.17 Scheme of the hydraulic installation of a bulldozer

The positions of the distributor are marked: D, N, S, and P (N—*neutral position*, D—*raising the blade or ripper*, S—*lowering the blade or ripper*, P—*so-called floating position*). The safety valve protects the installation from overload. The floating position of the blade is used in cases of planning the soil surface with a dozer blade. If the dozer is equipped with a *B-type* mechanism, the hydraulic system usually contains two lifting cylinders, one ripper cylinder, and a tilting cylinder. In this case, a piston distributor for tilting is installed.

Hydraulic installation uses a *load sensing pump*. This variable flow pump continuously monitors the pressure due to the load on the working tool. The oil flow is converted into useful work, and not into safety valve losses. When there is no load, when the piston distributors are in neutral position, then the pump is unloaded and it produces enough flow to maintain a lower standby pressure. If there is no standby pressure, the system will run sluggishly and slowly. *LS line*: the sensitivity regulation compares the pressure of the load at point B with the water pressure of the pump at point A, which regulates the operation of the pump in order to maintain a constant pressure difference, for example from 5 to 10 bar.

Hydraulic oil is prescribed by the machine manufacturer, according to the conditions of use. The viscosity marks of hydraulic oil are prescribed by the ISO 3448 standard. Commonly used viscosity oils are winter VG 22, summer VG 46 and VG 68. The basic characteristics of hydraulic oil for construction machines are:

ISO marks	Viscosity at 40 °C (cSt)		
	Medium	The lowest	The highest
ISO VG 10	10	9	11
ISO VG 15	15	13.5	16.5
ISO VG 22	22	19.8	24.2
ISO VG 32	32	28.8	35.2
ISO VG 46	46	41.4	50.6
ISO VG 68	68	61.2	74.8

Joint working area of the machine and the dozer blade

In bulldozer operation, two movements are required: of the basic machine and of the dozer blade. It is necessary to harmonize the operation of the machine and the working device and determine the area of optimal operation, Fig. 5.18. This is the result of the movement of the machine and the digging resistance of the blade, which is recognized on the torque and power curve of the engine. The depth of digging the soil, the cutting angle and the resistance of pushing the earth prism in front of the blade are the parameters of digging the soil that can be controlled by a microprocessor. Automatic gear change allows the operator of the machine to preselect the range of revolutions, so that changes in speed and driving direction are easy and efficient. There are several variants of *joysticks*, from simple controls with two axes (forward/backward, left/right) to those that may include a third axis (clockwise/counterclockwise rotation).

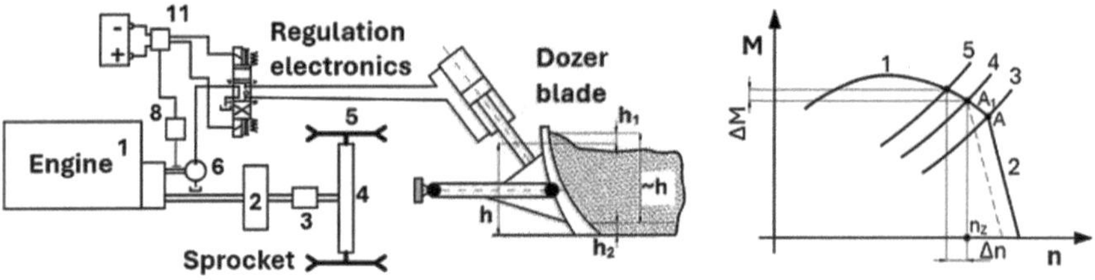

Fig. 5.18 Area of operation of the bulldozer driving system with the excavation depth regulation (h) [6]

5.4 Dozer Blade Parameters

When the dozer blade is in the maximum digging position, the maximum load on the machine occurs, Fig. 5.19. The vertical machine load ($G + R_y$) causes the ground reaction force (***N***). In order to eliminate the harmful effect of the sprocket load, the position of the normal ground reaction (***x***) must not be close to the sprocket. The position of the ground reaction should be at least one third of the track contact (L_0), which ensures the safety of operation (a). The position of the normal reaction (b) can be checked from the condition of the largest soil digging angle ($\varkappa \approx 20°$) and the resulting surface pressure force. In order to transmit the traction force, it would be ideal if the surface pressure under the tracks on the ground was even. However, when loading the dozer blade or the ripper, a trapezoidal load schedule (b) with a normal reaction (N) occurs.

The position of normal soil reaction:

$$x = \left(G\,d + R_y d_1 - h_R\right)/N, \quad N = G + R_y$$

G—adhesion weight of the machine.
R_x, R_y—components of digging resistance ($R_x = R_1$).

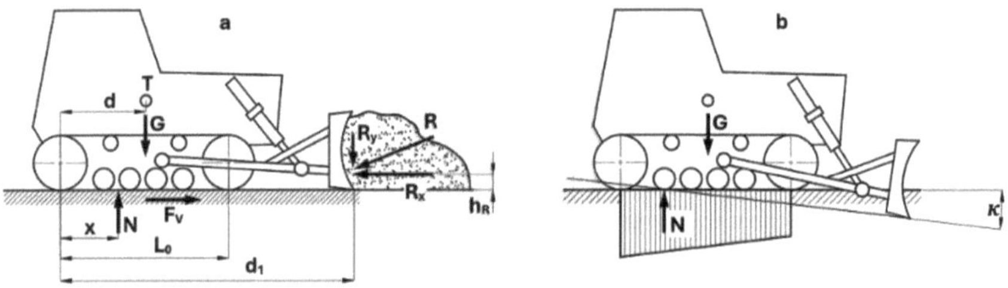

Fig. 5.19 Position of normal soil reaction (N). L_0-track contact, d_1-blade distance, $\varkappa$-blade digging angle

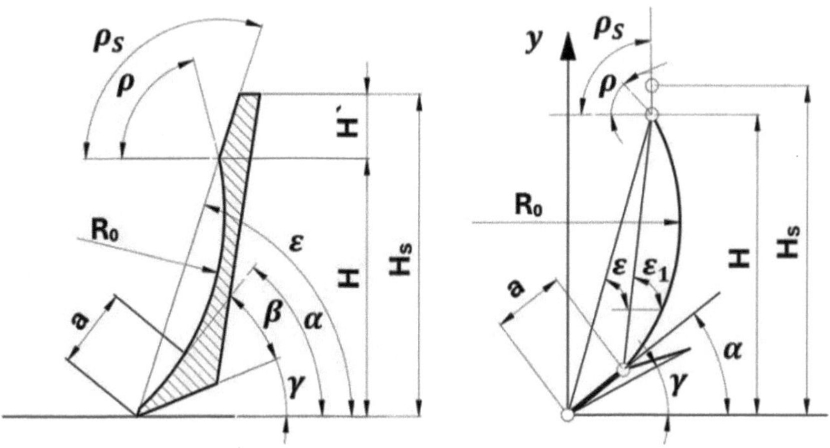

Fig. 5.20 Dozer knife profile parameters [4]

The general power parameters of the dozer blade are:

- ratio of nominal power and volume of prism, kW/m^3;
- ratio of nominal power and blade width, kW/m;
- ratio of traction force and blade width, kN/m.

The geometric parameters of the dozer blade profile are shown in Fig. 5.20. The digging resistance and the performance of the machine depend on the profile of the knife, its blade, and the condition of the soil. The blade of the knife is made of alloy steel, and the metal structure is made of sheet steel (8–20 mm).

Dozer blade profile parameters

R_0—radius of curvature of the knife, which forms an earthen prism in front of the blade.

$$R_0 = 0.75H\,[\mathrm{mm}]$$

H—height of the dozer knife $H = 250(P_n)^{1/3}$ [mm], P_n—rated engine power, $H = 800$–1500 mm;
L—length of the dozer knife, wider than the width of the machine, at least 100 mm on both sides. a—straight part of the knife blade, ensures the initial shaping of the cutting layer ($a \approx 200$ mm), the knife blade is performed along the blade in several parts, for maintenance or replacement.
α—angle of soil cutting ($55° \pm 10°$)
γ—back angle (30°–35°), β—blade angle
ε—angle of inclination of the blade profile (75°–80°)
ρ—angle of soil rejection (75°–80°) ensures the overturn of the soil in order to move it in front of the blade. Too large an angle pushes the soil over the blade,

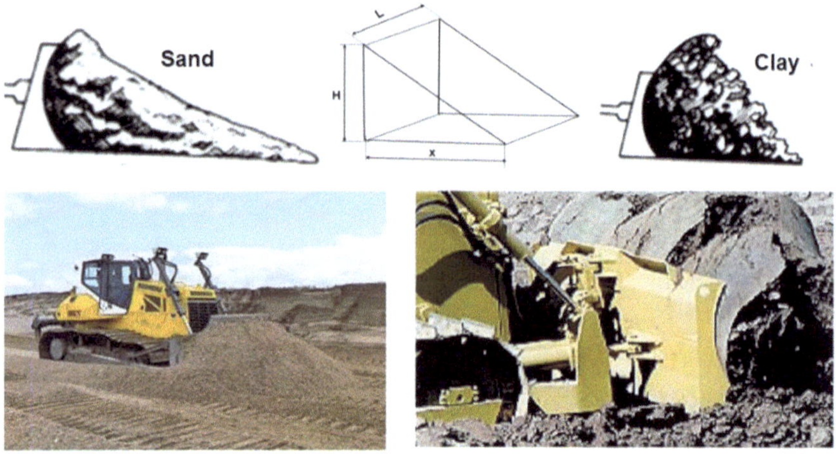

Fig. 5.21 The formation of an earthen prism when pushing sand and clay

and too small an angle—sticks the soil to the blade. A guard is added to the top of the blade, the height of which depends on the class of the dozer.

The volume of the earthen prism in front of the blade, approximately:

$$V_{zp} = HLx/2 \approx 0,5LH^2k_p[m^3]$$

The length of the blade depends on the application of the dozer, $L = 2–5$ m
k_p—blade loading factor (bound soil 1.4; loose soil 0.8)

Depending on the type of soil, the shape of the earthen prism differs, especially the length (x). Figure 5.21 shows the formation of the prism when pushing sand and clay.

Dozer blades, Fig. 5.16, are selected depending on the application of the dozer. The *U-blade* is obtained by adding side wings to the *S-straight blade*, the wings are set at an angle of 15°–30°, with the aim of directing the excavated soil towards the center of the blade. This reduces loose material losses. The performance of the machine can be increased up to 50% compared to a straight blade. *U-blade* is also called universal blade. The *SU-angle blade* is suitable for digging in the hard soil category, sharp scraping, dosing, and leveling. Modern bulldozers use a change in cutting angle. For example, CAT blade (VPAT).

5.5 Bulldozer Soil Digging Resistances

(a) Digging resistances on the dozer blade

Figure 5.22 shows the soil digging resistance on the dozer blade:

R_1—soil cutting resistance
R_2—blade pushing resistance
R_3—friction resistance of knife blade
R_4—resistance to movement of the soil in front of the blade
R_5—resistance to moving the soil by blade
R—resultant resistance.

Soil cutting resistance

$$R_1 = k_1 L h \text{ [N]}$$

k_1—specific soil resistance of a certain category,
L—length of the knife blade, h—depth of the cutting layer.

An example of calculating the nominal soil cutting resistance of a medium-class dozer.

I. category, light working conditions: $R_1 = 100$ kN gear III
II. category, medium working conditions: $R_1 = 150$ kN gear II
III. category, difficult working conditions: $R_1 = 200$ kN gear I
IV. category, very hard working conditions: $R_1 = 250$ kN gear I

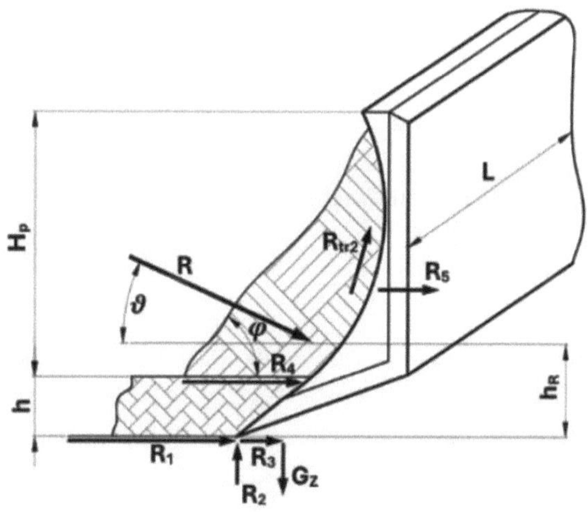

Fig. 5.22 Digging resistances on dozer blade

Fig. 5.23 Change in soil cutting resistance

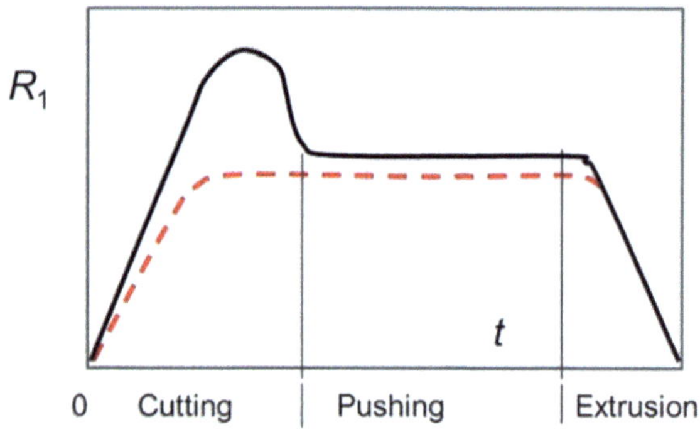

The change in cutting resistance during soil digging is shown in Fig. 5.23. At the beginning of the excavation and formation of the prism, the cutting resistance reaches the highest values, then falls to approximately constant values.

Blade thrust resistance

$$R_2 = \sigma A \quad [\text{N}]$$

σ—limit bearing capacity of soil for crushing, medium conditions 50–70 N/cm^2
$A = x\,L$, the surface of the blade perpendicular to the direction of force R_2
L—blade length, cm; $x = 1$–1.5 cm, bluntness of the tip of the knife blade.

Friction resistance of the knife blade

$R_3 = R_2\mu_1$—friction factor between the blade and the ground

Resistance to movement of the soil in front of the knife

$$R_4 = G_{zp}\mu_2 \quad [\text{N}]$$

μ_2—friction factor of loose soil on the ground, bound soil $\mu_2 \approx 0.6$; loose soil $\mu_2 \approx 0.9$

Resistance to moving the soil by blade

$$R_5 = R_{tr2} \cos\alpha \quad [\text{N}]$$

$$R_5 = \mu_1 G_{zp} \cos^2\alpha \quad [\text{N}]$$

R_{tr2}—force of soil friction in front of the blade
$R_{tr2} = \mu_1 G_{zp} \cos\alpha$; G_{zp}—weight of part of the earthen prism.

Resultant resistance

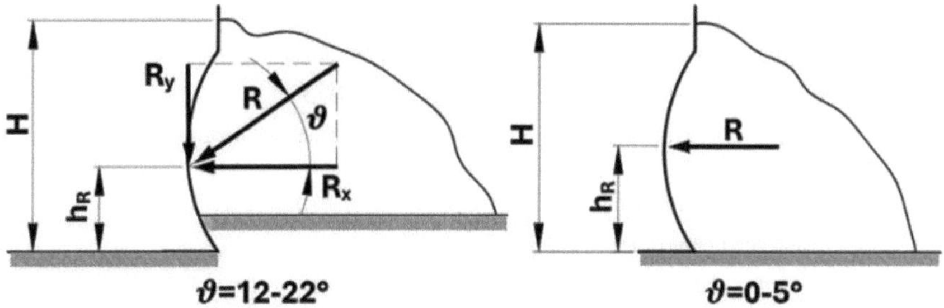

Fig. 5.24 The position of the resultant of the digging resistance

$$R = \sqrt{R_x^2 + R_y^2} \ \text{[N]}$$

The position of the digging resistance force, during excavation and soil dosing (h_R, ϑ), is shown in Figure 5.24. The angle of the resultant resistance (ϑ) depends also on the type of soil.

$$R_x = R_1 + R_4 + R_5; \quad R_y = R_x \text{tg}\vartheta$$

Ground digging resistances in the angle dozer variant.

Since the excavated soil moves laterally, and due to the smaller width of the excavation, the components of resistance to the blade in the direction of movement are:

- soil cutting resistance, $R_1' = R_1 \sin\varphi$ [N]
- movement resistance, $R_4' = R_4 \sin\varphi$ [N]
- resistance due to the upward movement of the soil, $R_5' = R_5 \sin\varphi$ [N]

Due to the movement of the soil on the blade and along the blade, an additional resistance R_6 follows:

$$R_6 = R_{tr3} \cos\varphi \ \text{[N]}$$

$$R_6 = \mu_1 \mu_2 G_{zp} \cos\varphi \ \text{[N]}$$

(b) Bulldozer motion resistance

Tracked bulldozer	Wheeled bulldozer
Movement resistance (rolling and climbing)	Movement resistance (rolling and climbing)
$R_f = R_k \pm R_\alpha$	$R_f = R_k \pm R_\alpha$
$R_f = G\,(f \cos\alpha \pm \sin\alpha)$	$R_f = G\,(f_k \cos\alpha \pm \sin\alpha)$

(continued)

(continued)

Tracked bulldozer	Wheeled bulldozer
$\alpha_{maks} \approx 70\%\ (35°)$	$\alpha_{maks} \approx 70\%\ (35°)$
f—rolling resistance factor, $f = f_{t1} + f_{gm}$	f_k—rolling resistance factor of the wheel
f_{t1}—soil deformation resistance factor	
f_{gm}—resistance factor of the crawler mechanism	
• dry dirt road: $f \approx 0.07$	• dry dirt road: $f_k \approx 0.06$
• wet dirt road, sand: $f \approx 0.15$	• wet dirt road: $f_k \approx 0.09$

(c) Total soil digging resistance and required power

The total resistance of the bulldozer R_u, is equal to the sum of the digging resistance on the blade R_x and the movement resistance (R_f, R_α) and inertia resistance R_i:

$$R_u = R_x + R_f + R_i\ [\mathrm{N}]$$
$$R_u = R_1 + R_k \pm R_\alpha + R_i\ [\mathrm{N}]$$

Power required to overcome resistance:

$$P_e = R_u v_r\ [\mathrm{W}],$$

v_r—machine operation speed

5.6 Dozer Device Design

The calculation of the dozer mechanism is based on the strength, working life (fatigue) and wear. The calculation schemes of the dozer device are: determining of the minimum weight of the working device, the beginning of excavation, excavation and pushing, raising of the blade, tipping of the dozer, and the load on the edge of the blade.

a—Determining of the weight of the working device

From the condition of balance of moments of forces at point C of the mechanism joint and the basic machine, the approximate weight of the working device follows, Fig. 5.25.

$$\sum Mc = 0;\ R_2 l_1 - R_3 m - G_{du\ min} l_o = 0;\ \ R_1 = 0$$

$$G_{du} \geq G_{du\,min} = (R_2 l_1 - R_3 m)/l_o \quad [N]$$

$$G_{du} = (180 - 260)P_e[N]$$

P_e [kW]—machine power.

b—Excavation and pushing, Fig. 5.26.

Soil digging resistances

$$R_x = R_1 + R_4 + R_5 \text{ [N]}$$

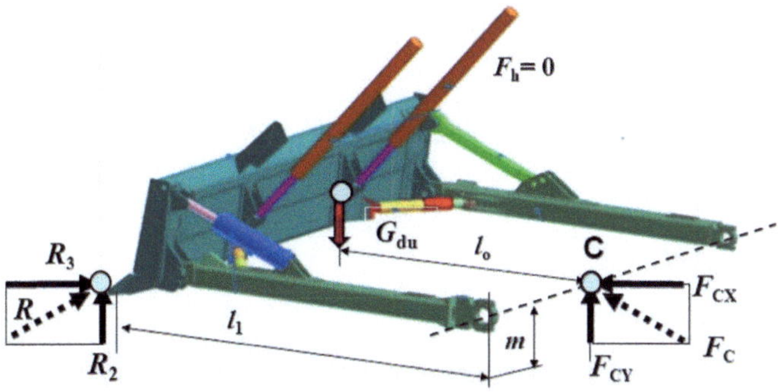

Fig. 5.25 Bulldozer loading scheme at the beginning of excavation. G_{du}—weight of the dozer device, F_{cx}—horizontal force component in the C joint. R_1—soil cutting resistance, F_{cy}—vertical force component in C joint. R_2—blade pushing resistance, F_h—hydraulic cylinder force. R_3—friction resistance of the knife blade

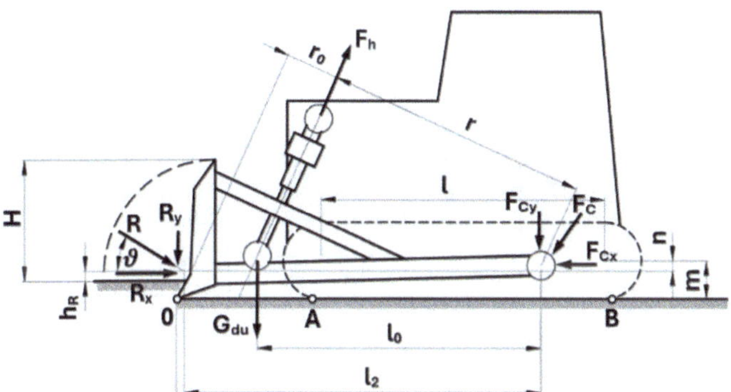

Fig. 5.26 Bulldozer load diagram at the end of pushing. R_x—horizontal component of total digging resistance, R_y—vertical component, F_n—force of hydraulic cylinders

$R_{xp} = k_d R_x, \quad k_d = 1.35 - 1.5$, dynamic inertia factor of hitting an obstacle

$$R_{yp} = R_{xp} \mathrm{tg}\vartheta$$

$$\sum M_c = 0; \ 2F_h r - G_{du} l_o - R_{xp} n - R_{yp} l_2 = 0$$

Hydraulic cylinder force

$$F_h = \left(G_{du} l_o + R_{xp} n + R_{yp} l_2\right)/2r \ [\mathrm{N}]$$

Loading of the C joint, components of forces

$$F_{cx} = R_{xp} + 2F_h \cos\varphi \ [\mathrm{N}]$$

$$F_{cy} = F_{cx} n + 2F_h r_o - G_{du}(l_2 - l_o)/l_2 \ [\mathrm{N}]$$

Reaction in the C joint: $F^2\mathrm{c} = F^2\mathrm{cx} + F^2\mathrm{cy}$.

c—Raising the blade, Fig. 5.27.

Soil shear resistance

The shear force between the soil raised by the blade and the ground depends on the sliding surface:

$$R_S = 0.65 G'_{zp} \mathrm{tg}^2\varphi \ [\mathrm{N}]$$

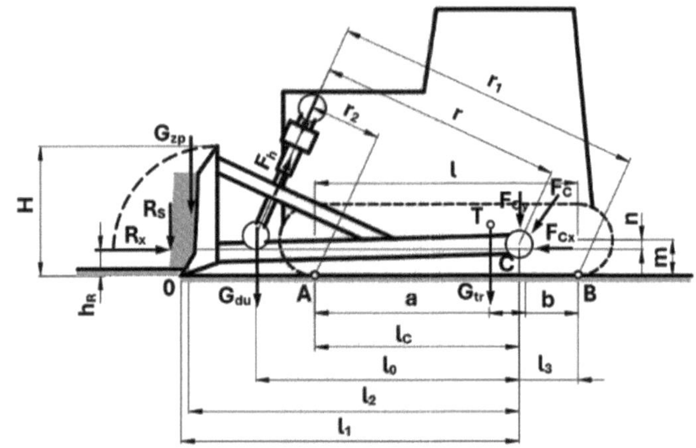

Fig. 5.27 Bulldozer loading scheme when extracting the blade

Weight of the prism that is lifted together with the blade: $G'_{zp} \approx 0.35\ G_{zp}$.
φ—internal friction/shear angle, 20°–45°, $tg\varphi \approx \mu$

Components of digging resistance

$$R_x = R_1 + R_4 + R_5\ [\text{N}]$$

$$R_y \approx 0,3R_x\ [\text{N}]$$

The force of hydraulic cylinders

$$2F_h = \left(G_{du}l_0 + R_S l_1 + G'_{zp}\,l_2 + R_{xp}n + R_{yp}l_1\right)/r\quad [\text{N}]$$

d—Dozer tipping

By pushing the knife into the hard soil or base with hydraulic cylinders, the front end of the dozer is raised around point B, Fig. 5.28a. The lifting of the rear part of the dozer around point A (b) is the result of blocking the blade and turning on the hydraulic cylinders.

Component of digging resistance, $\sum M_B = 0,\ R_y = (Ga_1 - 2F_h r_1)/b + b_1$
Force in hydraulic cylinders, $\sum M_c = 0,\ 2F_h = R_y(b/r) - R_x(\text{m/r})$
Component of digging resistance, $\sum M_A = 0,\ R_y = (Ga_2 - 2F_h r_2)/l_c$
Force in hydraulic cylinders, $\sum M\text{c} = 0,\ 2F_h = R_y(b/r) + R_x(m/r)$.

e—Force on the blade edge

The action of the lateral force at the end of the blade causes the turning moment of the bulldozer ($M_o = R_y\ l$), Fig. 5.29a.

The adhesion moment of resistance to turning is (a):

$$M_a = \mu_z GL/4\ [\text{Nm}]$$

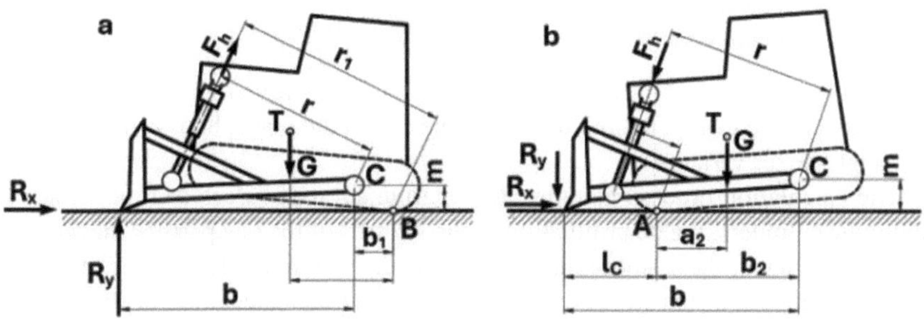

Fig. 5.28 Schemes of tracked bulldozer tipping [4]

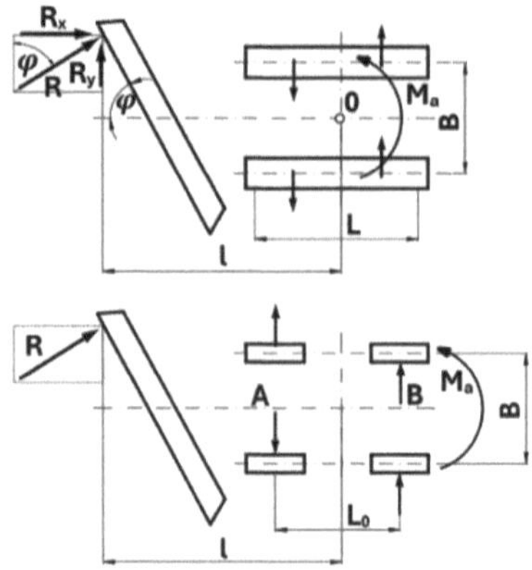

Fig. 5.29 Load on the edge of the blade. **a** tracked bulldozer, **b** wheeled bulldozer

μ_z—turning resistance factor ($R_T = B$; loam $\mu_z = 0.8$–0.9)
R_T—theoretical turning radius
$M_o < M_a$—working conditions
Force on the blade: $R = R_x / \sin\varphi$
Side component: $R_y = R_x \operatorname{ctg}\varphi$.

In case of a wheeled bulldozer (b)

- when turning around point B: $M_B = R_y\, l$
- when turning around point A: $M_A = R_y\,(l - L_o)$

Adhesion moments of resistance to turning

$$M_{aI} = G_A L_o \varphi_b \quad [\text{Nm}]$$

$$M_{aII} = G_B L_o \varphi_b \quad [\text{Nm}]$$

G_A and G_B—front and rear axle reactions.
φ_b—lateral adhesion factor.

Dozer mechanism load

To calculate the stress of the dozer mechanism, it is necessary to analyze the influence of the concentric load in the middle and at the end of the blade, both in the bulldozer and angle-dozer variants.

The calculation process includes the determination of forces and the choice of hydraulic cylinders, the determination of the stresses of the key parts of the working device, the blade, the frame, joints, and their dimensioning. The center of gravity of the calculation of the working device is based on the calculation of the C-frame as the key part of the dozer mechanism. It is made of square or rectangular profile,

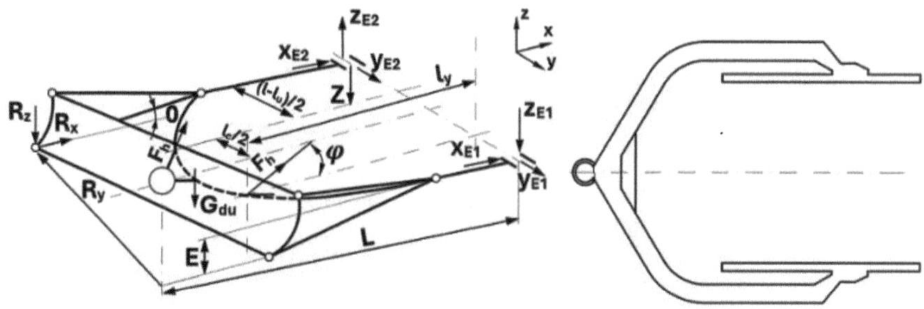

Fig. 5.30 Load scheme of the dozer mechanism. The components of the resistance force on the blade (R_x, R_y, R_z), the weight of the dozer mechanism (G_{du}), the forces in hydraulic cylinders (F_h), components in the left and right joints (F_{cxi}, F_{cyi})

of the welded structure. Attention should be paid to the joints and the frame at the points of connection with the hydraulic cylinders, Fig. 5.30.

The C-frame is made from a solid rectangular profile, usually of constant width and variable height, made of structural steel, with increased strength, high impact toughness values, as well as resistance to brittle fracture. The forces and moments of resistance in the joints are set for cases of the highest symmetrical and asymmetrical loads. The calculation procedure requires the application of the finite element method (for example, *software package, Solidworks*). To determine the working life, the load spectrum is determined as well as the number of the working cycles.

5.7 Bulldozer Ripper

The ripper is placed at the back of the bulldozer, Fig. 5.31. Rippers are used for the preparation of soil categories III and IV, hard surfaces of damaged roads, concrete and others, for the purpose of easier excavation with a dozer blade. The teeth of the rippers penetrate the soil with the movement of the machine and with the thrust of the hydraulic cylinders, and thus break the tough soil. The mechanism for ripping and breaking up hard soil is pivot (rotating), parallelogram, or combined. The pivot device gives the tooth a variable angle of penetration into the soil.

With the 4-joint mechanism, the teeth rise and fall vertically, i.e. the angle of penetration of the teeth remains the same. It is used for deep ripping because it retains the depth well. The parallelogram of the mechanism maintains the parallelism of the sides, and thus the constant angle of the teeth.

The combined ripper combines the good properties of the two previous mechanisms; light teeth penetration, deep penetration, and agility. To increase the performance in hard soil, frozen soil, soft rock and others, rippers reinforced with vibration devices are used.

The working performance of the rippers

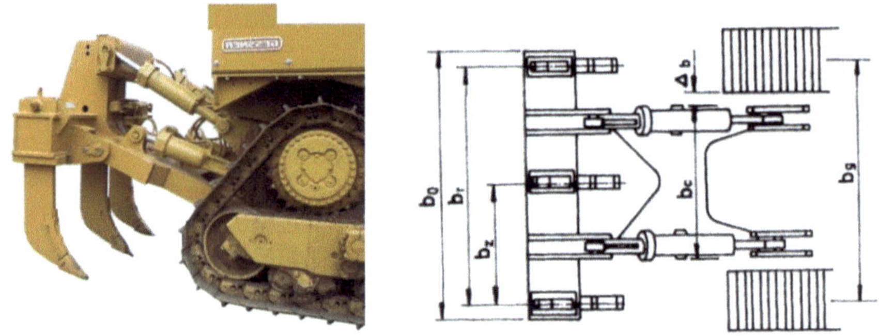

Fig. 5.31 Parameters of the combined ripper [2]

$$U = Bhv/z\ \left[\mathrm{m}^3/\mathrm{h}\right]$$

B, *h*—width and depth of ripping, *v*—ripping speed, *z*—number of passes.

5.8 Ripper Parameters

The teeth of the ripper are at a distance from the dozer tracks, $d = (1.5\text{–}2.0)h_{\text{maks}}$, so that the interspace is not blocked. Depending on the power of the dozer and the hardness of the layer, the ripping depth is 0.5–1.0 m (max. 2.0 m). According to the type of soil to be ripped, the type of ripper is determined. The dozer must have sufficient mass so that the teeth of the dozer can penetrate the soil, and engine power and traction force have to be sufficient to pull the teeth through the soil. The shapes of the teeth are straight, arched (bent), and combined, Fig. 5.32. A straight tooth is good in all types of soil, especially in stony soil and conglomerate, and reaches a ripping depth of 0.8–2.0 m. The arched tooth is good for hard compacted soil and stone deposits and a shallow ripping depth of up to 0.8 m, and has a higher penetration speed, breaks the material and shreds it. A combined tooth features the advantages of a straight and an arched tooth.

The tips of the teeth are the most exposed to wear, so their service life is limited. Therefore, the teeth have replaceable teeth tips, shoes, caps (*d*). When choosing the teeth, attention should be paid to the shape and length of the tooth tip, as well as the possibility of so-called self-sharpening, which means that the designed shape of the tooth wears evenly and maintains the degree of sharpness in operation. Sometimes the teeth are double-sided, so they can be worn on both sides before they are replaced. The tooth tip must be long enough to last longer, hard enough to resist rapid wear, tough enough and short enough to resist fracture. In relatively soft material that wears the tooth quickly, a longer tooth tip is a better solution, while very hard materials require short teeth.

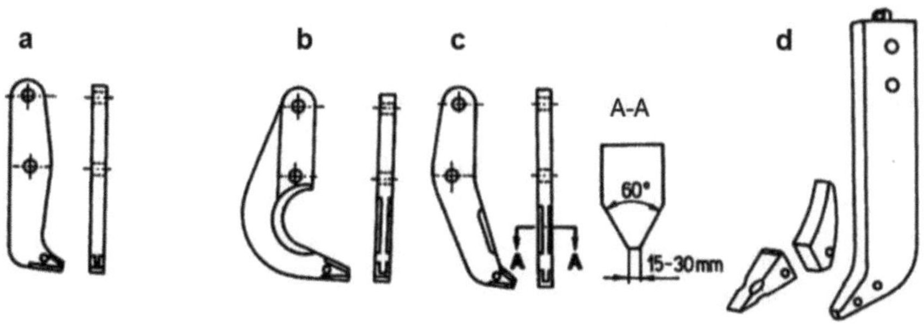

Fig. 5.32 The shape of the teeth of the rippers

Depending on the hardness of the soil, the depth of ripping and the traction force of the dozer, the number of teeth is 1–10, according to the traction force of the machine, from 30 to 250 kN, the most common number of teeth is 1 to 3. In ripping of frozen soil, usually one tooth is installed. The length of the teeth is $L = (1.1–1.3)$ h_{maks}, and the distance between them is 0.3–0.5 m with 5 teeth, and 0.8–1.2 m with 3 teeth. In rocky soil, tooth wear is more intense. Rolled and forged Mn-Mo steels, with strength of 1400–800 N/mm^2, are recommended. The tips of the teeth can be protected with an exchangeable shoe (hardness greater than 500 HB). The service life of shoes in rocky soil is about 10 working hours. The shoes can be regenerated several times. The parameters of the teeth of rippers are given in Table 5.1.

Tooth cutting angles

Ripping at the maximum depth
$\alpha = 40°–50°$, straight teeth, $\alpha = 30°–40°$, arched teeth
Back angle: $\gamma = 5°–7°$ (10°).

Other data

Most mining-class bulldozers are equipped with a giant ripper or multiple rippers. Ripping with a dozer is a method to loosen soil when blasting can be avoided as extremely expensive. The minimum digging depth should be 20–30% greater than the thickness of the layer removed with a bulldozer or any other machine. The working speed is 2.5-3.0 km/h. The mean pressure on the cutting edge of the tooth: $q_r = R_x/b\ n$; b—tooth width, n—number of teeth. Vertical pressure at the tip of the tooth: $q_v = R_y/A\ n$; A—support surface of the tooth.

Table 5.1 Parameters of the ripper teeth

Bulldozers	Ripping depth h_{maks}, mm	Tooth length L, mm	Tooth thickness b, mm	Tooth width b_1, mm
Light	200–500	450–800	25–27	75–300
Medium	600–800	9000–1400	75–90	250–350
Heavy	900–1100	1200–1800	75–100	285–380
Very heavy	1200–2100	1600–2500	70–110	305–420

5.9 Soil Ripping Resistance

Cutting resistance on the ripper tooth, Fig. 5.33.

$$R_x = k_r B h \quad [\mathrm{N}]$$

B—width of ripping
h—depth of ripping
k_r—specific resistance of ripping, 60–120 kN/m^2.

Tooth penetration resistance and tooth extraction resistance

$$R_z = R_x \mathrm{tg}\vartheta \quad [\mathrm{N}]$$

$$R_z \approx 0.25 R_x \quad [\mathrm{N}]$$

The highest force R_z at the beginning of penetration, Fig. 5.34.

$$R_{z\,max} = G_r(L - d_1)/L + d_2$$

Fig. 5.33 Soil ripping resistances on the ripper teeth [4]

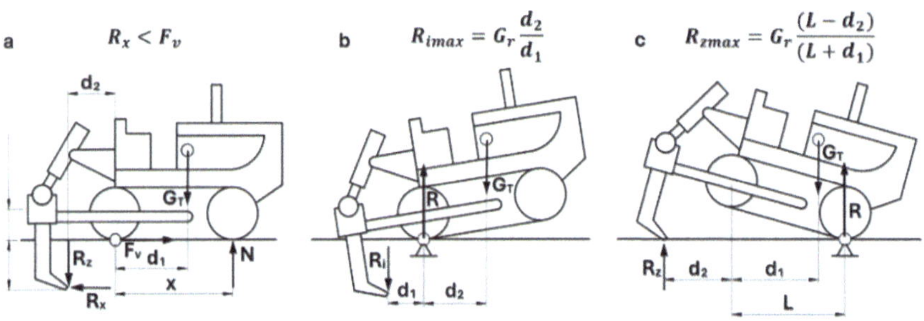

Fig. 5.34 Ripper tipping [4]. **a** ripping, **b** tipping around point B, **c** tipping around point A

The biggest force R_{iz} at the beginning of the extraction

$$R_{i\,max} = G_r d_2 / d_1$$

Total resistance

$$R_u = R_x + R_k \pm R_\alpha \text{ [N]}$$

Required power for ripping

$$P_r = R_u v_r / \eta_m \eta_{gm} \text{ [W]}$$

Safety of operation of the ripper

$$Nl_2 - R_x h - R_z l_1 = 0$$

$$x = l_2 = [(R_x h + R_z l_1)/N] < l_3$$

$x < l_3$, otherwise the teeth are extracted and the dozer turns around point A (N—normal reaction). Apart from the horizontal component R_x, there is a vertical component of the extraction resistance R_i (opposite R_z), with the condition: $l_2 = (R_x\, h - R_i\, l_1) < l_3$. If the condition $l_2 < l_3$ is not met, the dozer tips, from which the maximum values of the penetration force $R_{z\,maks}$ and the extraction force $R_{i\,max}$ are determined.

Other features

Hard ground should be ripped, especially if one wants to reduce the maintenance costs of the machine. The rule of ripping: if the soil can be ripped then it should be ripped. The soil is ripped twice in one direction or once in one direction and once in the other (longitudinally and transversely). In rocky soil, soil blasting is often carried out in order to prepare it for the use of rippers, because cracks are created in the loose soil that make ripping easier. For less wear on the machine and the teeth, it is better to carry out ripping with greater depth and at lower speed, and not the other way round.

5.10 Advanced Bulldozer Control System

Leveling the ground with bulldozers requires complex human skills and efforts. Automated machine control systems reduce the demands for strenuous labor and help the machine operator in planning the surface and reducing costs. This has contributed to the efficiency of construction works.

The *Global Positioning System* (GPS) is enabled by the *Global Navigation Satellite Systems* (GNSS), which consist of a configuration of multiple satellites in Earth orbit. The techniques such as Real-Time Kinematics positioning (RTK) provide precise coordinates of the object on the ground. The coordinates are determined based on the distance between the satellite and the receiver on the object. The leading construction machinery management platforms are *AccuGrade* and *SITECH-Trimble*, which use GPS and laser technology for more complex surface profiling work.

Management of machine operation is based on a GPS system for precise determination of the object position on the ground [7], Fig. 5.35. For example, the GPS *AccuGrade system* consists of a GPS base station, sensor equipment mounted on the machine and blade, Fig. 5.36. Settings and data displays can be configured on the display according to the operator's preferences. The blade guidance control system determines the exact position of the blade and the machine in 2D or 3D space, with accuracy in cm and mm. The blade movements such as lifting, tilting and turning are performed automatically during the machine movement. The dozer blade is controlled by the hydraulic system, while the operator pays attention to the machine speed.

Automated dozer blade control ensures precision and increased productivity. Based on the correction signals, hydraulic control distributors automatically raise or lower the blade in order to maintain the cutting edge position. This machine guidance technology eliminates the need for rework and surface ground repairs. One base station can support a larger number of machines, and the range can cover an area within 20 km, while the radio range is limited. The project productivity, such as land leveling and slope shaping, is increased by up to 50% compared to conventional joystick control.

The *AccuGrade GPS* blade tilt control system is suitable for surface leveling jobs where linear guidance is required. It is suitable for fine leveling of surfaces with one or two blade slopes. The *AccuGrade* function adjusts the position of the blade according to a three-dimensional computer plan of the construction site and signals the hydraulic system to raise or lower the blade.

The mounting brackets enable the installation of posts on the dozer blade. The display in the cabin provides the operator with real-time work information and communication with the system. The slope indicator shows the position of the blade

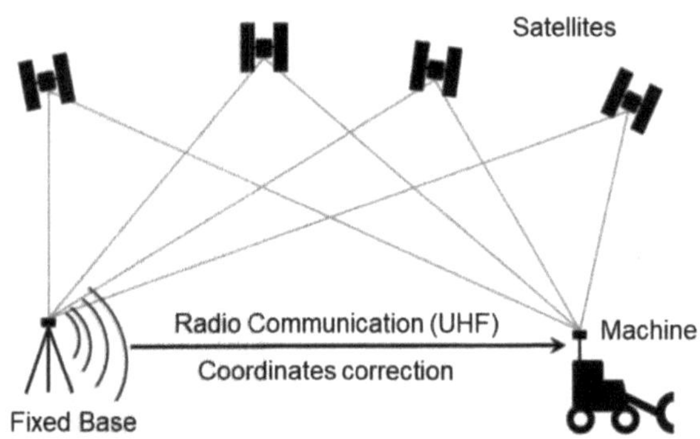

Fig. 5.35 GPS-RTK satellite radio navigation system

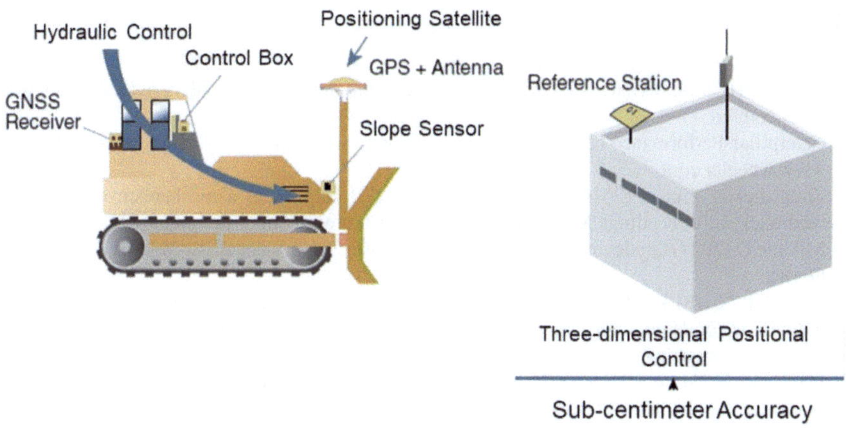

Fig. 5.36 Concept of GPS equipment (*CAT*)

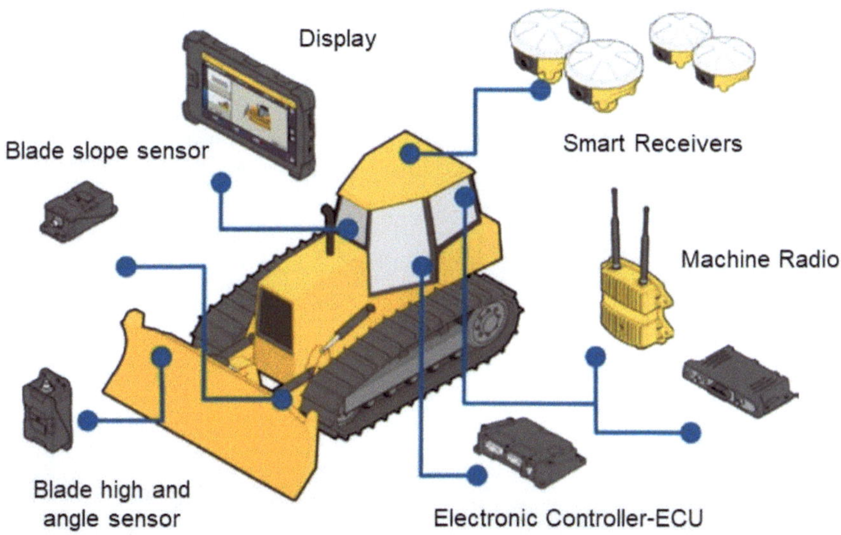

Fig. 5.37 GPS system configuration [8]

in relation to the required height of the blade, the angle and inclination, and the necessary cuts or backfills on the work surface. Settings and data displays can be easily configured according to the wishes of the operator/machinist. Switching from manual mode to automatic fine alignment mode is easy.

Advanced GPS systems use dual receivers on the cab roof (*SITECH-Trimble*), thus eliminating poles and cables from the dozer blade, Fig. 5.37.

References

1. Liebherr Hydraulic systems (2020) Product range, brochure. Image gallery available. www.liebherr.com/en
2. Caterpillar Performance (2020) Handbook, Product range, brochure. Image gallery available. www.wheeler.com
3. Zhunisbekov P et al (2014) Transition process of Bulldozer, as a mechanism with positive feed-back. In: Procedings XXII scientific-techn. Conference "trans & motauto'14"
4. Mikulic D (2024) Gradevinski strojevi. Teorija i konstrukcija. Veleuciliste Velika Gorica, Velika Gorica
5. Stewart L (2018) Track chain tension is your key to cutting undercarriage costs. www.forconstructionpros.com/equipment-management/article/21001484/track-chain-tension-is-your-key-to-cutting-undercarriage-costs
6. Durkovic R. Simovic S (2007) Regulation of working regimes of digging–transportation machines by changing the digging depth. Faculty of Mechanical Engineering University of Montenegro, Podgorica. FME Trans 35(2)
7. Lopesa JM, Trabancoa JLA (2019) Automated control systems for civil construction machinery using RTK-GNSS: an implementation review. Revista de Engenharia Civil 56. www.civil.uminho.pt/revista
8. SITECH Earthworks Grade Control System (2023) Comprehensive construction technology system. Trimble, Product catalog. www.sitechsw.com

Chapter 6
Loaders

Abbreviations

U [m^3/h]	Loader efficiency
V_n [m^3]	Loading shovel volume
T_c [s]	Working cycle time
G_o [N]	Weight of the basic machine
G_{ru} [N]	Weight of the working device
Q_n [N]	Nominal load of the loader
N_p [N]	Permitted front axle load
ρ [kg/m^3]	Soil density
v_r [km/h]	Working speed of the loader
P_{max} [W]	Engine power
F_v [N]	Traction force
F_μ [N]	Adhesion force
H_{is} [m]	Loader unloading height
B_{is} [m]	Unloading zone
k_s [-]	Rollover safety factor
R_o [m]	Radius of the loading shovel
l_n [m]	H-carrier length
H_n [m]	Length of the hydraulic cylinder for lifting
R_1 [N]	Shovel penetration resistance
R_{pI} [N]	Breakout force
R_{pII} [N]	Soil shear resistance
P_{zg} [W]	Power of the supported shovel
P_d [W]	Unsupported shovel power
F_1 [N]	Hydraulic cylinder force for turning the shovel
F_n [N]	Hydraulic cylinder force of an H-carrier

D. Mikulić and D. Mück, *Construction Machines*,
https://doi.org/10.1007/978-3-031-67845-5_6

6.1 Purpose and Division of Loaders

Loaders are intended for loading materials into transport vehicles and for digging soil, sand, gravel, etc. They can be wheeled or tracked. Most commonly used are wheeled loaders, Fig. 6.1.

Division of loaders

a. according to mass, engine power, loading shovel volume

   - very light, up to 5 t / up to 50 kW / 0.6–1.5 m^3 *
   - light, up to 10 t / 50–100 kW / 1.8–2.8 m^3 (2.0 m^3 *
   - medium, 10–25 t / 100–200 kW / 2.4–3.1 (3.5) m^3 *
   - heavy, 25–35 t / 200–300 kW / 2.7-3.1 (3.5) / 4.0–5.0 m^3 *

   * depends on soil density: larger shovel 1.8 (2.0) t/ m^3—smaller shovel 0.8 (1.2) t/ m^3

b. according to undercarriage type

   - wheeled loaders (articulated loaders)
   - tracked loaders (rigid loaders)

c. according to power transmission

   - with hydrostatic drive, for light and medium loaders
   - with hydrodynamic drive, for medium and heavy loaders

Fig. 6.1 Wheeled loader [1]

d. backhoe loader, with universal shovel and deep shovel

For communal works, light loaders are equipped with a universal shovel and an excavator device of a deep shovel for digging a trench.

Loader efficiency

$$U = k_p k_r V_n / T_c \left[\mathrm{m}^3/\mathrm{h}\right]$$

V_n—loading shovel volume, T_c—working cycle time
k_p—shovel filling factor, k_r—soil looseness factor

For example: wheeled loader efficiency (20 t, loading shovel 1.5–3.5 m^3), realizes the effect of loading loose material about 150 m^3/h, and about 100 m^3/h when excavating soil.

Wheeled loaders are operated so that the full loading cycle is completed in 25–30 s, as shown in Fig. 6.2.

Working cycle of the loader

A typical working cycle of a loader loading a truck consists of 4 starts, 4 stops, and 4 changes in the direction of movement:

- The articulated loader moves along the shortest path, the truck is placed as close as possible, obliquely and in relation to the landfill, Fig. 6.2.
- Movement and articulated steering of the loader forward towards the landfill, in the 1st gear.

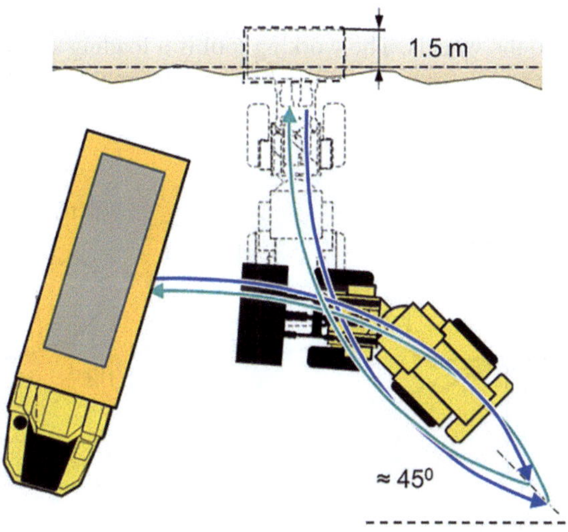

Fig. 6.2 Typical path of a wheeled loader when loading a truck (adapted [2])

- Penetration of the shovel into the material, pressing to the depth required to fill the shovel, which is up to 1.5 m for mid-range loaders, with a deceleration time of about 2 s.
- Closing the shovel, taking approximately 2.5 s, where it is assumed that the loader is braking even with the drive automatically switched off.
- Moving backwards in the 1st gear with simultaneous joint control and raising the shovel to the transport position (to a height of approximately 450 mm from the ground), taking about 1.0 s.
- Continuation of backward movement in the 1st gear, for a total of about 6 m, which can be considered an intermediate distance. Finally, the machine stops by braking.
- Moving forward—towards the vehicle in the 1st gear, articulated steering and slowly lifting the shovel, and braking the machine with the transmission off.
- Completion of lifting the shovel at the highest point, directly in front of the vehicle. The total duration of the lift, from the lowest point to the highest point, is about 7 s.
- Dumping the load into the vehicle takes about 3.0 s.
- Movement of the empty loader backwards to the same distance of about 6 m. Gear II is used. At the end of the movement, with the transmission switched off, the machine stops by braking.
- The new movement of the loader forward towards the landfill, in the 2nd or 1st gear.

6.2 Loader Construction

The structure of a modern loader is shown in Fig. 6.3. This defines their drive system, from the engine to the wheels. The working tool is a loading shovel, which can be replaced with other tools as needed. The basic loader machine can be wheeled or tracked, Fig. 6.4.

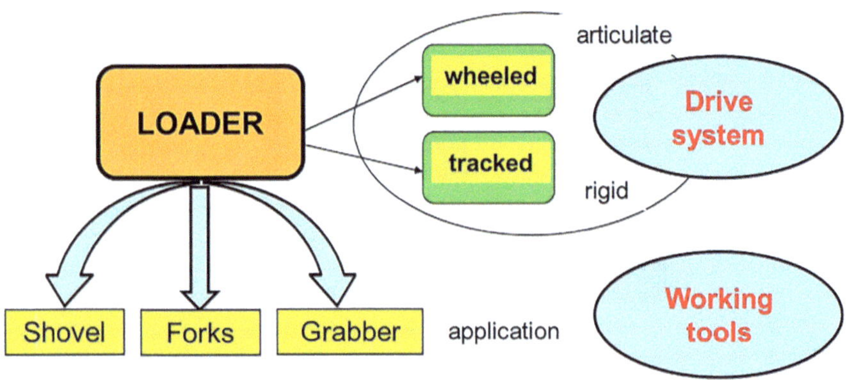

Fig. 6.3 Loader structure

Fig. 6.4 Wheeled and tracked loader

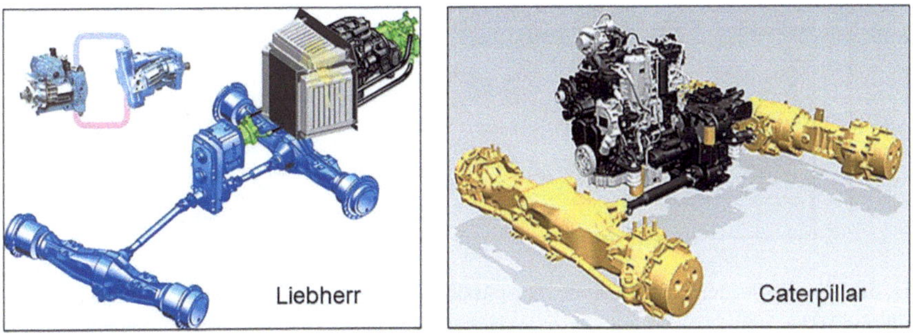

Fig. 6.5 Examples of wheeled loader transmissions [1, 3]

The main components of a loader are: a diesel engine, a hydrostatic or hydrodynamic transmission, a frame-housing, drive axles, and a cab. The hydrostatic (e.g. Liebherr) and hydrodynamic (e.g. Caterpillar) transmissions of a mid-range loader are shown in Fig. 6.5.

A hydrostatic transmission transmits power to the mechanical assemblies, from the power distributor to the drive wheels. Any change in the load on the wheels is balanced by the operation of the variable hydraulic pump. Joystick control provides a quick and easy change of the flow of the hydraulic transmission, i.e. adjustment of the required power and the machine speed.

The schematic of the hydrodynamic transmission is shown in Fig. 6.6. The transmission of the machine provides automatic adaptation of the tractive force to external resistances in a range of torque multiplication factors. Any change in the load on the wheels is balanced by the engine control characteristic.

The planetary gearbox consists of five planetary gear rows with two friction plate clutches for direction change (A, B) and four gear change clutches (C, D, E, F). By mechatronic engagement of the planetary rows, with the activation of the clutches, the required transmission ratios of the gearbox are achieved.

The traction characteristics of a medium-class loader with four forward-reverse transmission stages, with a hydro-dynamic transmission, are shown in Fig. 6.7. The towing force is the greatest in the 1st transmission stage, so it is mostly used for uphill digging and loading shovels (up to 250 kN), then the 2nd transmission stage (up to

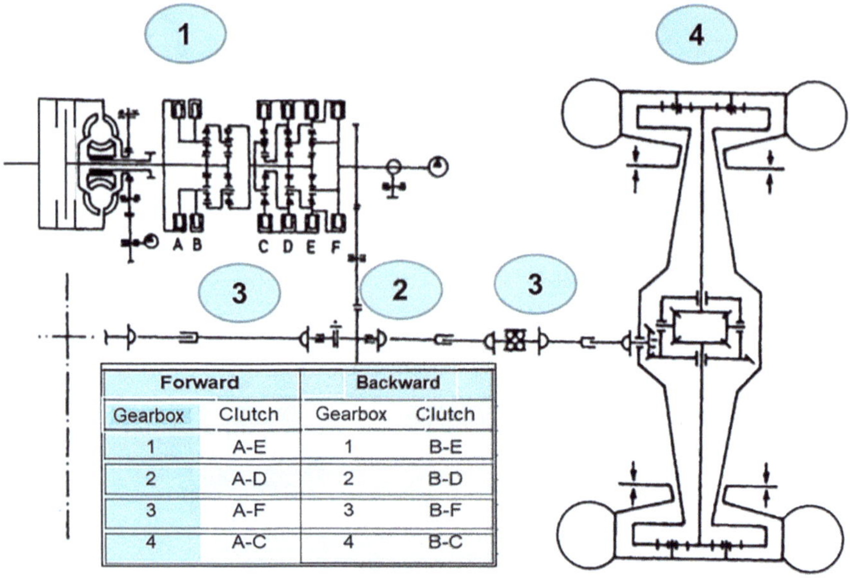

Fig. 6.6 Typical loader hydrodynamic transmission scheme 4 × 4. 1—hydromechanical transmission (converter and planetary gearbox), 2—power distributor, 3—cardan transmissions, 4—drive axles

150 kN). The 3rd and 4th transmission stages are used for machine relocation. In the first phase of the working cycle, it is important to obtain the necessary acceleration of the machine. Therefore, in order to shorten the working cycle, the time and the approach path are analyzed.

Modern loaders use a *Lock up* clutch to bridge the torque converter, for increased efficiency in transferring loads. The *lock up* clutch is installed between the engine flywheel and the converter turbine, thus achieving direct transmission of engine power to the planetary gearbox.

Wide off-road wheels and traction control system (TCS) ensure maneuverability on soft terrain. Optimal tire pressure (4–6 bar) ensures good traction characteristics and machine stability, so a tire pressure monitoring system is installed. The wheel that begins to slide is automatically braked, in order to achieve rolling and the realization of traction force on the ground. For the operation of the machine in the quarry, the wheels are equipped with chains.

The vehicle braking system is hydraulic dual-circuit, with lamellar wheel brakes. One brake pedal is used for normal braking, and the other (left one) is used for automatic disengagement of the transmission (*inch-brake pedal*), whereby all engine power is made available to the working device.

Articulated frame

The articulated frame of the loader consists of two parts, front and rear, Fig. 6.8. The complex structure of the articles is made of metal profiles. The rotation of the front

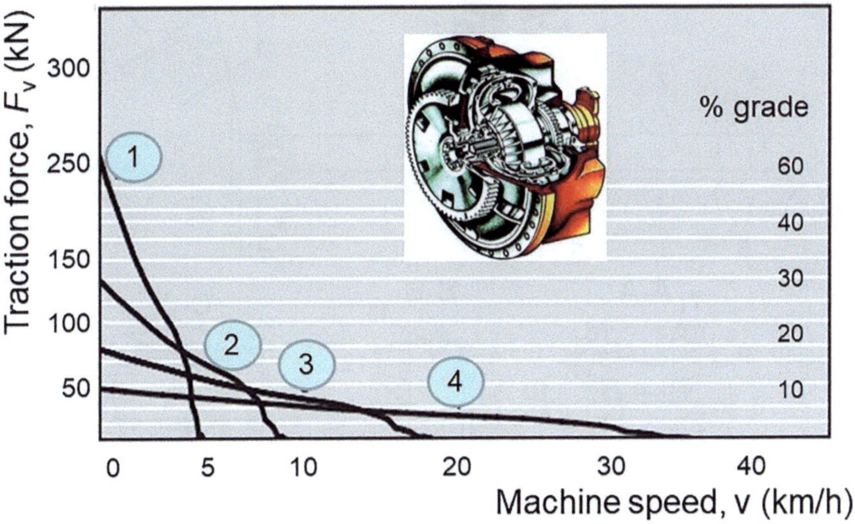

Fig. 6.7 Traction characteristics of mid-range loaders (adapted [1])

in relation to the rear article is performed by hydraulic power steering. The frame articles are mutually rotated around the central joint (*Pivot Joint*) ± 35° (45°), which provides maneuverability of the machine and shortens the working cycle time.

On the front part of the frame, on the upper side, there are reinforcements on which the loading shovel carrier is placed, and on the lower side there is a carrier on which the front drive axle is placed. On the rear part of the frame, on the upper side, a Diesel engine is installed, and on the lower side, the rear drive shaft (without classic suspension) is fixed on the support (6), but in such a way that it has the ability to adapt to the terrain (±15°). Loaders that achieve higher speeds use modern suspension, which is blocked when the machine is running. The central joint (*Pivot Joint*) with two bearings connects the front and the rear parts of the frame. The reinforced roller bearings (A–A) are installed between the bolt and the eyes of the upper bearing, and the lower bearing has a sliding sleeve (B–B).

Loader cabin

The loader cab is subject to the SAE and ISO safety standards for the protection of machine operators and occupational safety laws. A protective cabin has been designed, in case of overturning (ROPS) and falling objects (FOPS), Fig. 6.9. ROPS—Rollover Protection Structure, in accordance with SAE J394, SAE J1040 and ISO 3471 standards, and FOPS—Falling Object Protective Structure, in accordance with SAE J231 and ISO 3449.

The causes of overturning can be very different, from malfunctioning of the machine to collapse: faulty steering system, braking system, worn tires, gust of

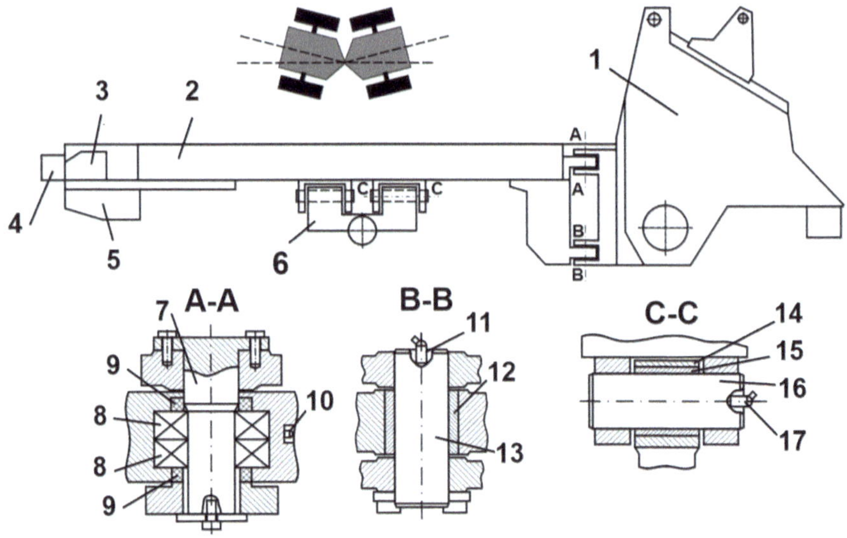

Fig. 6.8 Articulated loader frame. 1—front frame, 2—rear frame, 3 and 4—bumper, 5—counter-weight, 6—rear axle carrier, 7—bolt, 8—roller bearing, 9—seals, 10 and 11—grease nipples, 12—bushings bearing, 13—bolt, 14—ring, 15—bearing sleeve, 16—rear axle support bolt, 17—grease nipple

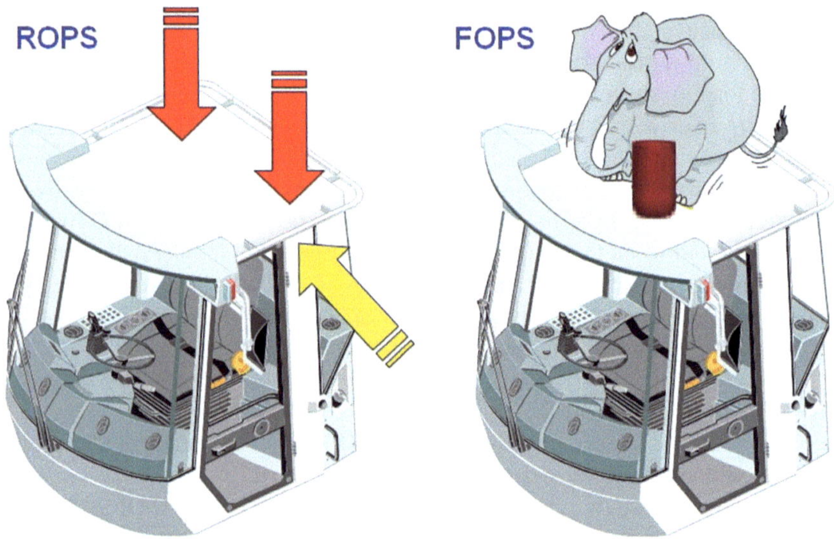

Fig. 6.9 Protective cabins ROPS and FOPS [3]

wind, and others. The result is a disruptive inertial force: $F = \Sigma m_i a_i$ (m_i, mass of cargo, mass of devices, articles, and corresponding accelerations a_i).

Besides protecting the human, the cab must provide ergonomic parameters and visibility of work with the loading shovel. Noise and engine vibrations must be at the level of the allowed values (ISO 6396). The entire equipment for the machine control is located inside the cabin. The electronic monitoring system of the machine parameters EMS—*Electronic Monitoring System*, keeps all the machine drive parameters under control.

6.3 Working Device Construction

Lever mechanism

The working device of the loader consists of a *lever mechanism*, a *loading shovel* and a *hydraulic installation*. According to the way of turning the loading shovel, there are three types of lever mechanism: *Z-kinematics*, *P-kinematics*, and *K-combined kinematics*. The lever mechanism ensures the realization of the highest soil shearing force during the lifting phase of the loading shovel. For medium and heavy class loaders, the *Z-lever mechanism* is most often used, Fig. 6.10.

In addition to the shearing force (*breakout force*), the task of the lever mechanism is to ensure the working positions of the loading shovel (during filling, lifting and emptying-shaking). It consists of the mechanism for lifting the H-carrier of the shovel (boom) and the mechanism for turning the loading shovel, which are connected by

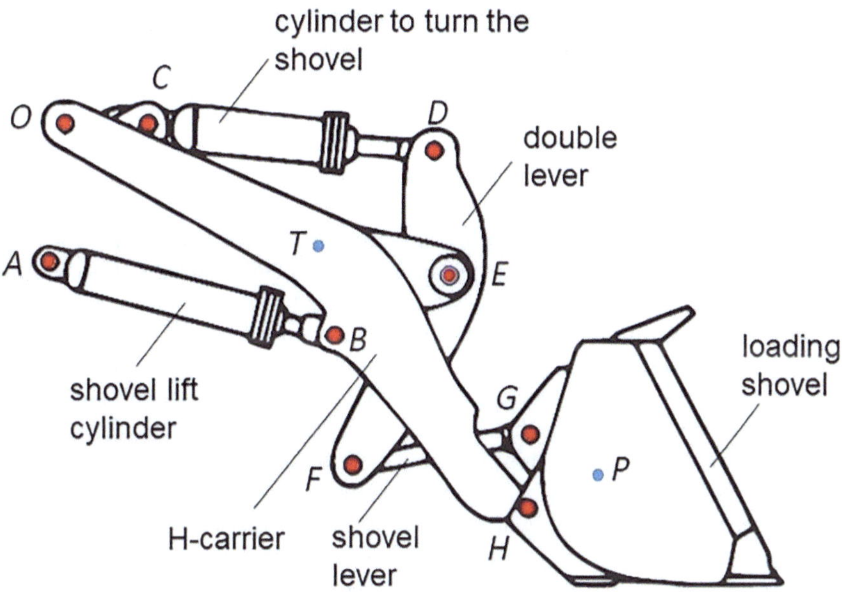

Fig. 6.10 *Z-lever* mechanism of the working device of the loader (adapted [4])

Fig. 6.11 Characteristics of the *Z-lever* mechanism

joints (O, A, B, C, D, E, F, G, H). The shovel carrier is connected by bearings to the undercarriage of the basic machine and on the other part to the loading shovel. It is usually made in the shape of the letter H. It is raised and lowered using hydraulic cylinders. The shovel turning cylinder turns the shovel via a two-arm *Z-lever*.

The *Z-kinematics* of the lever provides a large soil shearing moment in the shovel loading positions, which is suitable for earth excavations, Fig. 6.11. With *P-parallel kinematics*, a higher lifting moment in the upper positions is suitable for industrial drives, for example when working with logs, Fig. 6.12. A K-combined lever can also be used for lifting positions in industrial plants, ensuring greater torque at the lifting height, Fig. 6.13.

The comparison of *Z, P and K* mechanisms according to the change in soil shearing force, is shown in Fig. 6.14. *Z-lever* mechanism shows the smallest necessary shear force to overcome the same soil resistance in all load lifting positions. This is the reason why it is mostly used in loaders. For less loaded machines, such as backhoe loaders with a universal loading shovel, the *P or K lever* mechanism is used.

Considering the different types of work, two types of loading shovels are usually used; without teeth and with teeth, Fig. 6.15. A shovel with teeth is used for various excavations of hard soil, and a shovel without teeth is used for loading loose material. A knife blade is fixed at the front end of the shovel. The shape of the blade can be flat or V-shaped. For working in hard and stony soil, teeth are installed on the front of the shovel. The other type of the loading shovel is the universal shovel, which is used for lighter various works, as a loading shovel, as a forklift, then for loading logs, and others.

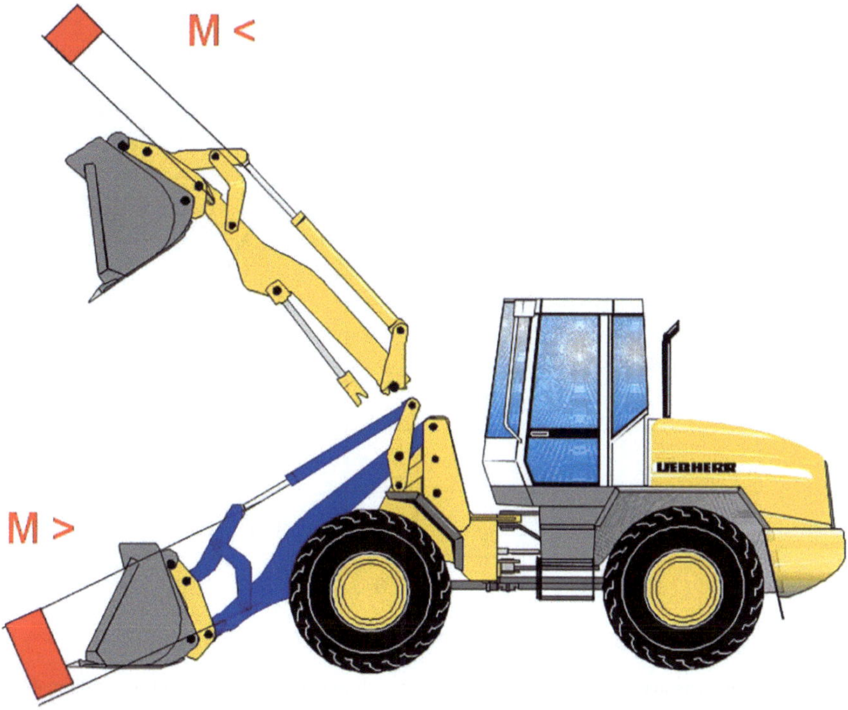

Fig. 6.12 Characteristics of the *P-lever* mechanism

Fig. 6.13 Characteristics of the *K-lever* mechanism (adapted [3])

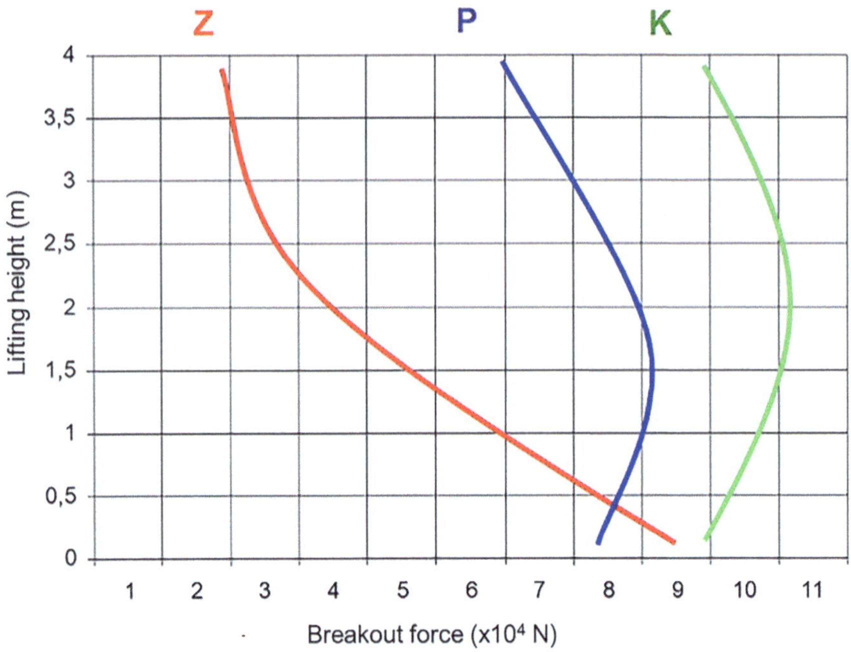

Fig. 6.14 Comparison of Z, P and K-mechanism according to the soil breakout force (adapted [3])

Loading shovels are made from steel sheets of welded construction. Adapters are attached or welded to the front side of the blade, on which the teeth are placed and locked, as well as the interdental blades. Depending on the shear force of the loader and the type of hard soil, the teeth of suitable dimensions and shape are installed, which reduce the cutting resistance. An insufficient distance between the teeth increases the cutting resistance, while an optimal distance reduces and protects the knife blades from rapid wear. The knife blades between the teeth serve to complete the cutting phase.

The teeth are made single or multi-part. Multi-part teeth are often performed in combination with an adapter and a fastener, Fig. 6.16. The rear part of the adapter is connected to the support of the front blade of the shovel by welding, and the front part is connected to the shape of the teeth and closed with a special fastener (*C-lock*). The connection between the teeth and the adapter is on two supports with a spring element that provides the necessary strength of the connection and stability of the load. Replacing teeth and interdental blades is performed using a simple tool.

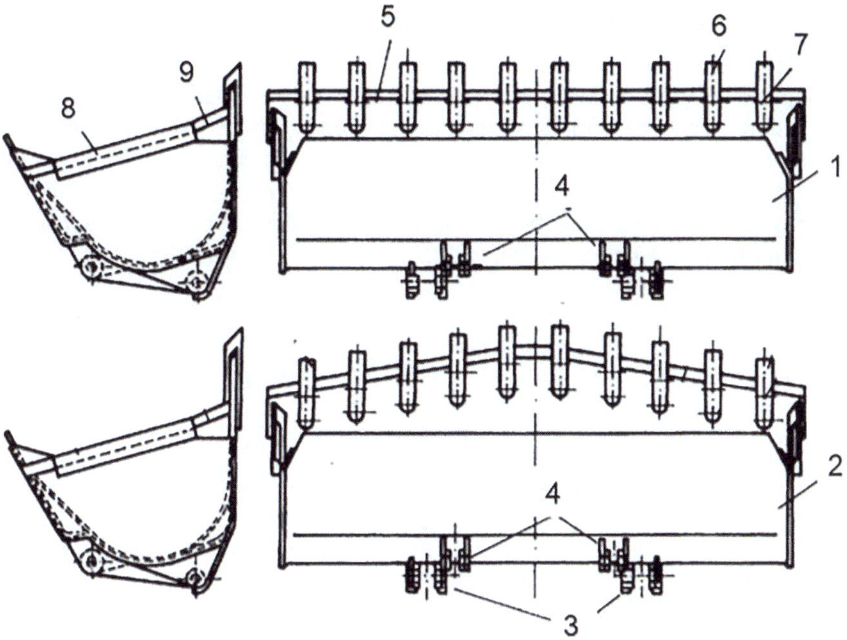

Fig. 6.15 Loading shovels with teeth. 1—standard loading shovel, 2—V-design shovel, 3—lugs for connecting with the shovel carrier, 4—lugs for connecting with the turning levers, 5—front blade, 6—shovel teeth, 7—bolt, 8 and 9—side blades

Fig. 6.16 Placing and securing of teeth (*C-lock*) [5]. 1—standard tooth, 2—tooth adapter, 3—C-fuse, 4—interdental blade, 5—side blade

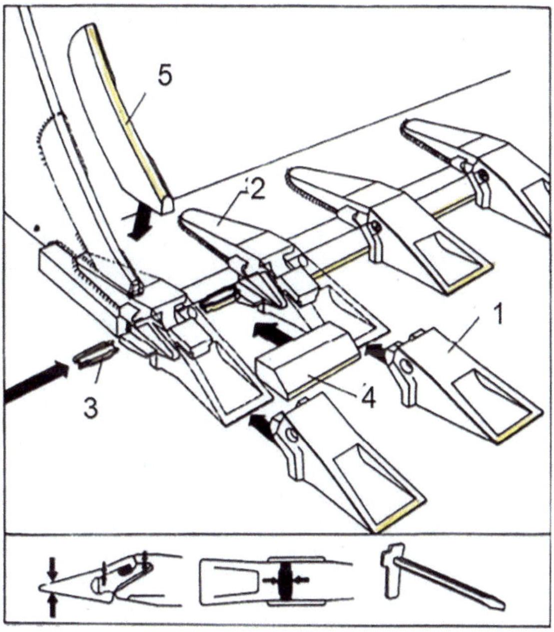

6.3.1 Hydraulic Loader Installation

Raising and lowering of the H-carrier, as well as filling and rotating of the shovel, is achieved by means of hydraulic system, as shown in Fig. 6.17. Using the distributor (4), the hydraulic oil is directed into the hydraulic cylinders (5, 8). The position of the distributor provides an additional so-called floating position of the shovel, which is used when planning the soil (instead of a grader). In middle-class loaders, a constant-flow gear pump is used, with an oil working pressure of about 150 bar, with a tank of about 200 L of oil.

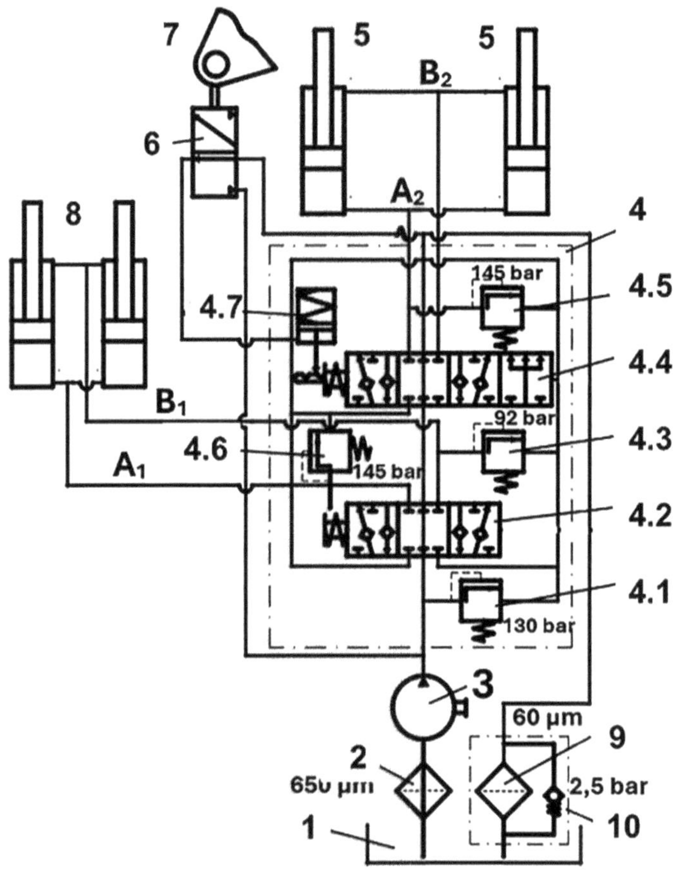

Fig. 6.17 Scheme of a simple hydraulic installation of a loader. (A1/B1) closing and shaking out the shovel; (A2/B2) raising or lowering the carrier. 1—oil tank, 2—suction coarse filter, 3—gear pump, 4—hydraulic distributor (manifold), 4.1—main safety valve, 4.2—distributor of shovel, 4.3—shovel safety valve when filling, 4.4—distributor of carrier, 4.5—safety carrier valve, 4.6—shovel safety valve during unloading, 4.7—lock cylinder, 5—hydraulic cylinders of the shovel carrier, 6—distribution valve, 7—travel stop cam, 8—hydraulic cylinders of shovel rotation, 9—fine oil filter, 10—non-return filter valve

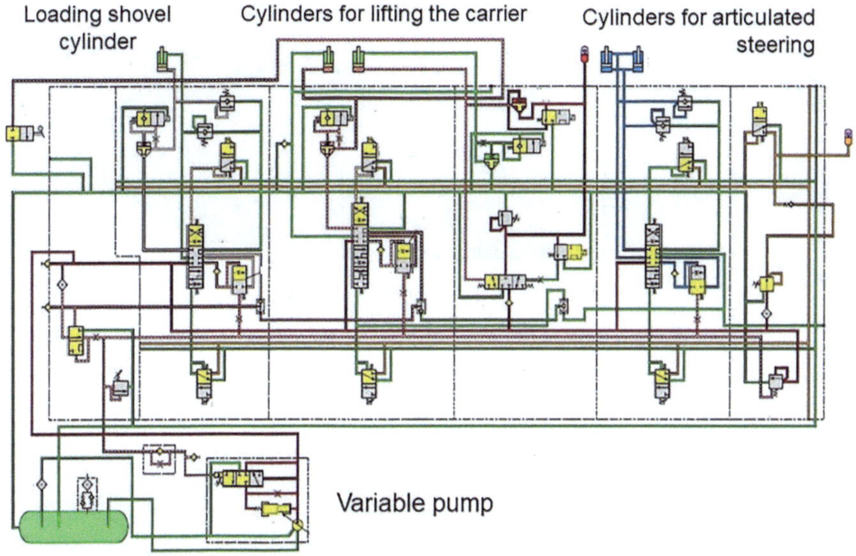

Fig. 6.18 Scheme of the hydraulic installation of the loader (*CAT 950*) [1]

The scheme of a modern hydraulic installation of the loader is shown in Fig. 6.18. Two hydraulic cylinders are used for lifting the H-carrier of the shovel, one cylinder for turning the loading shovel and two cylinders for articulated turning of the machine.

Compared to a constant flow pump, the use of a high-pressure variable pump made it possible to reduce the cross-section of pipes, reduce the dimensions of hydraulic cylinders and other parts of the installation. High pressure variable axial-piston load sensing pump (250–300 bar) produces enough flow and maintains a lower standby oil pressure, making the system quickly available. The type of hydraulic oil is prescribed by the machine manufacturer in accordance with the conditions of use of the machine. Hydraulic oil viscosity marks are prescribed by the ISO standard (ISO 3448).

Handling the loading shovel is shown in Fig. 6.19. When excavating the soil, three movements are required: the movement of the basic machine, the lifting of the H-carrier and the rotation of the loading shovel. The result of the combination of these movements, the movement of the machine and the resistance of the loading shovel control, is recognized first by the power transmission to the loading shovel. In conditions of the highest load on the shovel and soil shear, the highest breakout force is required on the blade of the shovel.

Using the inch-brake pedal, the machine operator transfers all power to shear the soil, Fig. 6.20. *Inching* is used to stop the machine and simultaneously raise the loading shovel, when all the power of the engine and hydraulic pumps is used to increase the shear force (*breakout force*). To achieve this, the *inch brake* pedal gradually disengages the transmission, at the same time as the brakes are gently applied, similar to the "clutch slip" of manual transmission. This is why it is called

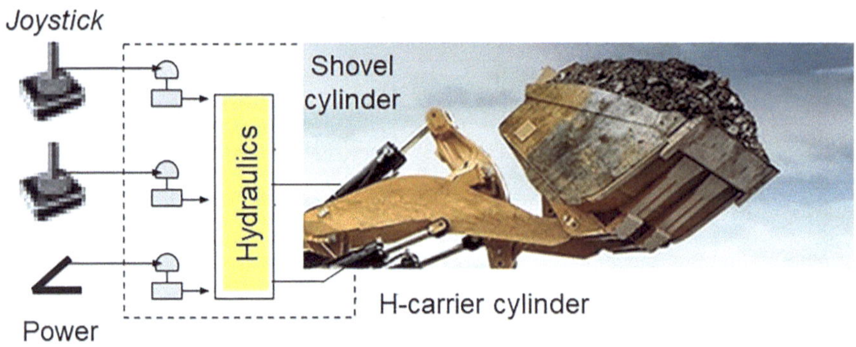

Fig. 6.19 Handling a loading shovel

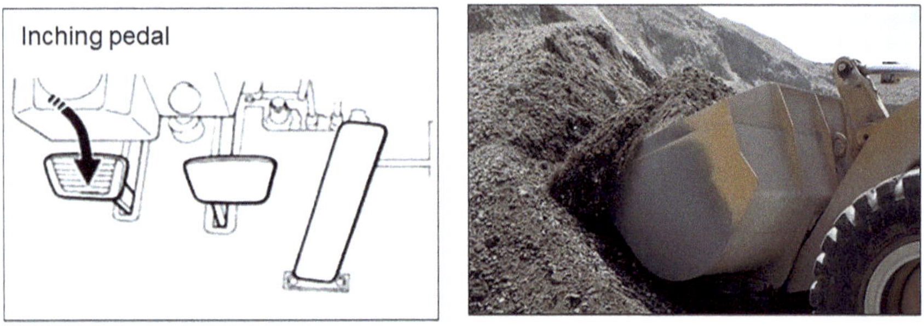

Fig. 6.20 *Inch brake* pedal

inching. This enables carefully controlled movement of the machine towards the truck with the simultaneous operation of the *inch brake* pedal and gas pedal.

6.4 Loader Parameters

Loading capacity

From the condition of static tipping of the loader about the front axle (which separates the rear wheels from the ground), in the position of the largest length of the mechanism and maximally rotated shovel, the nominal load follows, Fig. 6.21.

$$Q_\mathrm{n} = \frac{N_p \cdot x - G_{ru} \cdot (b + x)}{a + x} \text{ [N]}$$

Q_n—nominal load of the loader (N, kN),
Q_n—nominal loading capacity of the loader (kg, or t—ton),

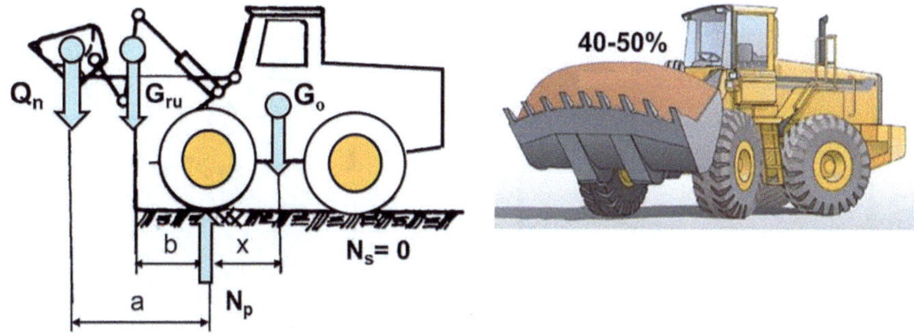

Fig. 6.21 Determining the load capacity of wheeled loaders (adapted [6])

N_p—permitted front axle load,
G_{ru}—weight of the working device (lever mechanism and loading shovel—without load),
$G_{ru} \approx (0.25\text{–}0.35)\, G_o$; G_o—weight of the basic machine,
X—coordinate of the center of gravity of the machine.

Maximum permissible capacity

Practically, for safety reasons, the maximum permissible payload of the loader is the useful payload of the machine that can be transported at operating speeds, subject to the following limitations:

$Q_k \leq 0.5Q_n$, with a wheeled loader (40–50%)
$Q_k \leq 0.35Q_n$, with a tracked loader.

The volume of the loading shovel

$$V_n = \frac{Q_n}{\rho \cdot k_n} \quad [\mathrm{m}^3]$$

Q_n—nominal load capacity/$0.5Q_n$
ρ—soil density, recommendation 1780 kg/m^3
k_n—shovel loading factor (25%, $k_n = 1.25$)
Nominal and geometric volume of the shovel, Fig. 6.22
a: $V_n = V_o + (b^2W/8) - (b^2/6)\,(a + c)$; $V_o = AW - (2/3)\, a^2\, b$
b: $V_n = V_o + (b^2W/8) - b^3/24$; $V_o = AW$.

Figure 6.23 shows the choice of a loading shovel for a specific loader depending on soil density. The density of different materials is given in Table 6.1.

Traction force of the machine

$$F_v = \frac{\eta_{tr}\, P_{maks}}{v_r(1-\delta)} - G_{ad} \cdot f_k \quad [\mathrm{N}]$$

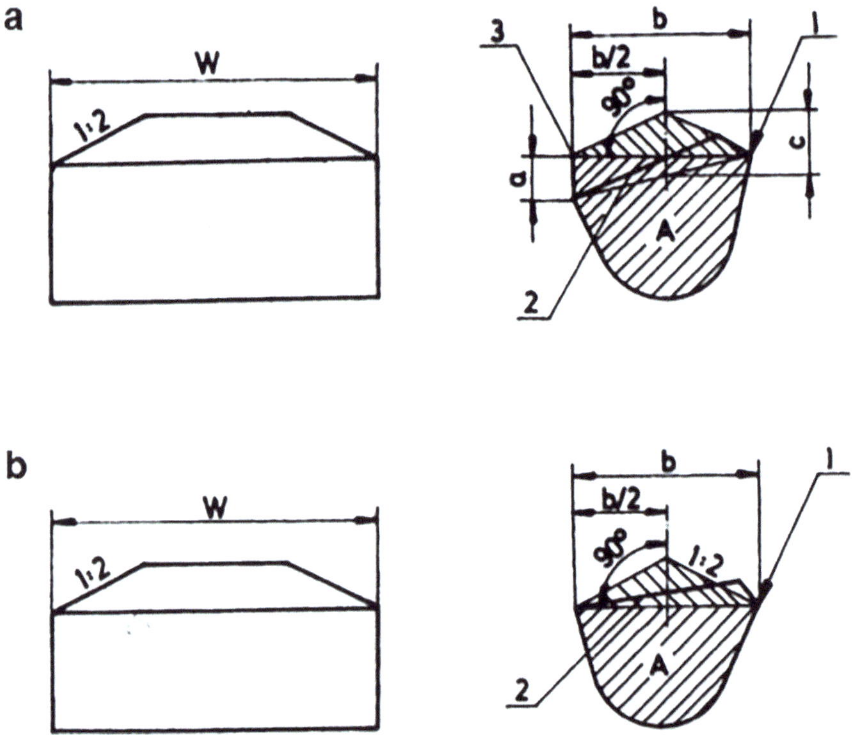

Fig. 6.22 The volume of the loading shovel (*SAE J742b*). **a** A shovel with a guard, without a protruding knife blade, **b** a shovel without a guard and without a protruding knife blade, 1—tip of the knife blade, 2—contour of the shovel, 3—tip of the guard, W—middle inner width of the shovel, A—surface-dashed, a—height of the guard, h—protruding tip of the blade, 1:2—slope

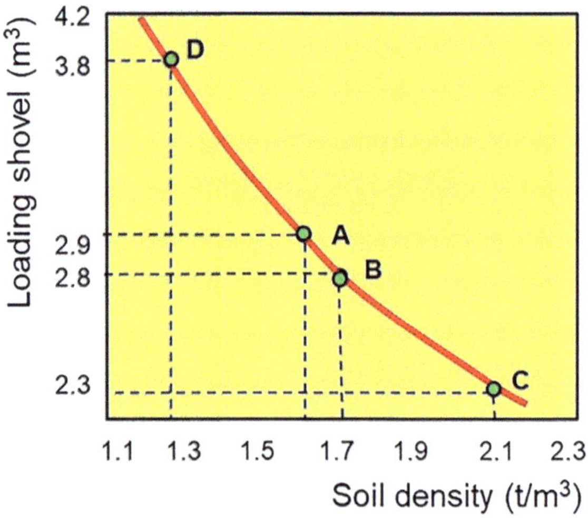

Fig. 6.23 Choice of loading shovel according to soil density. A—shovel without teeth volume 2.9 m^3, B—shovel with teeth volume 2.8 m^3, C—shovel volume 2.3 m^3 for broken stone after blasting, D—shovel volume 3.8 m^3 for lighter material

Table 6.1 Soil density

Soil, material	Density (kg/m^3)	Soil, material	Density (kg/m^3)
		Sands, dry	1550
Snow	200	Asphalt	1600
Sugar beet	530	Gravel	1650
Barley	600	Clay	1680
Wheat	730	Sand, wet	1890
Dry land	1150	Slate, limestone	2800

P_{maks}—engine power
v_{r}—working speed of the loader
η_{tr}—degree of utility of transmission
δ—slippage, tires: ≈ 0.2, (20%), tracks: 0.07 (7%)
f_{k}—rolling resistance factor, tires: 0.03–0.04, tracks: 0.06–0.1

$$F_{\text{V max}} = F_{\mu} = G_{\text{ad}}\mu$$

F_{μ} = maximum traction force that can be achieved from adhesion conditions,

$$G_{\text{ad}} = G\cos\alpha \ [\text{N}]$$

$$G_{\text{ad}} = (G_{\text{o}} + G_{\text{ru}} + \Delta G)\cos\alpha \ [\text{N}]$$

G—total weight of the machine, for $\alpha = 0$, $G_{ad} = G$, ΔG—additional weight of the equipment
$(G_{\text{o}} + G_{\text{ru}})$—weight of the machine and working device

Soil resistance to tool penetration

The greatest resistance to machine intrusion arises from the maximum traction force at the wheel slipping boundary and the force of inertia.

$$R_{\text{zt}} = F_{\text{v max}} + F_{\text{i}} \quad [\text{N}]$$

Axle load

The loading shovel is raised to the transport position (height 450–600 mm), the largest distance from the front part of the machine, Fig. 6.24a, b. The following is the static load of the wheeled loader, on the front and rear axles:

$$N_{\text{p}} = G(1 - x/L_{\text{o}}) + Q_{\text{n}}(1 + a/L_{\text{o}}) \ [\text{N}]$$

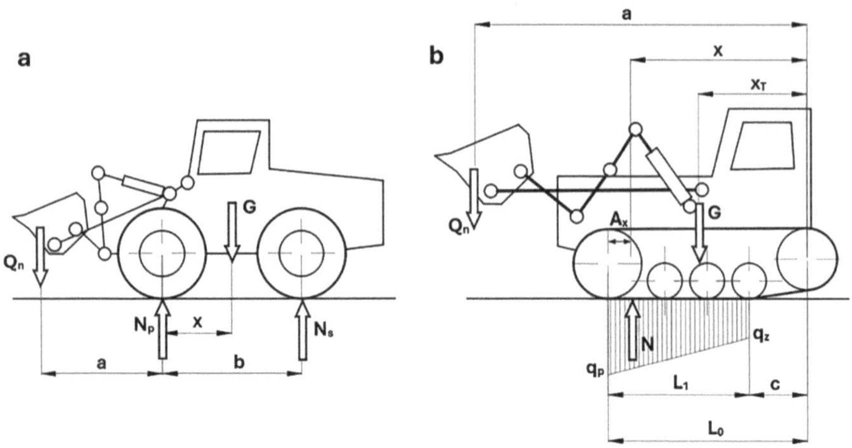

Fig. 6.24 Static load distribution in motion. **a** wheeled loader, **b** tracked loader

$$N_s = Gx/L - Q_n a/L_o \text{ [N]}$$

In the case of tracked loaders, with nominal capacity Q_n in the center of gravity of the shovel, the position of normal sol reaction $\boldsymbol{N}$:

$$x = (Q_n\, a + Gx_T)/N \text{ [mm]}$$

Stability condition, $\Delta x \geq L_o/6$, $(\Delta x = L_o - x)$.

For wheeled loaders, the pressure on the ground is approximately equal to the air pressure in the tires (p_s). By letting the air out of the tires, the pressure on the ground is reduced, which can increase the soil trafficability for a short time. For tracked loaders, the pressure on the base is approximately equal to the trapezoidal shape ($q_p > q_z$), Fig. 6.24b.

Loader unloading height

The unloading height is adopted according to the height of the transport trucks (Fig. 6.28). The highest unloading point is the height from the base to the top of the shovel blade in the unloading position.

$$H_{is} = h + \Delta h \quad \text{[mm]}$$

h—height of the transport truck
Δh—additional height (300–500 mm), ensures unhindered loading on an uneven surface, excludes the contact of the shovel with the vehicles sides on the return of the machine. $H_{is} = 2800$ mm.

Unloading zone

$$B_{\mathrm{is}} = B_{\mathrm{y}}/2 + \Delta b_{\mathrm{is}}\ [\mathrm{mm}]$$

B_{is}—the distance of the tip of the shovel tooth from the front of the machine (Fig. 6.28.)
$B_{\mathrm{is}} = 1.0$–1.5 m, B_{v}—width of cargo box.
Δb_{is}—safe distance between loader and vehicle (150–300 mm).

Shovel inclination during unloading at the highest point, $\varepsilon_1 = 40$–$50°$.
Back angle of rotation of the shovel at ground level, $\gamma = 40$–$45°$.
The lowest position of the shovel under the substrate, $H_{\mathrm{sp}} = 300$–500 mm.

Rollover safety factor

For construction machines, it is especially important to secure the machine from overturning. The safety factor against loader tipping k_{s}, is determined by the ratio of stability moments ΣM_{s} and overturning moments ΣM_{p} around the tipping axis, including moments of inertia, wind gusts, machine tilt, machine load in the highest position of the shovel and braking when turning the machine.

$$k_{\mathrm{s}} = \sum M_{\mathrm{s}} / \sum M_{\mathrm{p}} \quad k_{\mathrm{s}} \geq 1.5-2.0$$

Due to the change in the pivot points when turning the front axle relative to the rear, the coordinates of the machine center of gravity change, and hence its stability. Figure 6.25 shows the effect of the articulated chassis turning on the safety of the machine when turning on a side slope. A safer way of turning is shown in Figure b. The position of the machine center of gravity can be determined depending on the angle of the turn of the articles.

6.5 Parameters of the Lever Mechanism

When excavating the soil, the favorable angle of the loading shovel for digging is 10^0, which is marked with position meters, for example, mechanically with sufficiently aligned rapiers on the hydraulic cylinder of the shovel (8), Fig. 6.26. In the unloading position of the shovel, the limiter is activated to stop the rotation of the shovel.

Radius of the loading shovel

The radius of the shovel (R_{o}) is the distance between the axis of the joint around which the shovel rotates and the blade of the shovel, Fig. 6.27. The radius of the shovel is the basic parameter for the calculation of the geometric parameters of the loading shovel and the lever mechanism. It is determined according to the expression:

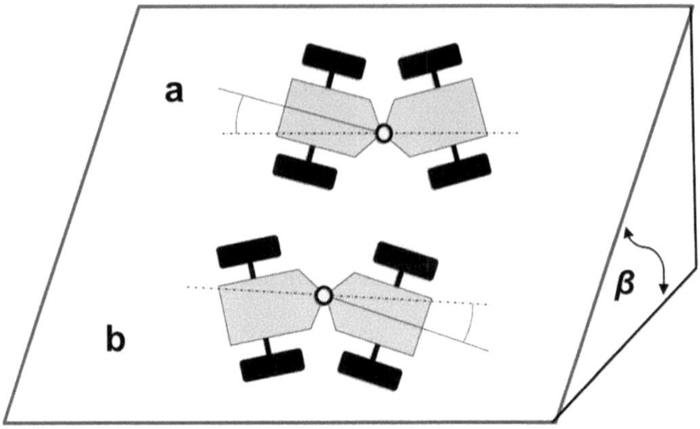

Fig. 6.25 Turning the articulated loader on a side slope. **a** Less safe mode, **b** safer mode

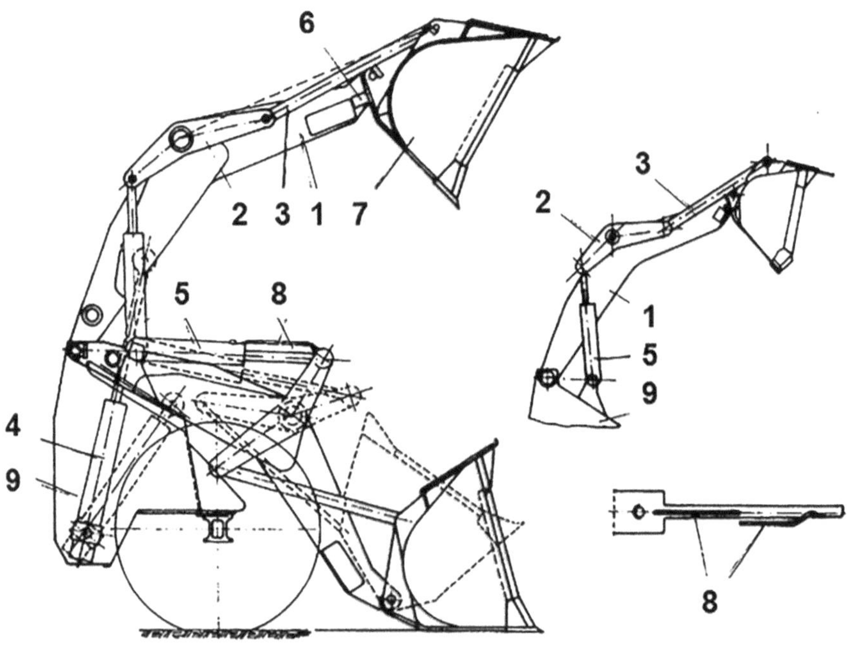

Fig. 6.26 Z-lever mechanism in unloading position. 1—shovel H-carrier, 2—two-armed lever, 3—shovel lever, 4—hydraulic cylinder of the H-carrier, 5—hydraulic cylinder for shovel rotation, 6—travel limiter, 7—loading shovel, 8—position bars, 9—front frame article

$$R_o = \sqrt{\frac{V_n}{b\lambda}} \quad [\mathrm{m}]$$

V_n—nominal volume of the loading shovel, b—width of the loading shovel, $b = B + \Delta b$; B—width of the basic machine, Δb = 50–200 mm, λ—relative parameter of the shovel design, $\lambda = 0.5\, c\, (c_1 + c_2 \cos\gamma_1) + \sin\gamma_0 - c_3{}^2[\mathrm{ctg}\, \gamma_0/2 - 0.5\, \pi\, (1-\gamma_0/180)]$; c—relative depth of the bottom of the shovel, $c = l_g/R_0$, (c = 1.4–1.5), c_1—relative depth of the rear part of the shovel, $c_1 = l_h/R_0$, (c_1 = 1.1–1.2), c_2—relative length of the guard, $c_2 = l_s/R_0$, (c_2 = 0.12–0.4), l_s—the length of the guard, c_3—relative radius ratio, $c_3 = r_0/R_0$, (c_3 = 0.35–0.40), r_0—internal radius of the shovel, γ_1—the angle of the guard in relation to the direction of the back side of the

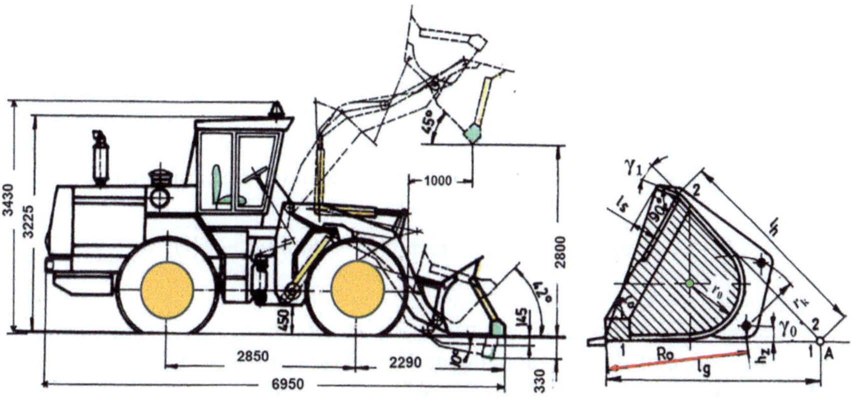

Fig. 6.27 Loader and loading shovel parameters (120 kW; 15 t; 1.5–3.5 m³) [5]

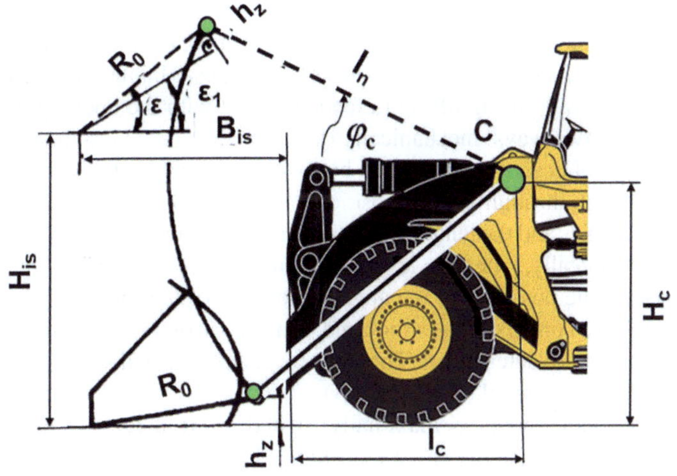

Fig. 6.28 Determining the length of the loading shovel carrier

shovel, $\gamma_1 = 5\text{–}10^0$ (from the condition of placing the guard at 90° on the edge of the shovel), γ_0—the angle between the back side of the shovel and the base, $\gamma_0 =$ 48–52°, r_k—turning radius of the shovel, h_z—height of the shovel joint.

Loader lever parameters

a. Shovel carrier

The parameters of the lever mechanism mostly depend on the selected point of the C-joint connection of the shovel carrier and the machine and the radius of the loading shovel, Fig. 6.28. The coordinates of joint C (H_c, l_c) should be determined. The height of joint H_c and its distance from the front part of the machine l_c is taken from the visibility conditions of the machine operator's work:

$$H_c = (1.2\text{–}2.0)R_0 \text{ [m]}$$

H-carrier length

Length of the loading shovel support is calculated based on the required unloading zone and the radius of the shovel. It is usually made in the form of letter H.

$$l_n = \sqrt{(Bis - R0\cos\varepsilon + lc)^2 + (His + R_0\sin\varepsilon - H_c)^2} \text{ [m]}$$

ε—angle of inclination of the shovel radius axis R_0, $\varepsilon = \varepsilon_1 + \text{arc sin}\,(h_z/R_0)$
ε_1—back angle of shovel unloading at the highest point (45°–50°)
B_{is}—unloading safety zone (1.0–1.5 m).

Angle of rotation of the H-carrier, $\varphi_c = 85°\text{–}90°$

b. Lever for turning the shovel

When lifting the loading shovel, the soil must not fall out of the shovel. The problem of leveling the loading shovel is solved by lever kinematics. Maintaining the horizontal position of the shovel during lifting (a loaded shovel with the opening upwards) is solved mainly in two ways: mechanically (kinematically) or hydraulically.

The mechanical design is performed by computer graphics based on the path of the loading shovel, through several positions, Fig. 6.29. The length of the lever for turning the shovel d_k and the length of the two-arm lever (leverage arms c_k and b_k) are determined graphically, when the lever mechanism is observed in several reference positions, i.e. in the positions of the angle of rotation of the shovel carrier φ_c (I. II. III. IV. V…). Based on this, the position and dimensions of the hydraulic cylinder for turning the shovel are determined. The lever mechanism usually uses one hydraulic cylinder to rotate the shovel (*Z-kinematics*).

In position I, the shovel is maximally rotated at an angle γ, and in the final unloading position at an angle ε_{max} (V-position). For safe shaking out of the shovel, it is necessary to ensure distance Δk:

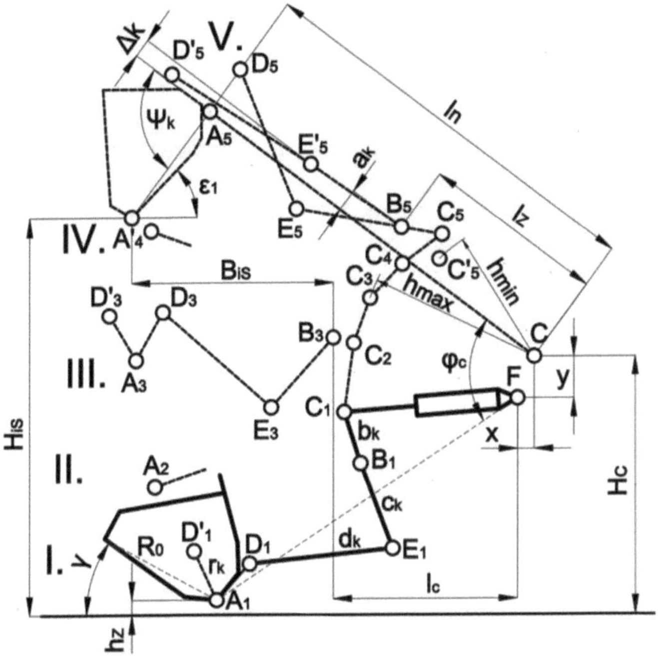

Fig. 6.29 Determining the dimensions of the levers for turning the shovel

$$\Delta k = (0.125 - 0.135)\, R_0 \quad [\text{mm}]$$

In the shovel shaking position, the following is the length of the lever for turning the shovel:

$$d_k = B_5 D'_5 - c_k \; [\text{mm}]$$

$$c_k = (0.27–0.3) l_n \; [\text{mm}]$$

Other parameters of the lever mechanism of the shovel are derived depending on the length of the support l_n, or similar designs: $l_z = (0.48–0.50) l_n$, distance from joint C to the joint of the two-arm lever, $a_k = (0.11–0.12) l_n$, the height of the joint of the two-arm lever, $b_k = (0.22–0.24) l_n$, the second step of the two-arm lever, $r_k = (0.13–0.14) l_n$, the turning radius of the shovel, $\psi_k = 10°–125°$, the angle between directions A_5 D'_5 and the length parallel to CA'_5, $\gamma = 40°–45°$, back angle of the shovel turn at ground level, D'_i, C'_i, E'_i—position of the shovel joints.

Hydro cylinder support for turning the shovel F (x, y)

For the considered positions of the shovel support (I–V), the comparative lengths A_iD_i according to the lengths $A_1D'_1$ (r_k) are determined. These lengths make the position of the shovel unchanged during lifting, which is necessary to maintain the leveling of the shovel. For each position, point C_i is determined, which forms the connection of the hydraulic cylinder with the lever mechanism. Through points C_1, C_2, C_3, C_4, and C_5, an arc h_{maks} is drawn, the center of which is the place of support of the hydraulic cylinder. Thus, the support point of cylinder F(x, y) on the machine, and the relationship between the coordinates of the support point of the support on the machine C (H_c, l_c) were determined. If not all points can be found on the arc, then they should be located inside the arc, so when lifting the H-carrier, the loading shovel will be turned more towards the front.

c. Length of the hydraulic cylinder to turn the shovel

A circular arc through points C_i represents in fact the length of the cylinder with the maximally extended piston rod. The required pullout of the cylinder is determined according to point C'_5, which is the position of point C_5 in the unloading position of the shovel. The drawn arc through point C'_5 (h_{min}) marks the location of the retracted connecting rod. The length of the hydraulic cylinder for turning the loading shovel:

$$H_1 = h_{maks} - h_{min} \text{ [mm]}$$

The unloading angle of the shovel ε_1 in the upper position is not less than 45°. To increase the unloading angle, the lengths: l_z and A_1D_1 (r_K) should be reduced.

Support of hydro cylinders for lifting the shovel H-carrier

The location of the articulated joint is determined at the points of reinforcement of the front frame. The extension of the hydraulic cylinder for lifting carrier H_n is determined by the position of the loading shovel in the lower position (maximally retracted piston rod) and in the upper position (maximally extended piston rod). With the obtained radii through the points of the articulated joint with the carrier, the length of the cylinder for lifting the H-carrier is obtained.

d. Length of the hydraulic cylinder for lifting the loading shovel

$$H_n = h_{n\,maks} - h_{n\,min} \quad \text{[mm]}$$

Computer graphics, methods of mechanism synthesis and mechanism load simulation contribute to the speed and accuracy of calculations of the working device mechanisms.

6.6 Loading Resistances of the Loading Shovel

Filling the loading shovel

Figure 6.30a, b shows two typical ways of loading a loading shovel. *The first method* of loading (a) consists of the penetration of the shovel to the depth of penetration L_r

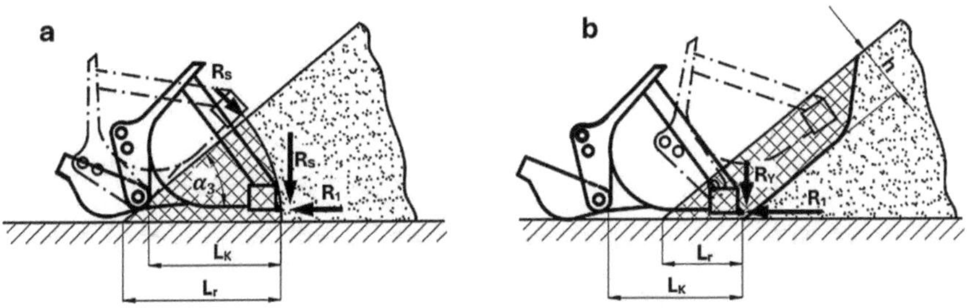

Fig. 6.30 Filling the loading shovel, **a** first method, **b** second method

= (1.1–1.15)L_k, (L_k—shovel depth), then turning the shovel (α_3). This method of loading the shovel is used for lighter and loose materials

Using the inertia of the loader mass and the traction force at the wheel slipping limit, the shovel is filled. The lever mechanism is exposed to impact loads, but due to the rigidity of the metal structure, it keeps the machine relatively stable. In order to reduce the frictional resistance of the shovel with the ground, the bottom of the shovel is raised to the path of the knife blade. Turning the shovel follows the shear resistance of the soil on the knife blade (*breakout force*).

The second method of loading the shovel (*b*) is more complex for the machine operator. The penetration of the shovel is shorter $L_r = L_k/3$, and with the turning, the shovel carrier is simultaneously lifted. At the same time, the path of the knife blade is parallel to the slope of the ground. The loader performs gradual loading of the shovel at the limit of wheel slippage. This loading method is less favorable from the point of view of the tire wear and shovel filling time; therefore, the wheel slip restriction system (ASR) has been introduced. The resistances to penetration and rotation of the shovel are lower than the resistance of filling the shovel using the first method.

6.6.1 Shovel Penetration Resistance

$$R_1 = bk_zk \quad [\text{N}]$$

b—width of the loading shovel
k_z—specific penetration resistance, L_r, Fig. 6.31.
k—size factor, 1.05–1.25, Table 6.2.

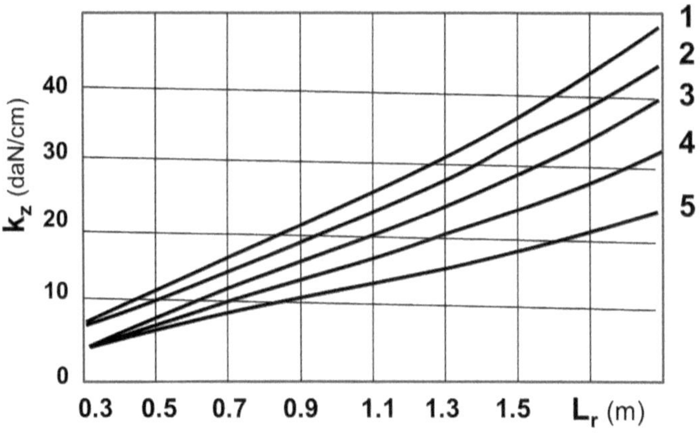

Fig. 6.31 Diagram of the specific cutting resistance k_z from penetration of shovel L_r. 1—fine gravel, 2—gravel, loose, 3—loose sand and soil, 4—slag, loose, 5—small pieces of coal

Table 6.2 Material bulk factor ($k = 1.05$–1.25)

	Material	Bigness (mm)	Factor (k)
1	Granite gravel	Up to 60 60–160 161–300	1 1.1–1.15 1.15–1.25
2	Grit, slag	Up to 60 61–160	1 1.05–1.1
3	Stone coal	Up to 60 61–160 161–300	1 1.05–1.1 1.10–1.25
4	Sand, loose soil	Normal humidity Increased humidity	1 –1.15

6.6.2 Soil Shear Resistance

The model of the soil shear prism with lateral planes DET and NMK and the soil shear to be overcome in the slip plane TKMD is shown in Fig. 6.32. Therefore, the reaction of the soil on the sides of the shovel (R_2) and the reaction of the soil on the front plane of soil shear (R_3) follow. These resistances depend on a large number of factors and are therefore not suitable for quick and accurate calculation. This practically allows determining the soil shear opening based on the case analysis of the supported and unsupported shovel, Fig. 6.33.

The shear resistance of the soil is the highest at the very beginning of the rotation of the shovel, then it decreases with a lower resistance in the shear planes and the correction of the weight of the earth prism Q_M. The highest value of soil shear

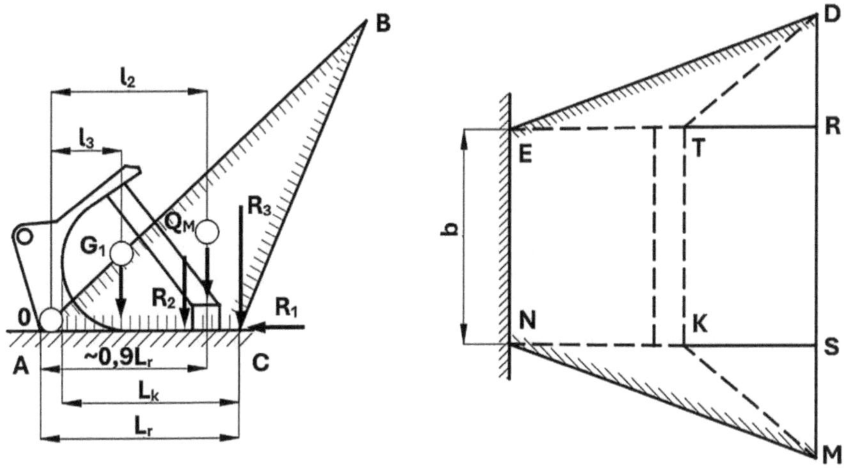

Fig. 6.32 Soil shear prism

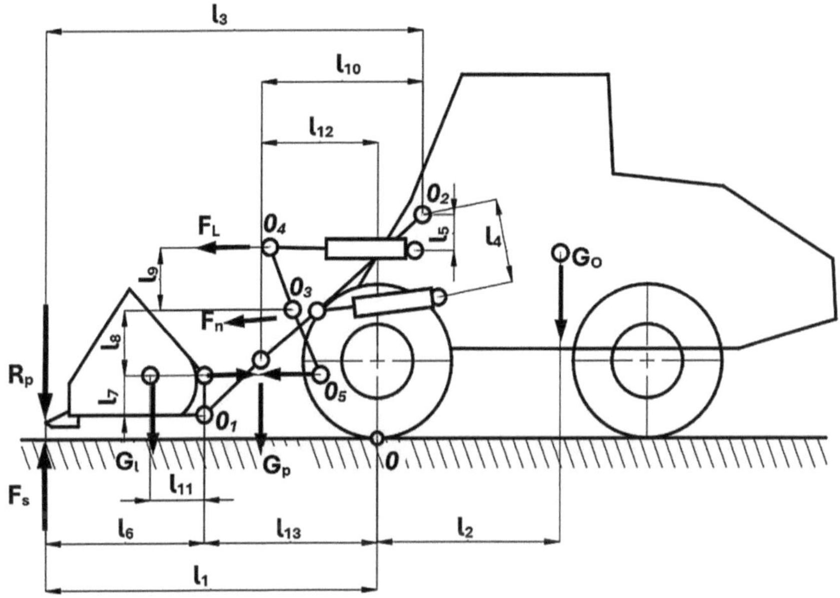

Fig. 6.33 Load scheme of the lever mechanism

resistance R_{pI} is determined for the beginning of turning around the shovel joint 0, the so-called case of the supported shovel.

AB–AC–CB–ED–NM–earthen prism in front of the shovel, G_1—weight of shovel without soil on arm l_3 from the pivot joint 0, G_{M}—weight of the earthen prism which is started by a loading shovel.

a. Case of a supported shovel

Loading shovel lowered to the ground (SAE J732c), Fig. 6.33. Shear resistance R_{pI}, assumes the calculated maximum load of the turning mechanism, with the limit condition of the permissible load of the lever mechanism. The entire engine power goes to shearing and lifting of the shovel (*use of inch-brake pedal*).

$$R_{\mathrm{pI}} = R_l \frac{l_7}{l_6} = F_l \frac{l_9}{l_8} \frac{l_7}{l_6} \ [\mathrm{N}]$$

$R_{\mathrm{pI}} = F_s$—*shear-breaking force, breakout force.*

Required power of the supported shovel

$$P_{\mathrm{zg}} = R_{\mathrm{pI}} v_{\mathrm{zg}} / \eta_{\mathrm{Zg}} \ [\mathrm{W}]$$

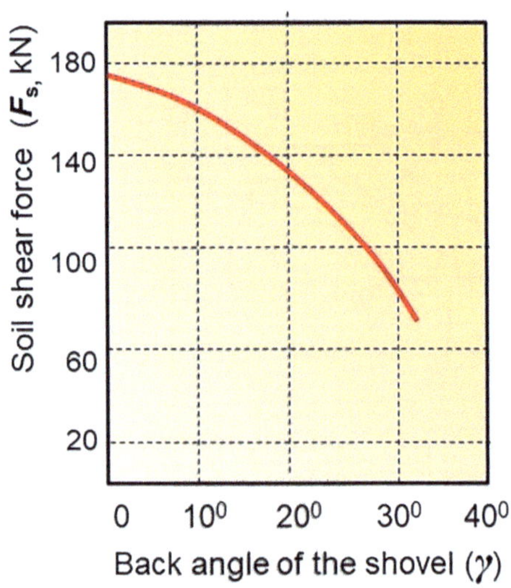

Fig. 6.34 Soil shear force when turning the shovel (middle class of loaders)

R_{pI}—shear resistance of the earth
v_{zg}—speed of rotation of the shovel edge upwards, $v_{zg} = R_0\,\omega_{lg}$
η_{zg}—efficiency rate of turning the shovel.

By rotating the shovel at the ground level, i.e. by changing the arms of the lever mechanism, the value of the shear force decreases, Fig. 6.34.

F_l—hydraulic cylinder force to turn the shovel

$$F_l = R_{pI}\frac{l_8}{l_9}\frac{l_6}{l_7}\ [\mathrm{N}]$$

b. Case of an unsupported shovel

Loading shovel raised above the ground, Fig. 6.33. Shear resistance assumes the design load of the turning mechanism ($R_{pII} = R_p$) with the limit operating condition of rollover stability about the front axle. Soil shearing is an operation that puts the maximum load on the front axle.

Soil shear resistance

$$R_p = R_{pII} = \left[G_0 l_2 - G_p l_{12} - G_l(l_{11} + l_{13})\right]/l_1 \quad [\mathrm{N}]$$

G_o—weight of the base machine
G_p—weight of the lever mechanism
G_l—weight of an empty shovel

(middle class of loaders: R_{pI} / R_{pII} > 1.5).

Unsupported shovel power required

$$P_d = R_{pII} v_{dn}/\eta_{dn}\ [\mathrm{W}]$$

R_{pII}—soil shear resistance
v_{dn}—shovel lifting speed, 0.3–0.7 m/s
η_{dn}—efficiency rate of lifting the shovel carrier.

The force of the cylinder to turn the shovel

From the conditions of balance of forces and the ratio of the lever mechanism, the following are the forces for turning shovel F_l and lifting support F_n:

$$\sum M_{03}:\ F_l l_9 = R_l l_8;\ F_l = R_l l_8/l_9$$

$$\sum M_{01}:\ R_p l_6 + G_l l_{11} = R_1 l_7;\ R_l = R_p l_6/l_7 + G_l l_{11}/l_7$$

$$F_1 = \left(R_p l_6/l_7 + G_l l_{11}/l_7\right) l_8/l_9$$

The force of one hydraulic cylinder for turning the shovel

$$F_1 = R_p i_p + G_l i_l \quad [\mathrm{N}]$$

i_p, i_l—leverage ratios

The force of the cylinders to lift the H-carrier

$$\sum M_{02} : 2F_n l_4 + 2F_l l_5 - G_p l_{10} - G_l(l_{10} + l_{11} + (l_3 - l_{10} - l_6)) - R_p l_3 = 0$$

$$2F_n = \left(R_p l_3 + G_p l_{10} + G_l l_k - 2F_l l_5\right)/l_4$$

The force of one hydraulic cylinder of an H-carrier

$$F_n = \left(R_p l_3 + G_p l_{10} + G_l l_k - 2F_l l_5\right)/2l_4 \ [\mathrm{N}]$$

6.7 Load Spectrum of the Working Device

First, unfavorable cases of loading on the blade edge, penetration, shearing/supported and unsupported shovel, and load transfer to the shovel, as shown in Fig. 6.35, are assumed.

a. soil resistance/penetration resistance R_1, for the most difficult working conditions, soil categories II and III.
b. vertical resistance R_p acts in the middle of the shovel blade, symmetrical load, for supported and unsupported shovels
c. the resultant resistance acts at the end of the shovel edge. The moments on the shovel from resistance components burden the shovel carrier, and the hydraulic cylinder loads. Unsymmetrical loading causes complex stress in the shovel carrier
d. load due to load transfer in the shovel.

The load transfer from the blade of the shovel is transferred to the parts of the working device. The key part of the mechanism of the working device is the H-carrier of the shovel (at the cost of production). The support is made from a solid rectangular profile, usually of constant width and variable height, from structural steels (or increased strength, high values of impact toughness at low temperatures, as well as resistance to brittle fracture).

The load forces and moments of resistance in the joints are set for the cases of the highest symmetric and asymmetric load. The calculation procedure requires the

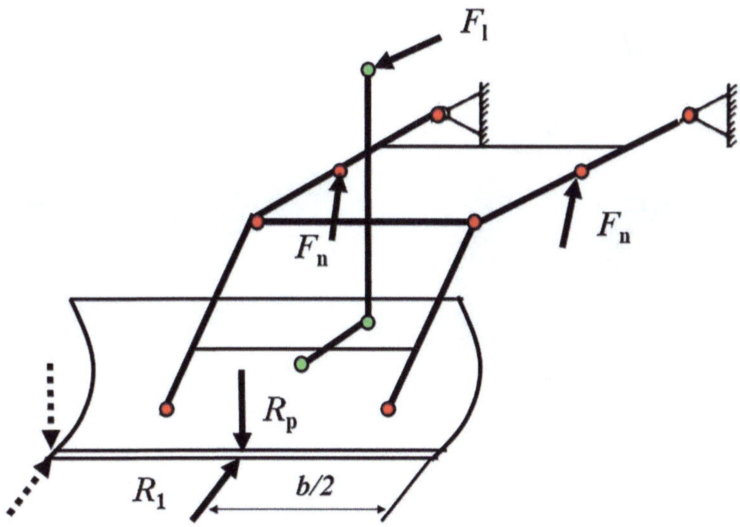

Fig. 6.35 Forces of the lever mechanism

application of the finite element method (software package, *Solidworks*), fatigue calculation of the critical section, and reliability calculation.

Practically, the cross-section of the support bears the greatest stress in the area of hinged joints with hydraulic cylinders for lifting, Fig. 6.36. In this zone, care is taken to eliminate stress concentration, profile reinforcements are performed by welding, and compound stress is checked. The stress of the transverse beam connections, at the beginning and at the end of the girder, are not significant. In addition to determining the dimensions of the hydraulic cylinders, it is necessary to determine the reliability of the hydraulic cylinders depending on the number of working cycles before overhaul or replacement.

Load spectrum

The working device of the loader is exposed to variable load, which can be represented by a random probability distribution function. For example, such dynamic loads can be computer simulated and given in the form of a load spectrum (collective).

Figure 6.37 shows the load spectrum using the law of normal distribution (s = 2, medium load regime). The relationship between variable and maximum load ($\mu = F_i/F_{maks}$) is shown as a function of the number of load changes (n) for the working life of the machine (10,000 h).

Practically, the unit spectrum is translated into a simulation model of the load spectrum (bar chart and tabular form) for quick tests on the test bench in the order

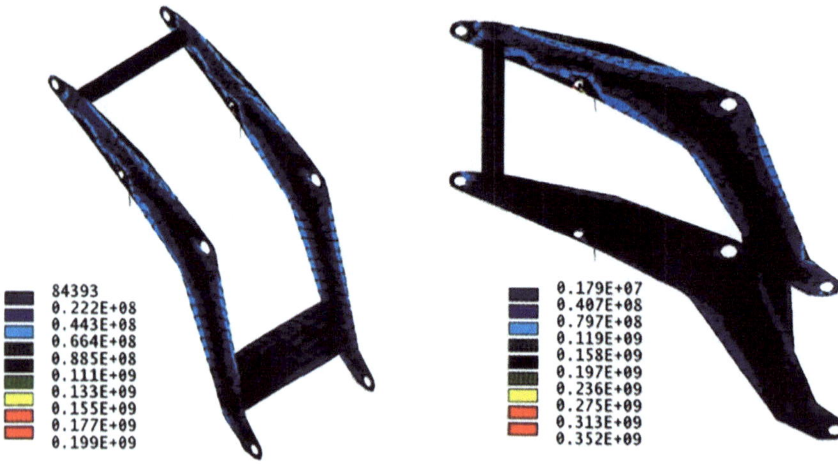

Fig. 6.36 Stresses of H-carriers under symmetric and asymmetric loading

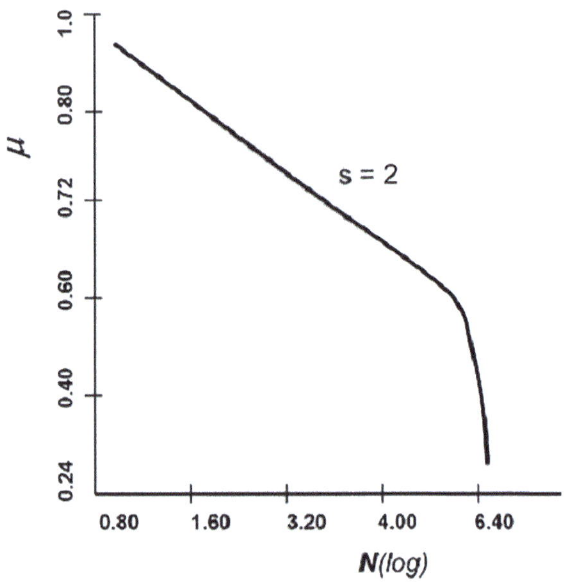

Fig. 6.37 Unit load spectrum of the working device [7]

of repetitions, Fig. 6.38, which includes device loads from penetration (categories I, II, III), shear and carrying loads in the shovel, all in 16 load blocks.

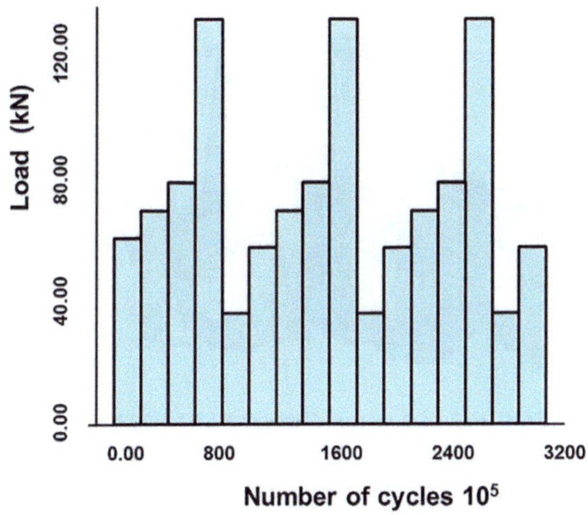

Fig. 6.38 Load spectrum for accelerated tests

Load simulation is performed with a vibration/servo-hydraulic device (load/number of cycles). The fatigue curve of parts is determined by monitoring the cracking in critical sections. By testing the critical and working load distribution functions, the conditions for predicting the reliability of parts and the working device are obtained. The starting parameters for shaping the load spectrum of the device are: the required working life of machine L, the frequency of work cycles f_i, the time of use by soil categories / working conditions T_i, and the load intensities R_i.

6.8 Universal Loaders - Backhoe Loader

Light wheeled loaders with a rigid frame are equipped at the back with an excavator device, Fig. 6.39. Such a general-purpose machine or universal loader (backhoe loader) is used to perform various communal works. A universal shovel is placed on the front of such a machine, for loading loose material, using forks, and more. In agriculture, the adapted tractor undercarriages are often used as the base for attaching the working device on the front part.

The excavator mechanism is connected to the digger frame using the so-called sleds, which provide transverse movement of the depth shovel. In order to achieve longer shovel reach, the arm is made with telescopic extension. When the machine is working, the wheels can be partially or completely unloaded by lifting the machine into a level position using hydraulic cylinders. Two hydraulic stabilizer feet are used to raise the rear part of the machine. The front part of the machine is lifted by the working device of the loading shovel, which serves as the front support during digging.

Fig. 6.39 Backhoe loader—universal loader (*CAT 434 F2*, 11 t, 70 kW)

The universality of the machine implies the possibility of connecting different tools, Fig. 6.40. These are the following tools, at the front: universal shovel, dozer knife and forks, and at the back: an excavator deep shovel for trench excavations, a hydraulic hammer for breaking hard objects, and a vibrating plate for soil compaction. A hook for lifting loads with a smaller load capacity, or a grapple, can be attached as well.

The maneuverability of the machine is an important feature of these machines. They can turn within a circle of 6 m (on the road), which is essential for working in an urban area. It is steered with 2 or 4 wheels and diagonally, using wheel sizes 24″ or 28″, with wide tires. In some machines, the track of the rear wheels is larger than the track of the front wheels, which provides a larger stabilization base. A comfortable cab of type ROPS and FOPS provides good visibility for driving and working with all tools.

Fig. 6.40 Backhoe loader–universal loader (*JCB 3CX*, 6 t, 55 kW)

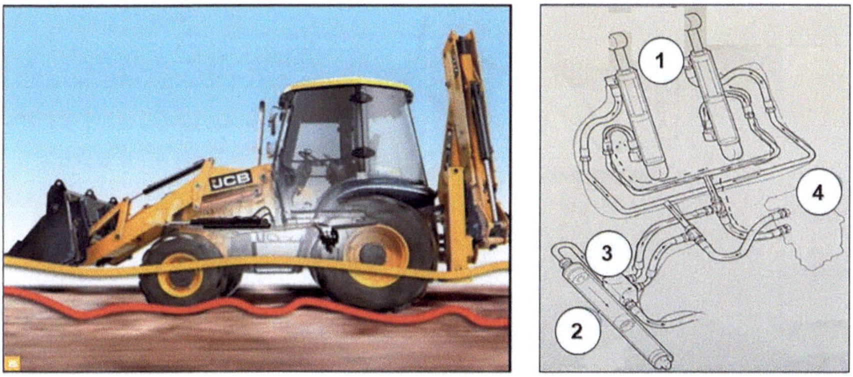

Fig. 6.41 Smooth Ride System [8]. 1—hydraulic cylinders of the carrier, 2—nitrogen tank, 3—control block, 4—oil tank

Three joysticks are used for control. One stick is used to work with the loader. The other two sticks are used to operate the excavator mechanism. By turning the seat, the excavator operation is started. The machines are equipped with a powershift automatic gearbox (4 gears forward and backward), or manual changes, and reach a speed of 40 km/h. The joysticks take over some of the pedal functions, such as the *inch brake* pedal for automatic transmission disengagement, with the full power of the engine being made available to the work device for shearing the soil and lifting the loader shovel.

The oil tank holds about 130 L. The working pressure of the oil is about 250 bar. The flow rate of the main constant flow pump is 75 l/min, which supplies the operation of the loader or excavator, and the auxiliary 50 l/min for steering the wheels. Each of the pumps supplies one hydraulic circuit. Pump regulation ensures that the pressure changes within a certain range at constant engine power.

Advanced machines use the function of smooth driving (shock absorbing, damping of inertia - leveling the shovel), Fig. 6.41. This reduces the bouncing of the machine when transporting load, increases the efficiency by reducing material losses and increases the comfort of driving. The hydraulic cylinders of the working device are connected to a gas shock absorber (nitrogen).

Figure 6.42 shows the handling of the excavation device of the deep shovel.

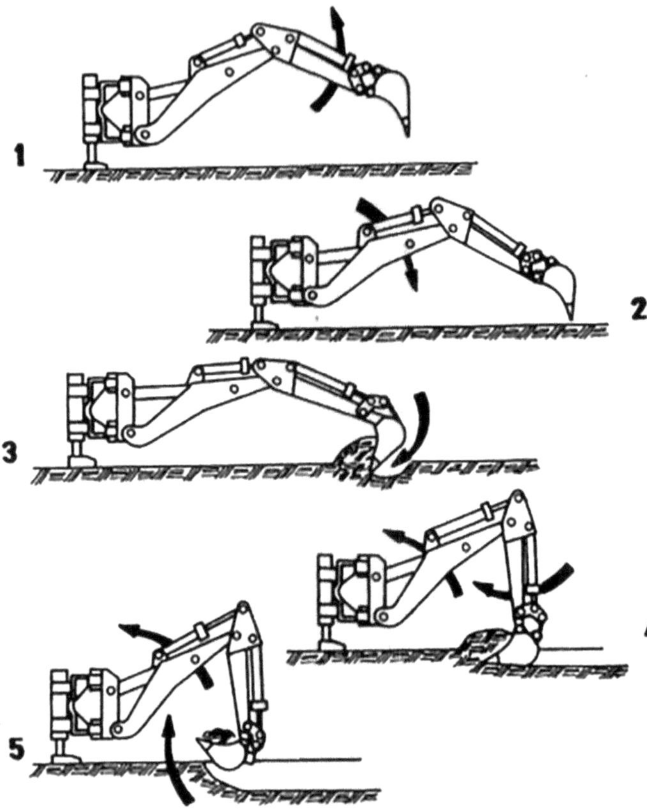

Fig. 6.42 Operation of the excavation device of the deep shovel. 1—reaching the excavation with a shovel hand, 2—lowering the branch (boom), 3—digging with the shovel, 4—digging and filling the shovel by hand, 5—closing the shovel, raising and turning the boom

References

1. Caterpillar Performance (2020) Handbook, images available. www.wheeler.com
2. Filla R (2003) Anläggningsmaskiner: Hydrauliksystem i multidomäna miljöer. Proceedings of Hydraulikdagar, Linköping, Sweden. https://gotoyard.com/en/factors-that-affect-the-efficiency-of-front-loaders
3. Liebherr Hydraulic systems (2020) Product range. www.liebherr.com/en
4. Addison A (2018) An electro-hydraulic traction control system for heavy duty off-road vehicles: formulation and implementation. A Dissertation, Purdue University, https://hammer.purdue.edu/articles/thesis/An_Electro-HydraulicTractionControl_System for_Heavy_DutyOff-Road_Vehicle
5. Mikulic D (2024) Gradevinski strojevi. Teorija i konstrukcija. Veleuciliste Velika Gorica, Velika Gorica
6. Ozdogan M, Ozdogan H (2019) Wheel loader buckets & digging forces. www.researchgate.net/publication/335618855

7. Mikulic D (1992) Doprinos modeliranju pouzdanosti gradevinskih strojeva. Dissertation (in Croatian). Fakultet strojarstva i brodogradnje Zagreb
8. JCB Brochure (2020) Images available. www.jcb.hae.org.uk

Chapter 7
Backhoes

Abbreviations

U [m^3/h]	Backhoe efficiency
V_b [m^3]	Loading shovel volume (bucket)
T_c [s]	Working cycle time
G_o [N]	Basic machine weight
G_g [N]	Boom weight
G_d [N]	Arm weight (stick)
G_l [N]	Shovel weight (bucket)
k [–]	Backhoe safety factor
R_1 [N]	Soil cutting resistance
k_1 [–]	Specific cutting resistance
b [m]	Shovel width
h [m]	Depth of the soil layer
R [N]	Digging resistance
F_d [N]	Arm (stick) hydraulic cylinder force
F_l [N]	Force of the hydraulic cylinder to turn the bucket
F_g [N]	Boom hydraulic cylinder force
S_E [J]	Hammer impact energy

7.1 Types and Purpose of Excavators

Excavators are used for the construction works of soil excavation, loading or transfer of soil. Depending on the place of work, they are divided into land excavators and floating excavators. Each group of excavators is classified into individual types,

D. Mikulić and D. Mück, *Construction Machines*,
https://doi.org/10.1007/978-3-031-67845-5_7

Fig. 7.1 Backhoe—tracked excavator [1]

depending on the use, mode of operation, and performance of the machine. In land-based excavators, the moving machine is a crawler or wheeled vehicle, and in floating dredges, the basic machine is the floating section.

Excavators with cyclic operation and excavators with continuous operation differ according to the mode of operation. Cyclic excavation and loading are performed with one tool in a specific working cycle. Continuous excavation has an even flow, and is performed with tools that have a series of loading buckets (rotor device, etc.). Floating dredges are used in the construction and maintenance of waterways, harbors, and piers, for excavations during hydrotechnical works on rivers, lakes, and the sea, for the extraction and preparation of gravel and sandy soil.

Tracked and wheeled excavators are intended for performing various works in all soil categories. With additional tools, such as hydraulic hammers, pliers, and demolition jaws, etc., their use has been expanded.

The largest number of excavators is carried out with a deep shovel (*backhoe*), Fig. 7.1. The loading shovel is filled by pressure from the front, it is then lifted together with the lever mechanism and turned to the place of shaking into the means of transport, or on the disposal site.

7.2 Backhoes with a Deep Bucket

Excavators with a deep bucket are used for excavating soil above and below the level of movement of the machine, with unloading to a truck or to a landfill. Usually, this variant of the hydraulic excavator is called a ***backhoe*** in professional jargon. It can be used in all types of soil, and replacing the shovel with additional tools (hammer, jaws) expands their area of use.

Tracked backhoes are produced in several classes. The largest number of operational excavators holds the class of range 10–30 t, power 50–150 kW, reach 7–10 m and digging depth 5–7 m, followed by the class 30–50 tons, power 150–250 kW, reach 9–2 m and depth excavation 6–9 m, followed by heavy excavators 50–100 t, power 250–300 kW, reach 12–16 m and excavation depth 9–12 m. The largest number of wheeled backhoes is produced in the 10–20 t class, with the power of 80–110 kW, reach of 8–12 m and excavation depth of 5–8 m. Medium and heavy excavators are carried out on tracks. Loading shovels are of different volumes, most often 100–1000 l (liters).

7.2.1 *Working Field and Working Cycle of the Backhoe*

The main parts of the lever mechanism of backhoe are: excavator boom (branch), shovel arm (stick), shovel (bucket), hydraulic cylinders (boom, stick and shovels), and hydraulic installation components. The working field of the backhoe is determined by the path of the bucket teeth, from the smallest (*position I*) to the largest values of angle change (*position II*), Fig. 7.2.

Backhoe efficiency:

$$U = k_p k_r V_b / T_c \,[\mathrm{m}^3/\mathrm{h}]$$

V_b—loading shovel volume (bucket), T_c—working cycle time.
k_p—bucket filling factor, k_r - soil looseness factor.

The working cycle of the excavator consists of the operation of filling the shovel and lifting the boom with turning to the unloading place. This is followed by the unloading of the material and the turning of the boom to the excavation site, which completes the working cycle, and it takes about 20 s, Fig. 7.3.

7.2.2 *Backhoe Undercarriage Construction*

The design of the backhoe consists of the lower and upper assembly with the working device, Fig. 7.4. The undercarriage can be tracked or wheeled. The upper assembly enables the backhoe to work. On the front part of the excavator on the left side, there is a cabin, next to which the excavator boom is connected to the undercarriage frame.

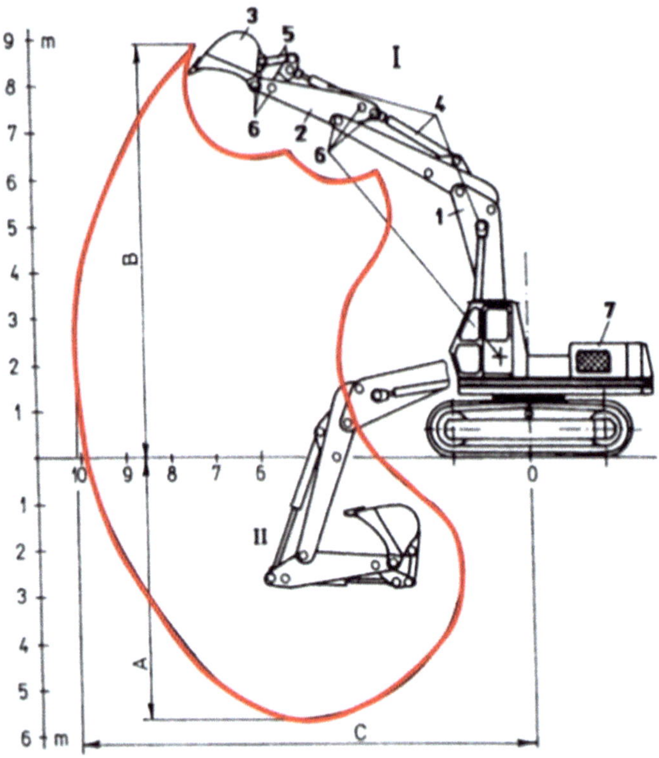

Fig. 7.2 Working field of backhoe. 1—boom, 2—stick/hand/arm, 3—shovel, 4—hydraulic cylinders (boom, arm, bucket), 5—bucket lever, 6—joints, 7—drive engine; **a** excavation depth, **b** excavation height, **c** excavator reach

Tracked undercarriage

The undercarriage consists of a rigid X-frame, a crawler mechanism and a power transmission device, Fig. 7.5. The performance of the machine movement is determined from the conditions of developing the necessary speed of the excavator (5–10 km/h) on a flat surface and from the conditions of overcoming the climb when working in pits, ramps and the like, up to 80%. There are several track widths (500/600/700/800 mm) for the use of excavators on low-bearing ground. A bulldozer blade is often placed in front of the excavator undercarriage to clean and level the approach terrain. Shock absorption on the guided wheel of the track is performed by a spring or by means of a special silicone oil—lubricant (PIT). Track tensioning is performed in one of the following ways, by tensioning the spring, using a lubricating chamber, or their combination.

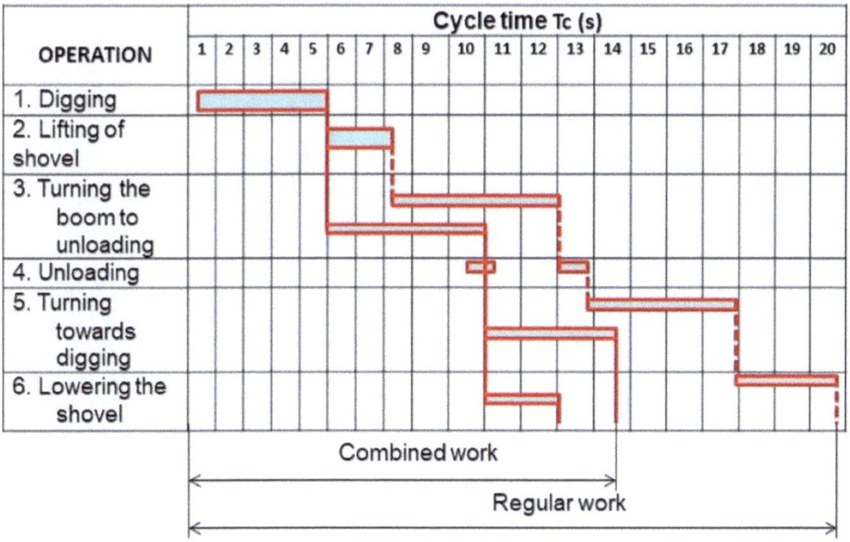

Fig. 7.3 Diagram of the backhoe working cycle

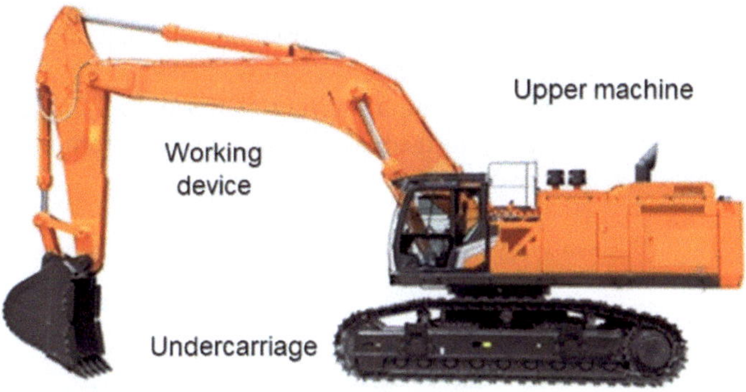

Fig. 7.4 Main units of the tracked backhoe

Ring gear

The connection of the upper machine and the undercarriage is realized with a gear ring and a large bearing, Fig. 7.6a. The excavator platform rotates using a hydraulic motor and a reducer. Axial-piston hydraulic motor. For heavier excavators, a planetary reducer with two gears is used or two reducers are installed, which ensures greater reliability in operation (b and c). The large bearing is a roller bearing, characterized by the method the upper and lower machine are supported. The bearings used are

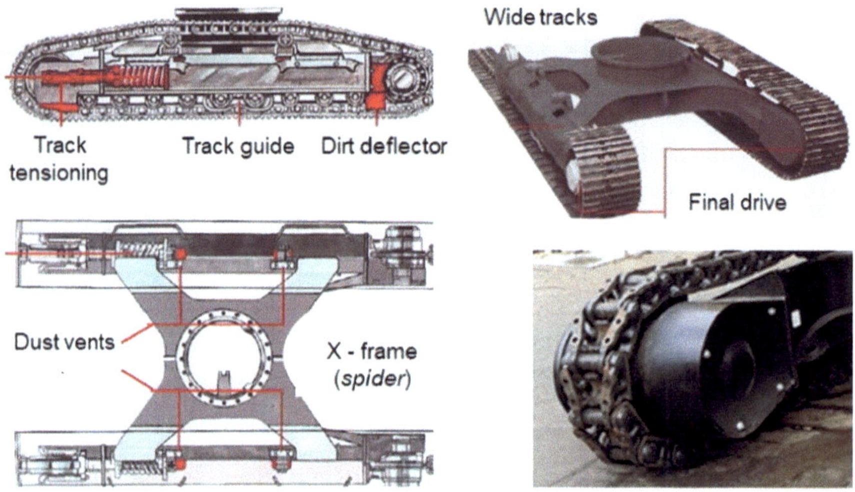

Fig. 7.5 Main undercarriage assemblies of a tracked backhoe [2]

single-row ball bearings, single-row roller bearings, and double-row ball bearings for higher load capacities. For special design requirements, bearings are used that require additional sealing to prevent water penetration. Small and large gears make up the assembly of the excavator ring gear.

Some excavators are equipped with a leveling platform, Fig. 7.7. Leveling or adjustment of the platform is performed at an angle, when performing additional works in a canal or on land leveling.

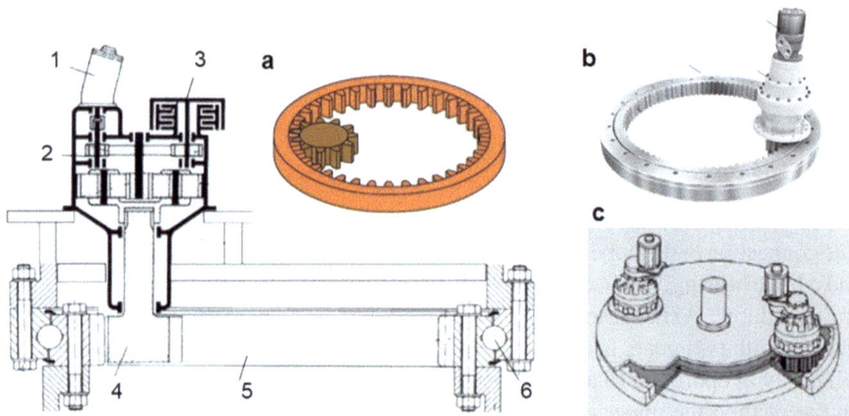

Fig. 7.6 Swivel assembly of the backhoe platform (adapted [3, 5]). **a** Section of the assembly for turning (1—hydraulic motor, 2—reducer, 3—brake, 4—small gear of the reducer, 5—large gear, 6—ball bearing); **b** hydromotor and brake with a reducer, **c** turning the platform with two reducers

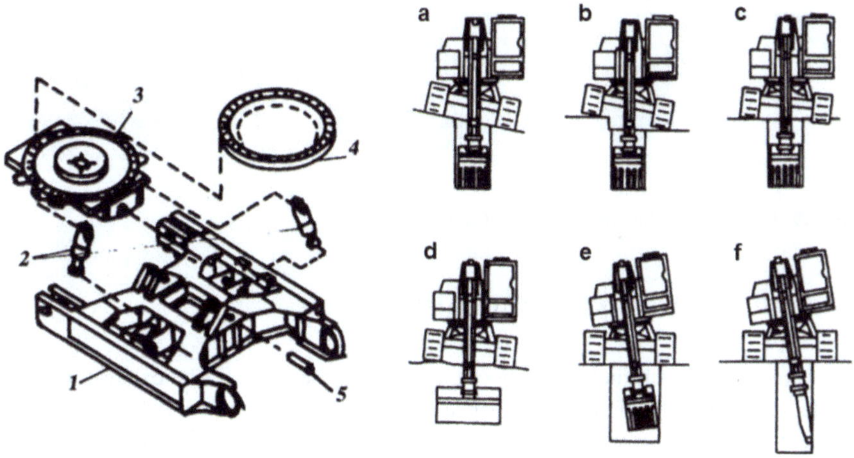

Fig. 7.7 Platform angle adjustment. 1—lower unit frame, 2— hydraulic cylinders, 3—lower ring, 4—upper ring, 5—pin; **a** construction of canals on a slope, **b** construction of canals on pavement, **c** on low-bearing soil, **d** land planing, **e** work below "cables", **f** arrangement of canals

Wheeled undercarriage

Wheeled backhoes reach a weight of 20–25 (30) tons, Fig. 7.8. The robust undercarriage ensures movement on uneven terrain. During operation, stabilizers are used as supports. A hydrostatic or hydromechanical transmission (*powershift*) is used to transmit the power of the diesel engine to the wheels, which achieves higher speeds, up to 30–40 km/h. Medium and heavy machines use double wheels 11.00–20 (*Dual Pneumatic*). The front axle offers lateral sway and hydraulic steering to adapt to the terrain. The height of the undercarriage is 350–400 mm.

Modern machines offer decrease of inertia and reduction of swaying, by adding a gas accumulator (nitrogen) to the hydraulic cylinder system, Fig. 7.9. This reduces the bouncing of the machine. Shocks are absorbed and the levelling of the loading shovel is maintained.

Fig. 7.8 Wheeled backhoe with two-part boom

Fig. 7.9 Calming backhoe rocking (*Ride Control*) [2]

7.2.3 Backhoe Lever Mechanism

The lever mechanism includes a boom, an arm, and a shovel. The excavator boom is the key assembly of the lever, since the type of boom mainly determines the application of the excavator, followed by the type of shovel. Different types are:

- backhoe with one-piece standard boom, Fig. 7.10a.

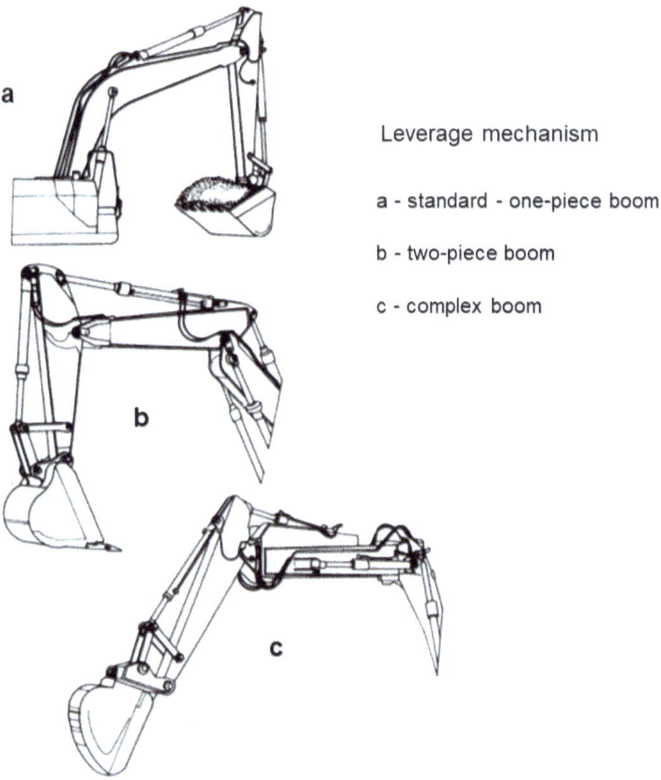

Fig. 7.10 Backhoe leverage mechanism **a** standard—one-piece boom, **b** variable boom, for digging near obstacles, **c** complex (parallel) boom for special spaces

- backhoe with a two-part (variable) boom, for digging near and under walls, in confined spaces, and the like (b).
- backhoe with a compound boom, provides lateral parallel movements of the arm to perform work in very narrow spaces, laterally close to walls and obstacles (c).

The main parts of the backhoe lever are shown in Fig. 7.11a–c.

The length of the arm depends on the depth of digging and the choice of the shovel.

The length of the boom determines the excavation field, for example, with a one-part boom, Fig. 7.12a, and a two-part boom (b),

The parameters of the working field of the backhoe (A, B, C, D, E, F, G) with one-piece and two-piece boom are shown in Fig. 7.12. The cutting force on the shovel teeth depends on the pushing force of the hydraulic cylinder of the arm and the length of the arm. Therefore, several arm lengths are offered, 2100 mm, 2500 mm, 3000 mm. The cutting force on the shovel teeth is:

- by pushing the hydraulic cylinder of the hand: 66 kN/2100 mm; 59 kN/2500 mm; 54 kN/3000 mm,
- by pushing the hydraulic cylinder of the shovel: 79 kN/2100 mm; 79 kN/2500 mm; 79 kN/3000 mm.

Backhoe shovel (bucket)

There are several types of backhoe shovels: *general purpose shovel, grid shovel, heavy duty bucket, canal cleaning shovel,* etc. The basic indicator of an excavator bucket is the volume (L, m^3), which primarily depends on the width (l_2) and the turning radius (X, R_0), as well as the filling factor (k_p), Fig. 7.13. General purpose excavator shovel data are shown in Table 7.1. For difficult conditions, they produce shovels reinforced in critical places.

The bucket filling factor (k_p) is defined as a correction coefficient that corrects the theoretical volume of the material, which causes a change in the shape of the filling (A, B, C) and the position of the center of gravity of the load in the shovel. The filling factor is, for example, for light excavation 0.95–1.25 (loose dry soil, fine gravel, sand), for medium excavation 0.8–0.9 (excavation of dry or less moist loam, coarse gravel, compacted soil and soft weathered rocks), for medium-hard excavation 0.65–0.8 (fractured and weathered rocks), for hard excavation 0.4–0.65 (excavation of rock, by blasting sandstone and conglomerate and very hard plastic clays).

The teeth of the shovel make it easier to cut hard layers of soil. The teeth are placed in front of the knife blade, so that they start cutting, and the knife blade finishes cutting. The shape of the teeth, Fig. 7.14, determines the purpose of the shovel: short teeth (for tough soil), long teeth for general purpose, long teeth for heavy soil (so-called standard teeth), teeth for deep soil penetration, sharp teeth at the ends of the shovel or in in the middle of the shovel (when penetration is more important than strength). With teeth, a lower cutting resistance is achieved in hard soil than a shovel blade without teeth. However, in compacted soil, the resistance to soil increases.

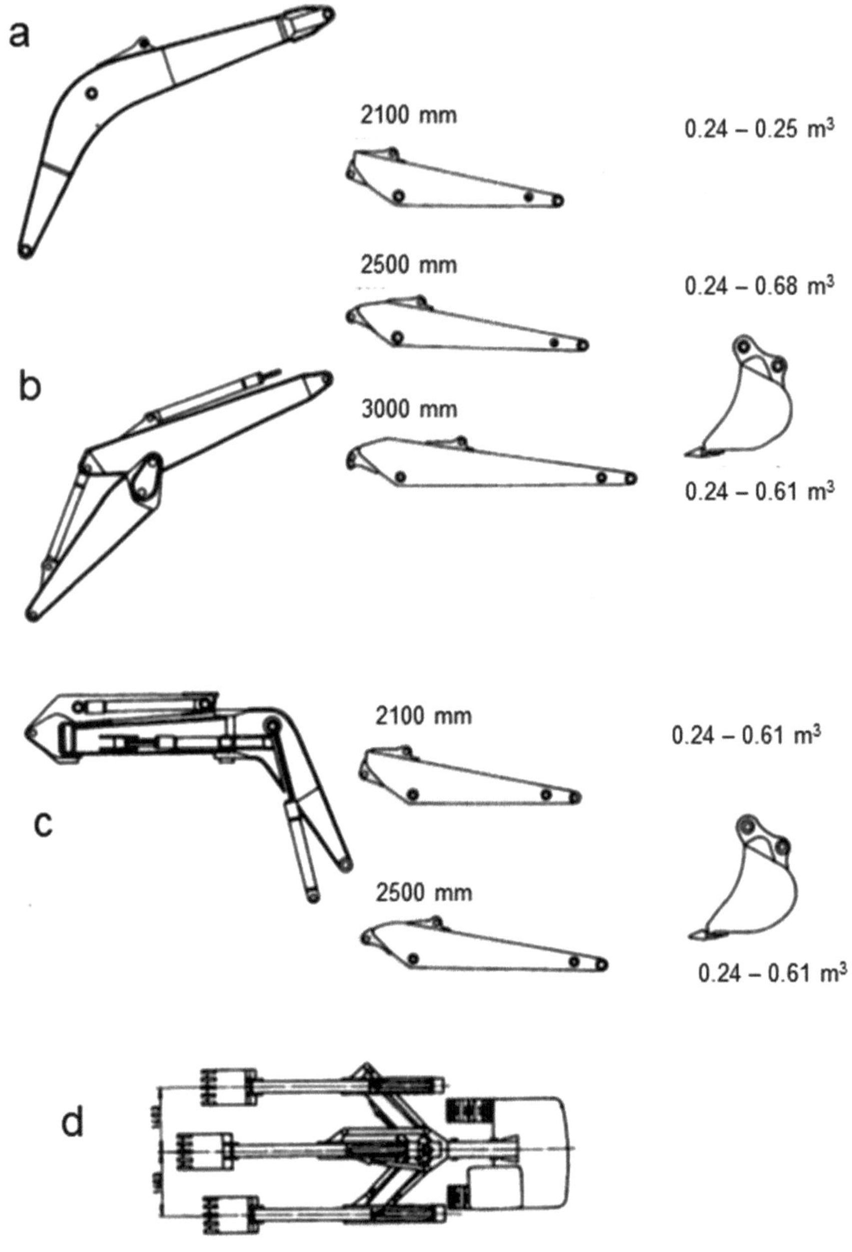

Fig. 7.11 Backhoe lever and choice of bucket (13 t, 63 kW) [2]. **a** one-piece boom, **b** parts of a two-piece boom, **c** parts of a parallel boom, **d** lateral movements of the complex boom (left, right)

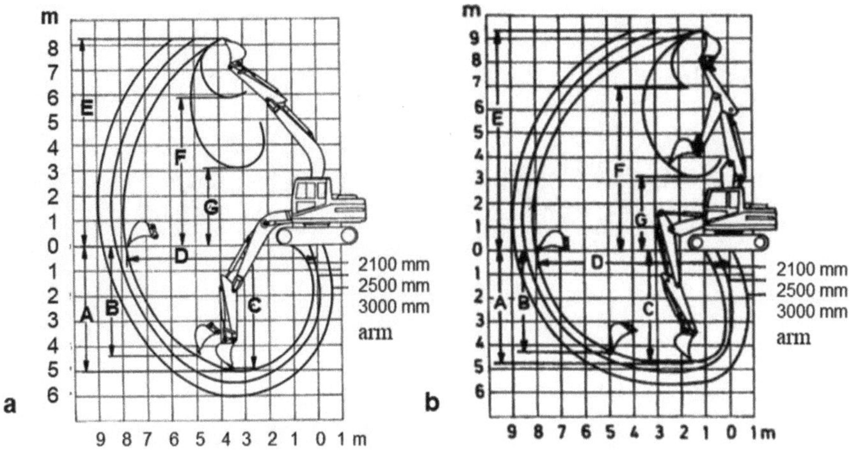

Fig. 7.12 Comparison of the backhoe working field (13 t, 63 kW, 0.54 m3) **a** one-piece boom, **b** two-piece boom

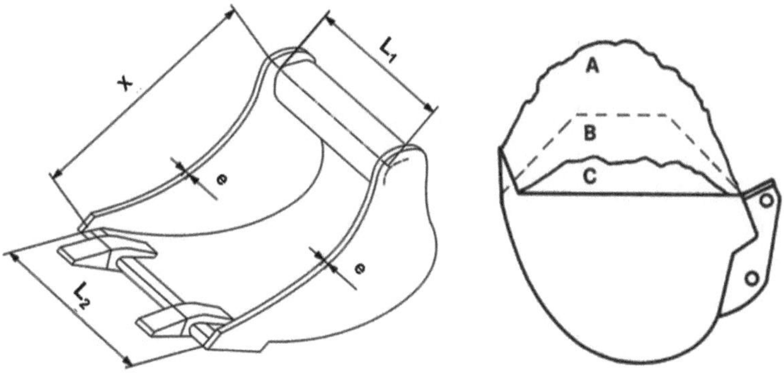

Fig. 7.13 Shovel (bucket) dimensions and different filling shapes (ISO 7451)

Table 7.1 General purpose buckets, radius $R_0 = X = 1220$ mm (*CAT*/13 t, 63 kW)

Bucket width l_2 (mm)	450	600	700	900	1000	1100	1200
Volume (*L*)	235	330	400	540	6100	680	750
Number of teeth	3	3	4	5	5	6	6
Mass (kg)	234	264	287	324	348	371	391

The resistance to cutting the soil depends on the condition of the teeth and the blade of the shovel, and the cutting method (blocking and semi-blocking cutting methods cause the greatest resistance). If the worn teeth are not replaced, the cutting resistance increases, so the wear intensity is transferred to the interdental blade. Since

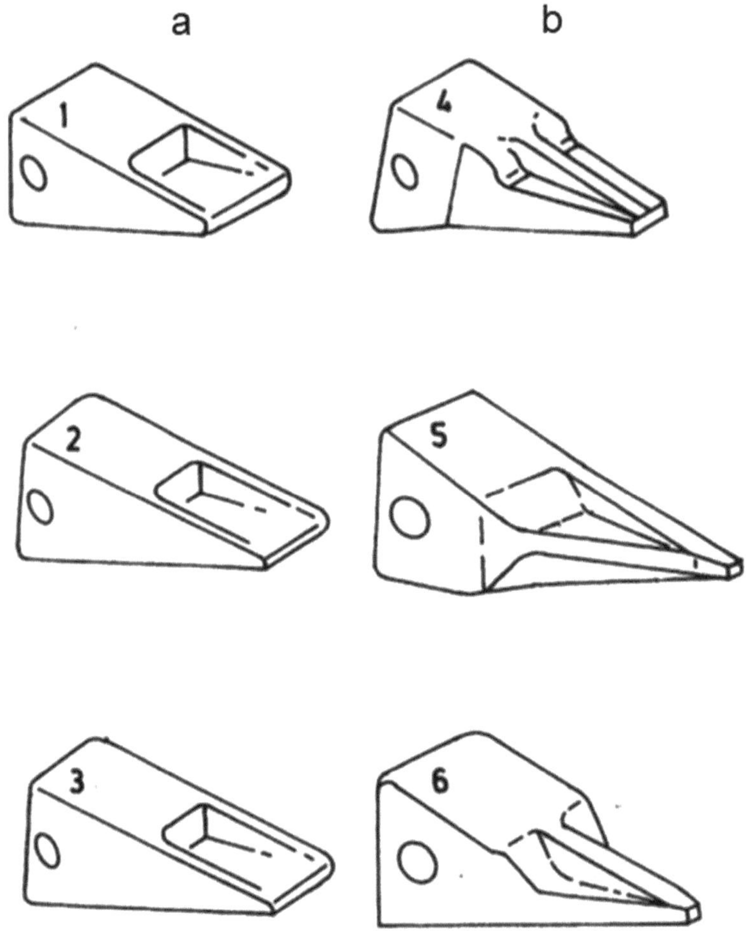

Fig. 7.14 Shapes of backhoe teeth [3]. **a** standard teeth (1—short tooth, 2—long general-purpose tooth, 3—standard tooth, heavy duty), **b** teeth for deeper penetration (4—basic tooth, 5—end tooth, 6—middle tooth, max penetration)

the shovel works with a variable cutting angle, the tip of the tooth blade and the knife blade wear more quickly. For example, the teeth are replaced after wear (5–7 mm).

The arrangement of the teeth along the knife blade is very important. The teeth are placed on the ends of the shovel in order to increase the looseness of the soil and prevent friction of the sides of the shovel against the side edges of the excavation. Proper tooth spacing protects the knife blade and reduces the cutting resistance. A larger distance between the teeth causes more wear on the knife blade than the teeth, and the cutting resistance increases. The ratio between the teeth and the width of the teeth is in the range of 2.0–2.5. The tooth width is calculated according to the metal strength criterion. The angle of sharpening the teeth depends on the hardness of the rock; 25–30% for soft, 30–35% for hard rocks.

7.2.4 Backhoe Hydraulic Installation

The hydrostatic system of the backhoe has two closed and independent circuits: one for the undercarriage—for the movement of the machine, and the other for the turntable and the functions of the working device, Fig. 7.15. The diesel engine drives a double variable pump that creates the required oil pressure and flow. The variable pump adjusts according to the pressure signal. The *joysticks* are used to control the hydraulic cylinders and hydraulic motors via the distributor.

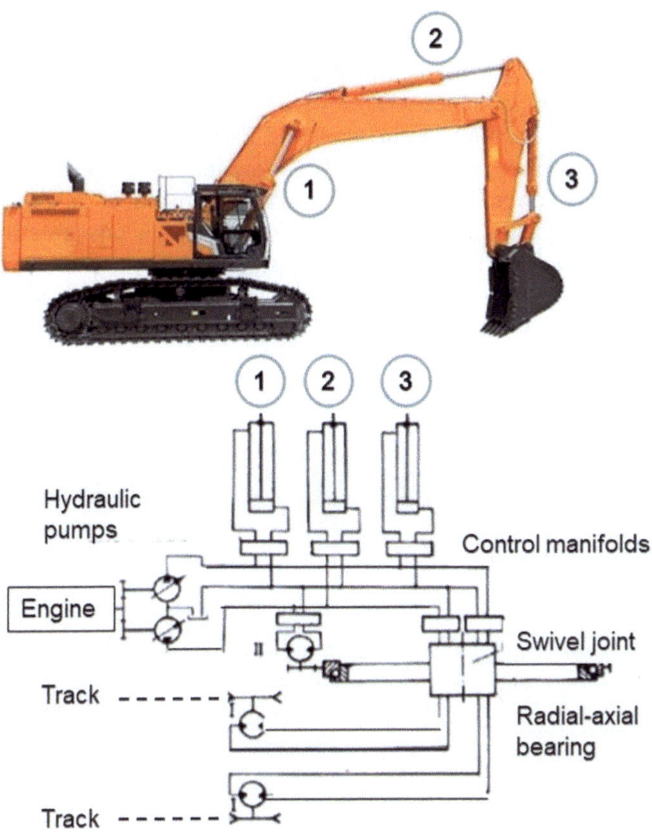

Fig. 7.15 Display of a simple backhoe hydraulic installation. 1—boom cylinder, 2—stick/hand/arm cylinder, 3—bucket cylinder

The working pressure of the oil when the machine is running is 250–350 bar, and when it is moving it is about 350 bar. The amount of hydraulic oil in the installation of medium excavators is about 60 l, and about twice as much in the oil tank (120 l). When moving the machine using the hydrostatic transmission, the total power of the driving diesel engine is made available, and the machine control (forward, backward, left, right, turn) is performed with independent drive of the left and right tracks.

The modern hydraulic installation of a backhoe is shown in Fig. 7.16. The machine operator first determines the movement of the machine or the work with the working device. When moving the machine, the operator uses lever arms and pedals, which are controlled by tracks, with left and right hydraulic motors. The working device is controlled with the joystick, raising and turning the bucket. Two cylinders are used to raise the boom: one cylinder for working with the hand, and one cylinder for working with the shovel.

The turning of the machine platform is done using a hydraulic motor. Two high-pressure variable axial-piston *load-sensing* pumps produce sufficient flow and maintain a lower standby oil pressure, making the system quickly available. An auxiliary supply pump is connected to the main pump to supply pilot oil for machine operation.

The machine operation process takes place with two simultaneous movements, and the possibility of combining them is multiple. Each of the hydraulic cylinders and hydraulic motors operating simultaneously with separate hydraulic circuits requires an adjustable drive power. The power depends on the amount of digging resistance and piston speed, which is adjusted automatically. Hydraulic oils prescribed by the machine manufacturer are used in the installation.

In order to reduce hydro shocks due to higher inertial forces and reduce vibrations when using a larger number of hydraulic cylinders, hydro accumulators are installed,

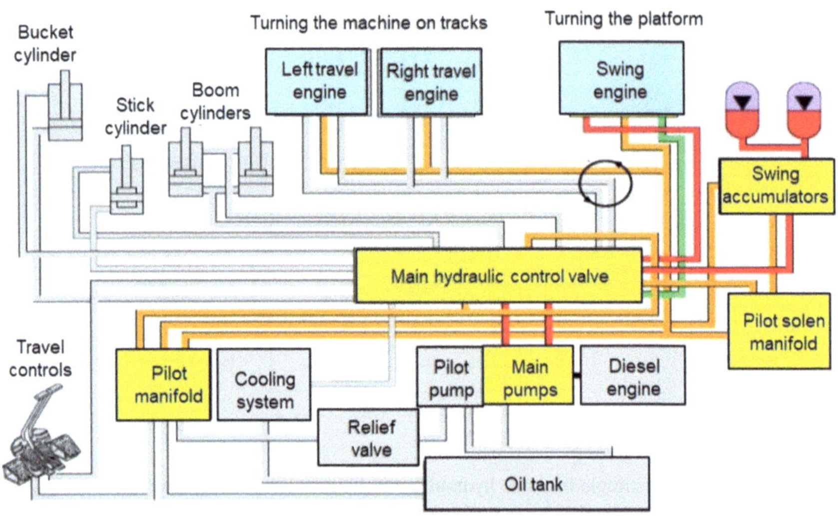

Fig. 7.16 Block diagram of the backhoe hydraulic installation (*35 t, 336E*) [2]

which accumulate energy when they are not needed and release energy when it is needed. When the pistons approach the end position, shock-free shock absorption, braking, oil heating, and battery charging follow. The result of all the losses in the system and the maintenance of the working temperature around 600 °C is maintained by the oil cooler, whereby the oil maintains a viscosity of 35 cST.

Backhoe control

To operate the excavator, it is necessary to move the control sticks. The machine operator can achieve three or more simultaneous movements of parts of the working device. However, following three simultaneous movements requires great stress—causing rapid fatigue and reduced working capacity. Practically, the number of simultaneous movements of parts is reduced to two. Combinations of movements of parts of the working device of the excavator, which are carried out simultaneously, during the operations of the work cycle, are shown in Table 7.2.

Carrying out the working cycle of a backhoe requires handling skills.

I. The first operation is filling the shovel. The shovel is most often filled with the simultaneous activation of the hydraulic cylinders for arm and shovel movement (3 and 4), or by the coordination (combination) of boom and arm movement (2 and 3).
II. The second operation is turning the machine to the place of unloading into the vehicle or to the landfill. At the same time as the machine rotates, the boom (1 and 2) is raised, so that the machine is ready for unloading after the rotation is completed.
III. The material is unloaded from the shovel by moving the arm and the shovel (3 and 4).
IV. By turning the platform towards the excavation site and starting the boom, the cycle ends.

By means of two manual controls (*joysticks*) on two consoles, the simultaneous starting of parts of the working device and the turning of the platform are managed (SAE model), Fig. 7.17.

Table 7.2 Combinations of moving the backhoe during the working cycle

Operation	I. loading the shovel	I. loading the shovel	II. turning for unloading	III. unloading of material	IV. Rotation towards to excavation
1. Turning the platform			■		■
2. Starting a boom	■				■
3. Starting an arm (stick)	■	■		■	
4. Starting a bucket		■		■	

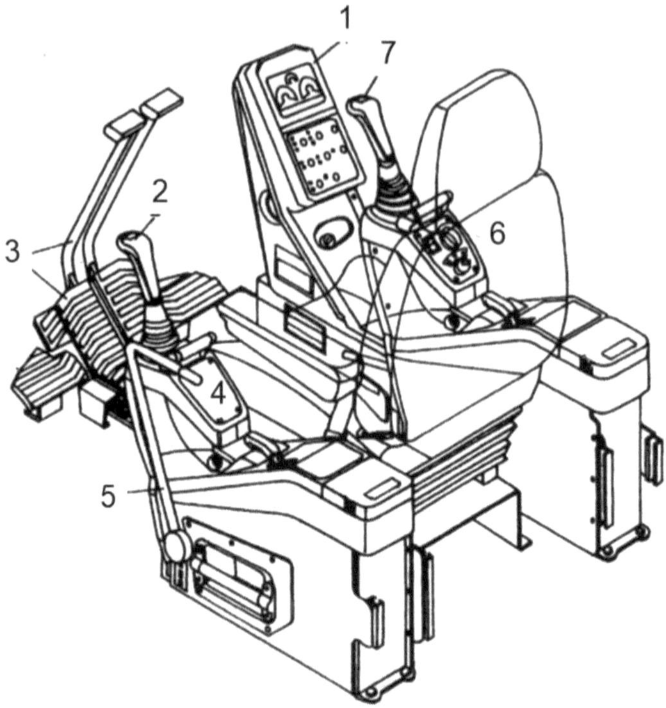

Fig. 7.17 Backhoe control console [2]. 1—right console, monitor, 2—manual control/joystick, 3—pedals and levers for driving and turning the machine, 4—left console, 5—hydraulic closing lever, 6—power mode and operation mode menu, 7—button to set the engine speed

Starting the engine, operating the excavator and turning off is on the left console. On the front of the console there is a monitor for monitoring electronic actions of the machine. Monitored are the engine speed, fuel level, engine temperature, hydraulic oil temperature, light indicators, automatic engine operation control, etc.. At the start, the power mode and mode of working with the working devices are determined.

Right joystick:

- forward–backward movements/lowering—raising of the boom (down-up)
- left–right movements/loading—unloading the shovel
- automatic engine speed control or manual/increase–decrease of number of revolutions (button on the top)

Left joystick:

- forward–backward movements / raising and lowering the shovel arm (up–down)
- left–right movements, turning the boom to one side or the other
- sound signalling (button on the top)

The diagonal movement of each joystick performs two functions at the same time.

The center of the electronic system is a computer that optimizes the tasks set by the operator. In addition to adjusting the required power, the operating parameters of the machine are continuously monitored, the current load is calculated and compared with the permitted one. By choosing the appropriate operating mode, the electronic system optimizes the working cycle. Correct choice of the power mode and operating mode results in fuel savings (1, 6—power mode and operating mode menus).

Three power modes—for three operating conditions

The operator can choose one of three engine power modes, with a choice of the engine speed.
Mode III—100%, for the regime of the heaviest use of the excavator and high productivity during excavation, truck loading, canal construction, high movement speed, and hydraulic hammer use.
Mode II—80%, for normal use of the backhoe when digging canals, utility services and more.
Mode I—60%, for lifting and laying pipes, shaping and cleaning banks, slopes, working in narrow spaces, etc.

Three working methods—mechanisms of the working device

Using the operating method menu, the machine operator can select the highest performance hydraulic circuit. The average working cycle can be shortened by up to 10%.
Method III, the hydro circuit of the boom, to achieve the highest working speeds using the boom, in heavy excavations, truck loading.
Method II, hydro circuit of turning the shovel, for work in canals, recessing, basement excavation, where the strength of the wall is important.
Method I, hydro circuit of the excavator arm, for fine control of work, shaping of banks, leveling, and others.

Automatic engine speed control or manual/7—button, three degrees:

Degree I, automatic reduction of 100 revolutions per minute,
Degree II, reduction to 1300 revolutions per minute,
Degree III, reduction to 1100 revolutions per minute, fuel saving, noise reduction.

Backhoe movement and turning

The tracked machine is controlled by pedals, Fig. 7.17. The left pedal and the left lever control the left track, and the right pedal and the right lever control the right track. Pressing both pedals or moving the levers forward moves the machine forward; otherwise, the machine moves backward. Moving one pedal or lever more than the other, the machine turns to one side or the other.

7.2.5 Calculation of the Working Device

Backhoe parameters.

The position of the backhoe working device is determined by three characteristic angles: boom, arm, and shovel (ξ_1, ξ_2, ξ_3), Fig. 7.18. The range of angle changes depends on the arrangement of the lever joints, the length of the boom (l_g), the length of the arm (l_d), the radius of the shovel (R_o), and the length of the hydraulic cylinders (h_x). The digging resistance on the shovel teeth (R) regarding intensity can reach several times the value of the weight of the soil in the shovel. The digging resistance can be divided into tangential (R_1) and normal force (R_2). Practically, it is sufficient to consider the digging resistance (R) within the limits of the ratio $R_2/R_1 \leq 0.6$. The corresponding soil shear force (breakout force) should be equal to or greater than ($F_s \geq R$). The stresses of the parts of the working device depend on the external load and the mutual position of the parts: digging resistance, weight of the soil in the shovel, own weight of the mechanism parts (boom, arm, shovel, cylinders), inertial forces, wind force, hydraulic cylinder force [6].

Soil cutting resistance (*penetration force*):

$$R_1 = k_1 b\, h\, [\mathrm{N}]$$

k_1—specific cutting resistance
b—shovel width
h—depth of the layer to be cut (Fig. 7.19).

Digging with a deep shovel is done in three ways:

- by pushing the hydraulic cylinder of the shovel (Fig. 7.19a)
- by pushing the hydraulic cylinder of the arm (b)
- their combination (Fig. 7.20).

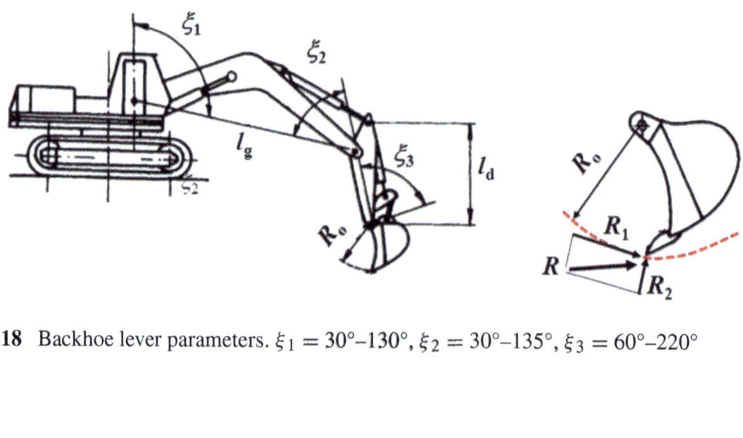

Fig. 7.18 Backhoe lever parameters. $\xi_1 = 30°–130°$, $\xi_2 = 30°–135°$, $\xi_3 = 60°–220°$

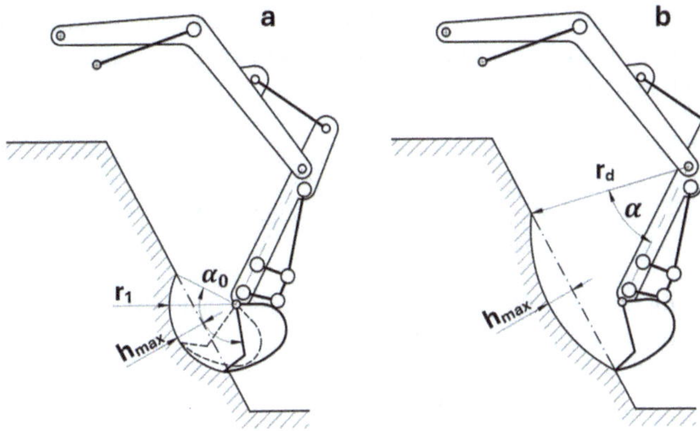

Fig. 7.19 Backhoe excavation method [4]. **a** by pushing the hydraulic cylinder of the shovel, **b** by pushing the hydraulic cylinder of the holder (stick)

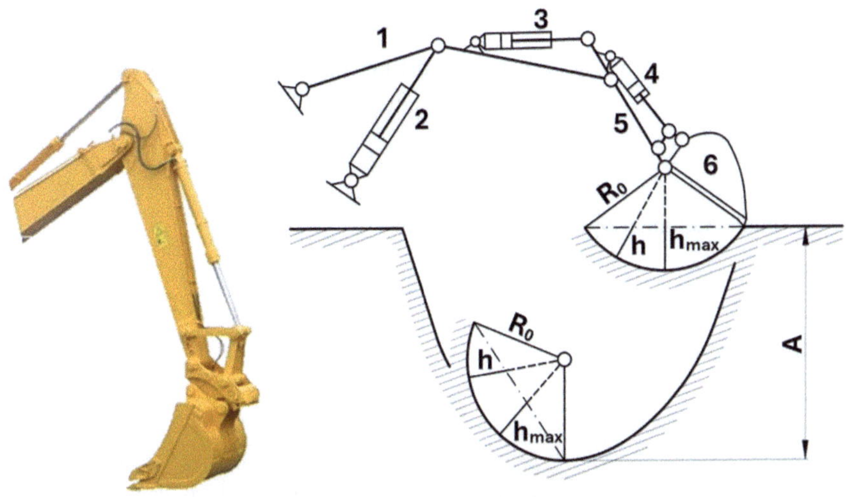

Fig. 7.20 Combined method of excavation. h_{max}—maximum thickness of the cutting layer, A—3, depth; 1—excavator boom, 2—boom cylinder, 3—arm/hand (stick) cylinder, 4—shovel cylinder, 5—arm/hand (stick), 6—shovel, R_o—shovel radius, R_1—cutting resistance

Digging the soil using hand pressure requires more power than digging with a shovel. Combined loading (starting with a shovel and continuing with the hand) reduces the total force required.

When teeth are placed on the blade of the shovel, the cutting resistance of hard soil is reduced by about 25%. If the wear of the shovel teeth is calculated, it can be calculated with less efficiency.

Maximum cutting layer depth

$$h_{\max} = r_{1/\mathrm{d}}[1 - \cos(a_0/2)]\ [\mathrm{m}]$$

$r_{1/d}$—digging radius of the bucket, Fig. 7.19. (a) or hand (b)
α_0—angle of engagement of the shovel or hand
$\alpha_{0\max} = 120°–130°$ when digging with a bucket
$\alpha_{0\max} = 60°–80°$ when digging by hand.

Practically, the depth of the soil cutting layer is required in order to optimally fill the bucket:

$$h_{\max} \approx 0.35R_o\ [\mathrm{m}]$$

Cutting resistance in the direction of the tangent to the trajectory of the bucket blade:

$$R_1 \approx 0.35k_1 b R_o\ [\mathrm{N}]$$

Normal component of digging resistance:

$$R_2 = R_1 \mathrm{tg}\alpha_0 = (0.2 - 0.6)R_1$$

Digging resistance:

$$R = \sqrt{R_1^2 + R_2^2}\ [\mathrm{N}]$$

Breakout (shear) force on the blade of the shovel ($F_s \geq R$) is generated by the hydraulic cylinders of the hand or bucket or combined.

Limiting digging forces

It is necessary to determine the limit calculation forces of the load of parts of the working device, from the conditions: the possibility of lifting the load in the shovel in different positions of the working field, achieving the maximum force of the hydraulic cylinders (R^l—shovel, R^d—hand/arm), stability of the excavator (overturning) R^{st}, and pulling the excavator R^{ad} (adhesion).

By activating the hydraulic cylinder of the shovel: $R_{HL} = \max\{R^l, R^d, R^{st}, R^{ad}\}$
By activating the hand hydraulic cylinder: $R_{HD} = \max\{R^l, R^d, R^{st}, R^{ad}\}$
Design force R_{max} that can be loaded on the working device:

$$R_{max} = \max\{R_{HL}, R_{HD}\}.$$

Digging resistance from machine safety conditions

Turning the shovel changes the direction of the digging resistance force. If the horizontal component of the force is greater than the adhesion force ($R^{ad} = \mu G\cos\alpha$, braked machine), the machine may move. The situation in which the shovel is empty and the teeth of the shovel encounter an obstacle is realistic enough and corresponds, for example, to an attempt to excavate a larger stone. The tipping axis of the excavator is located below the drive axis or tension wheel of the tracks. The highest loading force of the shovel is determined by the degree of safety (strength of the mechanism) and stability against the machine tipping over.

Around which point 0_1 or 0_2 the resistance force will be able to overturn the excavator depends on the direction of force R_1, Fig. 7.21. When the blade of the shovel is in the extension of the axis of the arm, the starting position for digging is determined. Since force R tends to overturn the excavator around point 0_1, the limit value of the force is determined from the balance of forces around same point (when digging below or above the movement level):

$$R^{st} = R_{max} = \left(G_0 l_0 + G_g l_g + G_d l_d + G_l l_l\right)/l_{ki}$$

$l_{ki}, l_0, l_g, l_d, l_l$—arm of forces
G_o—weight of the basic machine

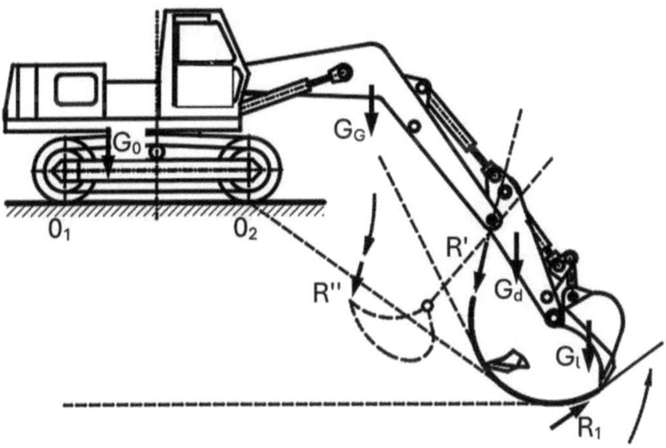

Fig. 7.21 Static stability of the backhoe [5]

G_g—boom weight
G_d—hand/arm weight (stick)
G_l—weight of the shovel with soil.

A certain position of the loaded shovel can overturn the machine around point O_2. When the piston rod of the hydraulic cylinder of the arm is most extended, load force R'' is:

$$R_p^{st} = R_{max} = \left(G_0 l_0 - G_g l_g - G_d l_d - G_l l_l\right) / l_{ki}$$

l_{ki}, l_0, l_g, l_d, l_l—arm of forces
Based on the case of rotary digging (lateral resistance R_b), when the force acts on the end tooth of the shovel, the lateral resistance is:

$$R_{b\,max} = M_v / L$$

M_v—the biggest moment on the gear ring
L—distance of the tip of the shovel teeth from the axis of platform rotation.

Lifting and transferring loads with a backhoe

Table 7.3. shows the nominal load-carrying capabilities with the standard machine boom, for reference heights of lifting the load above and below ground level (±), and for reference of the lever mechanism positions, front and side. These are the lifting capacities limited hydraulically, before overturning occurs, in accordance with the standard for excavators ISO/DIN 10567. The favorable characteristics of the excavator are in the range of 3–6 m of reach, 4.5 m of digging above ground and 3.5 m below ground.

The safety factor of the excavator from overturning conditions (DIN 15019) is determined by the ratio of stability moments and overturning moments around the overturning axis (including moments of disturbance, wind impact, etc.). Backhoe

Table 7.3 Nominal load capacity of the backhoe [2]

	1.5 m		3.0 m		4.5 m		6.0 m		7.5 m				
													m
7.5 m											*1250	*1250	5.78
6.0 m							*1950	*1950			*1050	*1050	7.19
4.5 m							*2350	2090			*1000	*1000	7.98
3.0 m					*3000	*3000	*2650	1990	*1500	1290	*1000	1090	8.38
1.5 m			*6150	5590	*3950	2990	2740	1890	1890	1290	*1100	1040	8.44
Soil			*7100	5040	4090	2740	2590	1790	1840	1240	*1250	1090	8.18
−1.5 m	*3600	*3600	7890	4890	3940	2640	2540	1740			*1500	1240	7.55
−3.0 m	*6000	*6000	*7500	4940	3940	2640	2540	1740			*2050	1590	6.45
−4.5 m			*5850	5140	*3700	2740					*3250	2490	4.83

safety factor: $k = \Sigma Ms/\Sigma Mp$, for example $k = 1.33$, i.e. 75% safety from the overturning limit.

Nominal load capacity of the backhoe (kg), backhoe 13 t, 63 kW; stick 3000 mm, shovel volume $V = 0.54$ m^3, track width 600 mm, ground pressure 0.33 bar, soil density, determine the load of the shovel ($V\,\gamma$); * lifting capacities limited hydraulically/before overturning.

Loads of hydraulic cylinders

a. by activating the hydraulic cylinder arm (stick)

In the position of maximum extension of the piston rod of the hydraulic cylinder of the arm, Figure 7.22. (stick in position II), the force of the hydraulic cylinder of the arm from the condition: $\Sigma\, M_E = 0$

$$F_d = (R_1 l_1 - G_d l_d - G_d l_d - G_l l_l)\,/l_{dk}$$

The force of the hydraulic cylinder to turn the shovel, with the transmission ratio of the lever:

$$F_l = R_1 i_p$$

Boom hydraulic cylinder force with respect to point C, shovel position I:

$$F_g = \left(R_1 l_i - G_d l_{di} - G_l l_{li} - G_g l_{di}\right)/l_{gi}$$

l_i—arm of forces

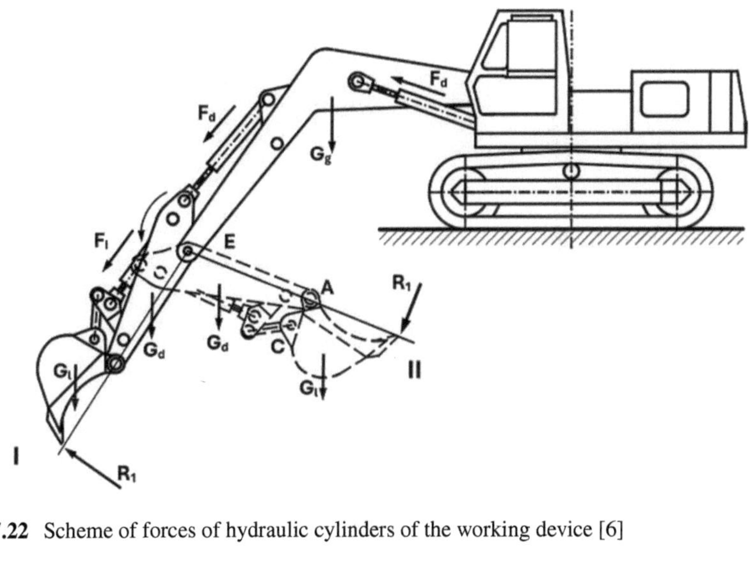

Fig. 7.22 Scheme of forces of hydraulic cylinders of the working device [6]

Diameter of hydraulic cylinders of the working device: arms (d), shovels (l), booms (g)

$$D_{\mathrm{d,\,l,\,g}} = \sqrt{\frac{4\,F_{d,l,g}}{p_{dop}\eta_{hc}\pi}}\ [\mathrm{mm}]$$

$F_{\mathrm{d,l,g}}$—corresponding hydraulic cylinder force ($\approx 50\% >$ calculated)
p_{dop}—permitted oil pressure, η_{hc}—degree of efficiency of the hydraulic cylinder.

Oil flow for supplying hydraulic cylinders:

- stick: $Q_{\mathrm{d}} = A_{\mathrm{d}}\ v_{\mathrm{min}}$, bucket: $Q_{\mathrm{l}} = A_{\mathrm{l}}\ v_{\mathrm{min}}$
- full flow: $Q_{\mathrm{u}} = Q_{\mathrm{d}} + Q_{\mathrm{l}}$
- A_{d}, A_{l}—hydraulic cylinder piston surface, v_{min}—piston speed

Based on flow Q_{u}, the number of revolutions of the drive engine n_{min} and the pressure in the hydraulic cylinders, the choice of the pump of the hydraulic device follows.

b. by activating the hydraulic cylinder of the bucket

The change of force in the hydraulic cylinder of the bucket with the angle of engagement of the shovel is shown in Fig. 7.23.

At the zero position of the bucket (0), i.e. when the piston rod of the hydraulic cylinder of the bucket is fully retracted, the required force F_{l} to overcome the cutting resistance R_{l}:

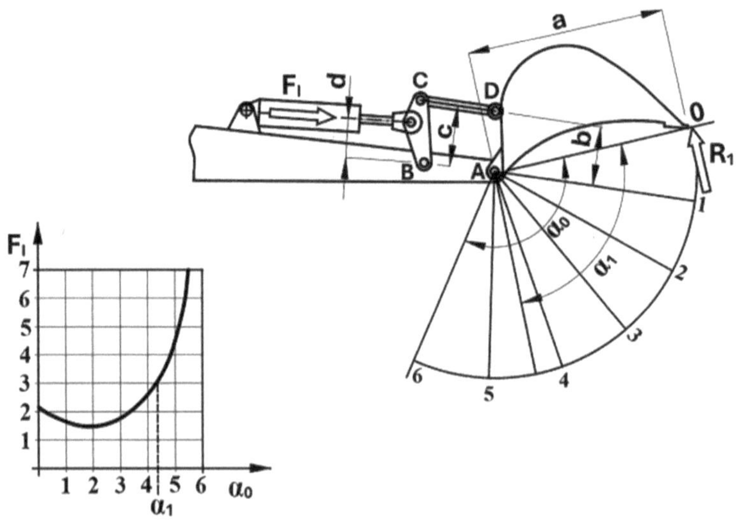

Fig. 7.23 Force change in the hydraulic cylinder of the bucket

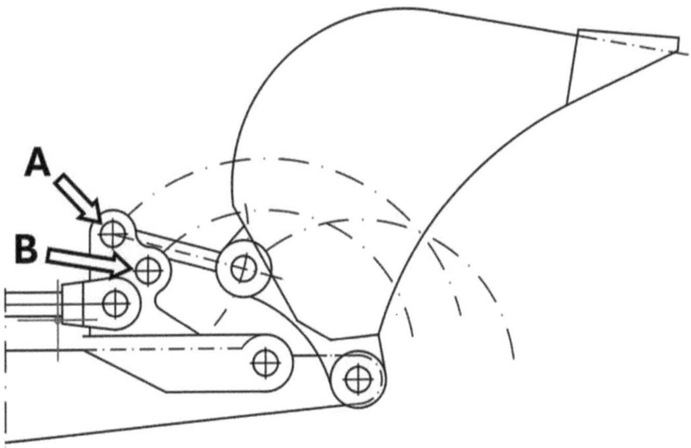

Fig. 7.24 Bucket joint change

$F_1 = R_1\, i_p$ [N]; i_p—shovel leverage ratio
$F_1 = R_1\, \frac{a}{b}\frac{c}{d}$ [N]; a, b, c, d—leverage arms.

By turning the bucket, from point 1 to point 2, the force curve drops, after that, by further turning the bucket, the necessary force in the hydraulic cylinder of the bucket increases progressively.

The highest force in the hydraulic cylinder of the bucket to overcome the same resistance is in fields 5 and 6. However, there is no need for large forces in this field of turning the bucket, since it is a field in which soil is rarely dug. Practically, the force F_1 (from the zero position) increased by 50% can be adopted as the design force (α_1).

In some machines, the grip of the bucket (α_i) can be adjusted to a greater or lesser force of digging the soil, by moving the lever joint, for greater or lesser force arm, Fig. 7.24.

Changing the position of the shovel joint provides:

- in position A: the largest rotation of the shovel, greater excavation depth, greater force of the hydraulic cylinder of the shovel (greater leverage ratio);
- in position B: less rotation of the shovel, less excavation depth, lower force of the hydraulic cylinder of the digging shovel.

Design of parts of the leverage mechanism

Parts of the leverage mechanism are made in the form of a beam, profile as in Fig. 7.25.

On the basis of box-like shapes, the parts are made mainly as beams of constant width and variable height, and the calculation using the finite element method (e.g. software package, *Solidworks*), includes:

- determining the optimal dimensions of critical sections (for the boom and for the arm—at the point of connection with the hydraulic cylinders), for the case of complex stress;

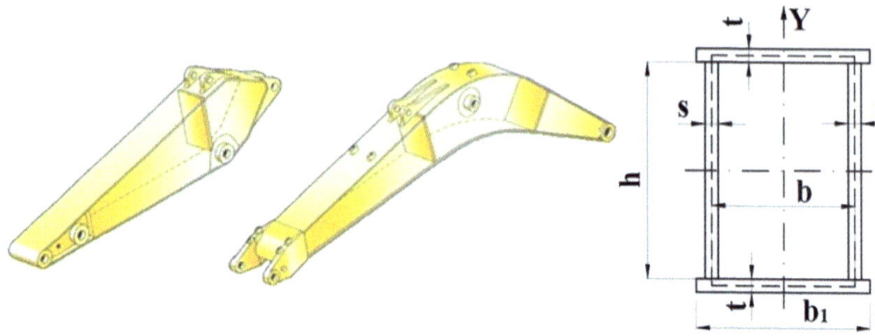

Fig. 7.25 Shape and profile of parts (arm and boom) [6]

- adopting the section width and sheet thickness, for constant lengths of the boom and arm, and determining the height of the other sections from the condition that the composite stress is less than allowed;
- determining the minimum weight of parts of the working devices, boom and arm.

Since the parts are exposed to complex stress (from axial force, bending moment and twisting moment, friction moment in the joint), the critical stress is located in one of the profile angles (*Huber-Hencky hypothesis*).

The probability distribution function of the working load and critical load is the basis for determining the reliability of parts for their working life. With the theory of the accumulation of fatigue damage, the prediction of the working life of parts is computationally accelerated in the project phase, without conducting expensive tests.

7.2.6 Hydraulic Hammer

Backhoes can use several types of executive tools: *hydraulic hammer for breaking, grid shovel, grapple, shovel for crushing concrete and stone, tongs for metal waste, jaws for crushing, chipper-miller for surface treatment, compactor-vibrating plate*, and others, Fig. 7.26. This diversity of tool usage reduces the overall cost of purchasing other machines.

On an excavator with a standard one-piece boom, a hydraulic hammer is most often used for breaking and pulverizing stone, concrete, asphalt, etc., Fig. 7.27.

The hammer is mounted on the excavator working directly on a bench to break away rock which can then be removed by truck. This is most commonly used where blasting is not allowed.

Working principle and effect of the hammer

A hydraulic hammer is connected to the existing hydraulic installation of the excavator. The hydraulic circuit of the hammer installation is shown in Fig. 7.28. The

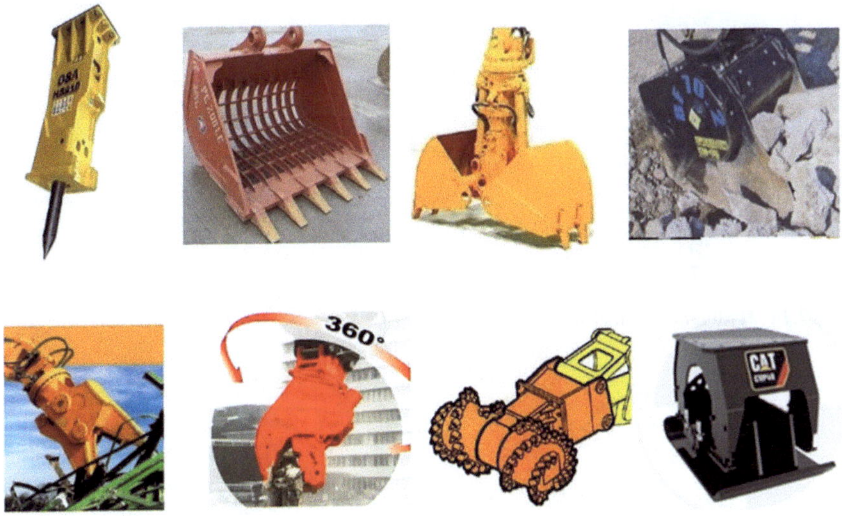

Fig. 7.26 Types of excavator tools

Fig. 7.27 Hydraulic hammers for crushing stone [7]

supply and return lines for hydraulic oil must be of adequate dimensions. Oil is supplied to the hammer by means of the hammer valve. The hammer is started from the cabin by pressing the pedal of the distributor, which opens the flow of oil to the hammer in the hydraulic distributor.

Figure 7.29 shows the working principle of a hydraulic hammer. The operation of the shock piston (1) is caused by the difference in oil pressure on surface A_1 ($A_1 > A_2$), which is achieved by the alternating connection of surface A_1 with the pressure line P (from the pump) and the return line T (oil tank). By lifting the distributor (3), the compressed oil gets access to surface A_1. The resulting force rapidly drives the impact piston (2) downwards.

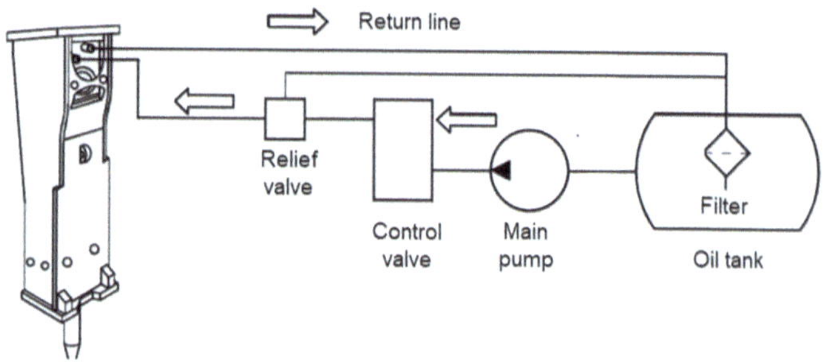

Fig. 7.28 Hydraulic hammer installation

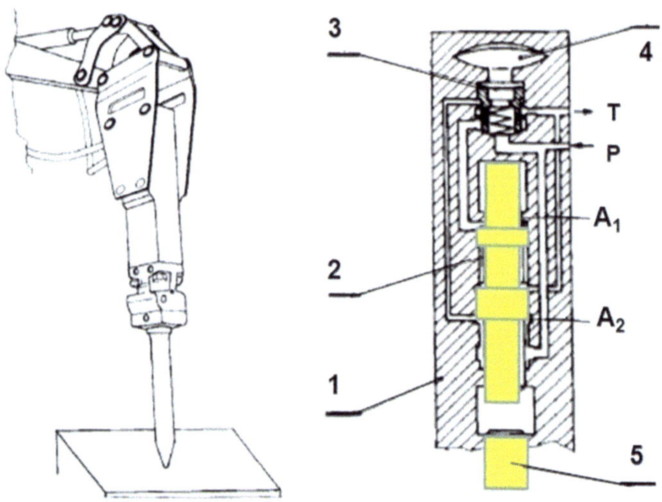

Fig. 7.29 Hydraulic light hammer (*HM 110*). 1—hammer housing, 2—impact piston, 3—distributor, 4—gas accumulator N_2, 5—socket tool–chisel

The resulting kinetic energy is transferred by the piston to the socket tool (5) at the end of the working stroke. A hydraulic accumulator (4) prevents rapid pressure drop in the pressure channel due to the rapid movement of the piston. During the return stroke, the membrane pressure accumulator is filled by compressing nitrogen N_2, from 50 bar to the working oil pressure of 150 bar.

Impact energy of hydraulic hammer

For a more powerful impact hammer, it is necessary to develop as much impact energy as possible, which depends on the speed of the impact piston and its mass. Impact energy increases linearly with the mass of the hammer.

Impact energy (*Tuncdemir hypothesis*) [7]:

$$S_E = k/\sqrt{E_i}[J]$$

E_i—energy of individual impact of the piston, k—constant
$E_i = m\,v^2/2$ [J]
m - piston mass (kg)
v - piston speed (m/s).

Hammer efficiency

$$H_E = P_{iz}/P_{ul}$$

$P_{iz} = n\,E_i$ [W], input power
$P_{ul} = Q\,p$ [W], output power
Q—oil flow (L/min)
P—oil pressure (bar)
n—impact frequency (min^{-1}).

With light backhoes, a series of light hydraulic hammers, weighing up to 250 kg, are attached. A hydraulic hammer with a mass of 120 kg, an oil flow of about 50 l/min, maximum oil pressure of 130 bar, reaches an impact rate of 1000 min^{-1}. The impact energy is 400–500 J (Nm). The working output is about 150 m^3/8 h day. The hammer can use several types of sockets, tips and chisels: a conical punch, a flat cutter and a wide flat cutter, which can be easily replaced.

With an increase in the mass of the hammer, its performance increases. This is followed by a series of medium heavy hydraulic hammers 400–1200 kg, with a number of impacts of 400–800 (1200) min^{-1}, and a series of very heavy hammers of 1500–7000 kg, with a number of impacts of 300–800 min^{-1}.

The working performance of hydraulic hammers, for example in the exploitation of technical stone in surface mines of sandstone and limestone, from light to heavy hammers, is from 100 to 1000 m^3/day.

The basic type of tools are shown in Table 7.4.

1. Moil point. The point-shaped tool directs the impact energy of the hammer smashing into a single point.
2. Cutter-chisel. When excavating rocks a tool in the form of a chisel (wedge) is effective because the rock is excavated by driving it into the existing cavities and cracks in the rock.
3. Blunt breaker. The blunt end achieves a strong transfer of the impact energy of the hammer to the surface, which is effective for breaking rock blocks.

A section of a modern hydraulic hammer and its parts are shown in Fig. 7.30.

Table 7.4 Basic types of socket tools

1	Moil point breaker	
2	Cutter-chisel point	
3	Blunt breaker	

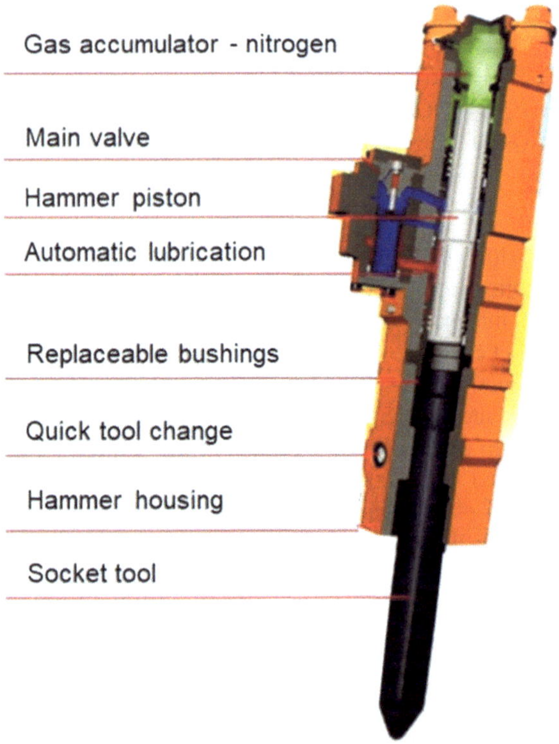

Fig. 7.30 Section of hydraulic hammer (*NPK*)

There is automatic lubrication of parts. These are the parts that wear out and develop temperature. The operating temperature should be maintained by lubrication, so the pump delivers high-viscosity lubricant to the lubrication holes.

Choice of hydraulic hammer

In order to choose a hydraulic hammer for a particular excavator model, it is necessary to determine the compatibility of the joint operation of the excavator and the hammer. The criteria for checking compliance [7]:

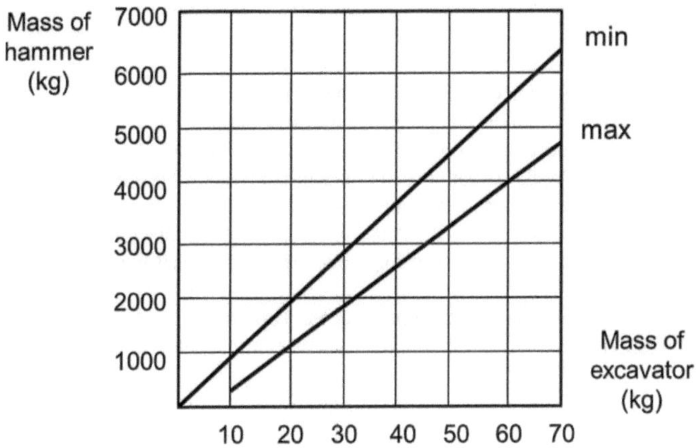

Fig. 7.31 Hammer and excavator selection diagram

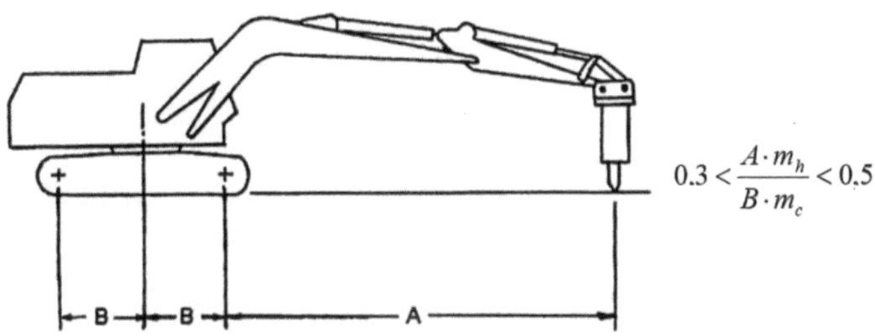

Fig. 7.32 Checking the alignment of the hammer and the excavator [7]: m_h—mass of the hammer, m_c—mass of the machine, A—arm of the leverage mechanism, B—arm of the track support

a. choice of hammer according to the weight of the excavator, Fig. 7.31.
b. matching according to the formula by *J. T. Anderson and W. N. Papineau*

Checking the alignment of the hammer and the machine *determines* the alignment interval, Fig. 7.32:

Practically, the hydraulic hammer can be mounted when the weight of the hammer is about 6–8% of the excavator weight. If the hammer and the machine are not well aligned, harmful vibrations of the working device appear and the efficiency of the hammer decreases. The longer the arm of the excavator—the weight of the machine must be greater. The hydraulic hammer is used on backhoes with one-piece boom. Due to the large number of joints, the two-piece and complex booms of the excavator are not suitable for the use of a hydraulic hammer.

In order for the machine to be reliable in operation, it is necessary to maintain the longitudinal symmetry of the boom and the undercarriage, and avoid the operation of the hydraulic cylinders in the extreme positions. The hammer is always placed perpendicularly to the work surface. Otherwise, the parts of the hydraulic hammer will be exposed to greater wear and deformation.

References

1. JCB Performance Handbook (2020) www.jcb.hae.org.uk
2. Caterpillar Performance Handbook (2020) www.wheeler.com
3. Mikulic D (2024) Gradevinski strojevi. Teorija i konstrukcija. VVG, Velika Gorica
4. Holodov AM (1986) Proektirovanie masin dlja zemljanih rabot. Masinostroenie. Harkov
5. Kelic VN (1989) Hidroprenosnici. Naucna knjiga, Beograd
6. Mikulic D, Stasic T, Grbac D (1999) Analysis of excavator's attachment for application on armoured engineer vehicle. In: Proceedings of 13th international conference of the ISTVS, Munich
7. Ruzic S (2021) Iskop stijena hidraulicnim cekicem. Diploma paper. Rudarsko-geolosko-naftni fakultet, Zavod za rudarstvo i geotehniku, Zagreb. www.aggbusiness.com/sites/ropl-ab/files/89005.jpg

Chapter 8
Soil Compaction Machines

Abbreviations

U [m^3/h]	Work output of roller
v [km/h]	Roller speed
h [m]	Layer thickness
b [m]	Width of the compaction layer
m_{AB} [kg]	Mass of vibrator
m_{VC} [kg]	Vibrating cylinder mass
R [mm]	Vibrator eccentricity
a [mm]	Vertical amplitude of the cylinder
n [Hz]	Vibrator frequency
D [m]	Roller diameter
M_u [Nm]	Vibrator torque
R_u [N]	Roller movement resistance
P_{vib} [W]	Vibrator power
P_{kr} [W]	Roller power for movement
P_v [W]	Total power of the roller
F_z [N]	Soil compaction force
F_{st}[N]	Static force of the roller
F_c [N]	Centrifugal force of the roller
F_i [N]	Roller inertia force
F_p [N]	Excitation force of vibro-plate
P_p [W]	Vibro plate power

D. Mikulić and D. Mück, *Construction Machines*,
https://doi.org/10.1007/978-3-031-67845-5_8

8.1 Purpose and Division of Machines for Soil Compaction

By compacting the soil layer, its density increases, and thus also the load capacity, which is very important for the load capacity and durability of road and other structures. Soil compaction depends on the type of material, moisture content, and method of compaction. Compaction reduces the pores between the soil particles; therefore, the mass of the soil per unit volume increases. Incoherent and coherent materials are compacted: gravel, soft stone, slag, sand, clay, greasy clay, asphalt, and others.

Mechanical soil compaction is an artificial method of soil subsidence in a short time. Machines of the roller and plate type are used for soil compaction, Fig. 8.1. Rollers compact the soil in contact—linearly, and the plates compact the surface.

Division of machines according to the principle of soil compaction, Fig. 8.2:

- Static machines compact the soil statically—with their own weight; smooth rollers, spiked rollers, and pneumatic rollers.
- Vibrating rollers and plates compact the soil dynamically using vibrations; vibrating rollers, oscillatory rollers, vibrating hedgehog rollers, vibrating plates, and groups of vibrating plates (compactors).
- Machines that compact soil by impact are divided into manipulators that use free fall and motorized rammers.

Each principle of compaction has its own field of application. The static compaction principle is better for compacting coherent materials, and the dynamic one for compacting incoherent materials. The soil is compacted linearly with rollers or with plates on the surface. Vibrating rollers are effective for deep soil compaction, and vibrating plates for surface compaction.

Traditional vibratory rollers operate by rapid repetition of the impact of the drum (cylinder) on the ground surface, Fig. 8.3. Impact centrifugal forces for soil compaction are created by a vibrator inside the drum according to the harmonic law. Under the influence of centrifugal force and circular vibrations—the drum of the

Fig. 8.1 Roller and plate for soil compaction

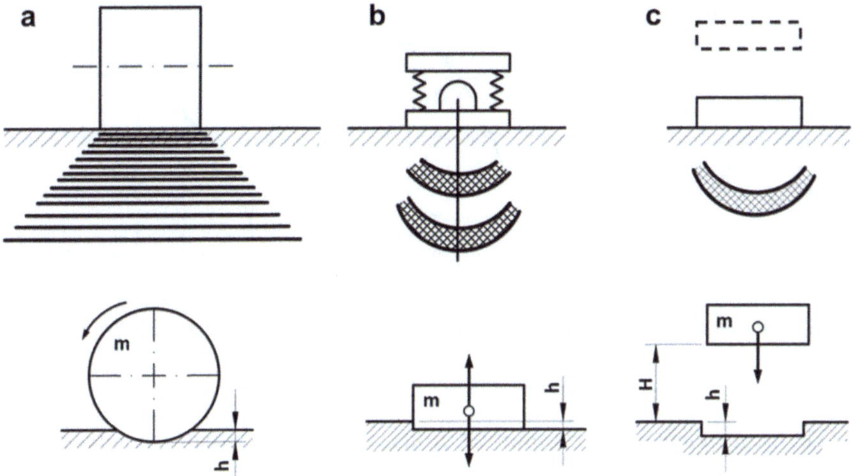

Fig. 8.2 Principles of soil compaction. **a** static compaction, **b** dynamic compaction, **c** impact compaction

roller (*bandage*) bounces and compacts the soil. The *centrifugal force* of the vibrator reaches 50% of the vibrating mass.

From the surface of the soil, pressure waves penetrate the soil one after the other and set the soil particles in motion at a greater depth, reducing soil porosity (by displacing air and excess water). A change in soil volume is possible if the excess water is removed from the pores and the forces of cohesion between the particles that oppose soil compaction are neutralized. Vibration frequency and impact force amplitude are the influential compaction parameters, so their choice is very important for the machine operator.

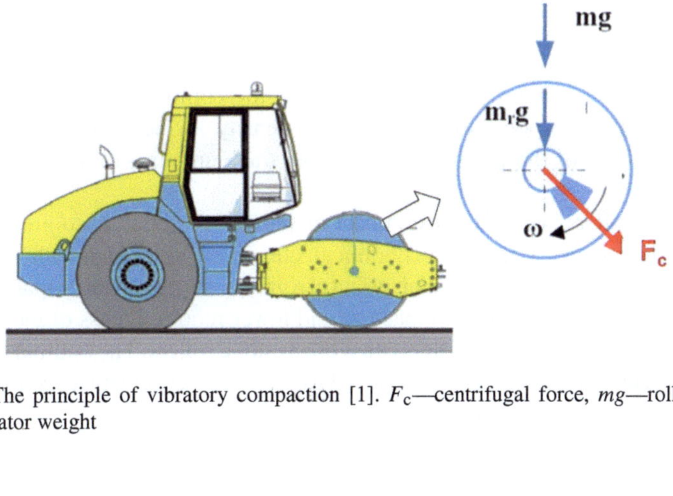

Fig. 8.3 The principle of vibratory compaction [1]. F_c—centrifugal force, *mg*—roller weight, m_rg—vibrator weight

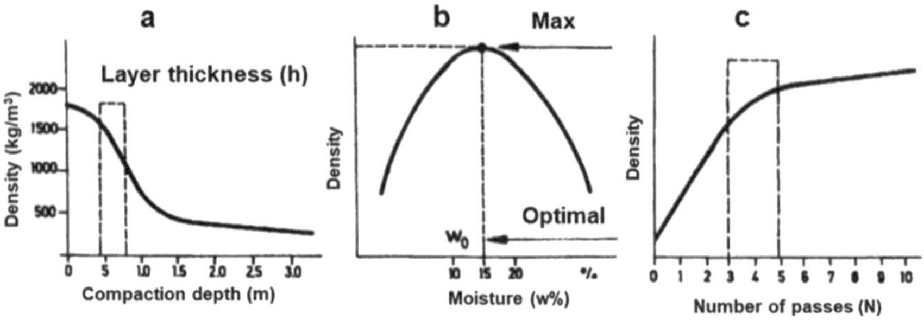

Fig. 8.4 Basic factors of soil compaction [2]. **a** layer thickness, **b** moisture content, **c** number of passes

The active element of compaction is the drum (steel cylinder, *bandage*). The pressure per length centimeter of the roller with the soil is called contact pressure (e.g. 500 N/cm). Under the influence of the vibrator, the rollers (or plates) bounce and compact the soil. Soil compaction depends mainly on the energy input and soil moisture.

The thickness of the compaction layer (h) is determined according to the grain size and cohesion (for asphalt up to 10 cm, for gravel up to 60/80 cm, etc.). The number of passes of the machine (N = 2–8 passes) depends on the type of soil, content, moisture, features of the roller and required compaction of the soil. Compaction is usually performed with several rollers, rarely with one roller. A significant working effect on compaction (m^3/h) is achieved with vibrating rollers.

In order to achieve the same density over the entire profile of the filled layer, it is necessary to fill the layer on a solid or compacted base. After compaction, the next layer is placed, etc. If the density of the substrate is not sufficient, compaction is continued (e.g. soil density $\rho_0 \approx 1400$–$1600\ kg/m^3$).

Figure 8.4a–c shows the influence of the key factors on the density, layer thickness, moisture, and the number of machine passes. A layer of a certain thickness (a) is compacted; the compaction density drops rapidly with a depth greater than 50 cm.

Moisture content is a very important factor; neither dry nor too wet soil is good. Optimal moisture reduces the friction between the compacting particles. Optimal moisture $\boldsymbol{w_0}$ (b) corresponds to the highest soil density. At the same time, the optimal number of passes of the machine on the same track amounts to, for example, 3–5 passes (c).

Vibrating rollers successfully compact incoherent material up to greater depths with several passes. They are economical compared to rollers with only static compaction. The advantage is in their effect at a greater depth and incomparably fewer passes to achieve the desired compaction and lower required mass of the roller. By turning off the vibrations, the rollers can work as static rollers. The optimal frequency value is 40 Hz, and the number of passes 3–6, because the density does not increase equally with the number of passes.

Improper use of the vibrating machine can cause so-called "leafing"—i.e. layers separation. Determination of compaction in the field is carried out by the method of extracting samples of compacted soil or automatically by using a device on the machine.

The work output of the compaction machine is the volume of soil that the machine can bring to the required density in one hour:

$$U = v h b / N \left[\mathrm{m}^3/\mathrm{h}\right]$$

V—roller speed, h—layer thickness after compaction $h = 0.65$hp, (h_p—layer thickness before compaction), b—useful width of the compaction layer without overlapping, N—number of passes of the roller along the same track

When and which type of the roller will be applied depends mostly on the material and soil moisture. The order of selection of machines for soil compaction, depending on the type of the fill layer and optimal moisture. The first transitions should be compacted with lighter machines, and when a certain compaction is achieved, the weight of the machine should be increased. Two machines, light and heavy, are most often used for optimal compaction. The number of passes per track of a light machine is 30–40% of the total number of passes. Working with two machines reduces the number of passes by about 25%, and hence also the cost of compaction.

8.2 Static Rollers

Static rollers compact the soil with their own weight. Modern static rollers are made with one or two cylinders for compaction, Fig. 8.5. They are used for surface compaction of a layer thickness of 10–20 cm, and especially for finishing the already compacted layers, as well as asphalt. They can be rigid or with articulated chassis, with one front roller and two rear ones, which are also powered, and with one front and one rear roller. Static tandem rollers for smoothing the final surfaces do not leave longitudinal marks, so they are suitable for rolling asphalt layers.

Roller cylinders can be driven or guided. Behind A guided cylinder leaves behind the undulation of the surface as result of the extrusion of the material in front of the roller that has to be crossed. For lower surface tolerances (airports), rollers with all drive roller cylinders are used, because the material is drawn under the cylinder and the surface remains flat. Static rollers are manufactured in the following classes: light up to 6 t, medium 6–12 (15) t, and heavy more than 15 t. To increase the static pressure, the rollers can be filled with water and moistened to prevent sticking.

According to linear pressure $q = G/b$ and the number of passes N, static rollers are divided into:

- light, $q < 400$ N/cm, $N = 4$–6, incoherent soil
- medium, $q = 400$–600 N/cm, $N = 10$–12, coherent soil

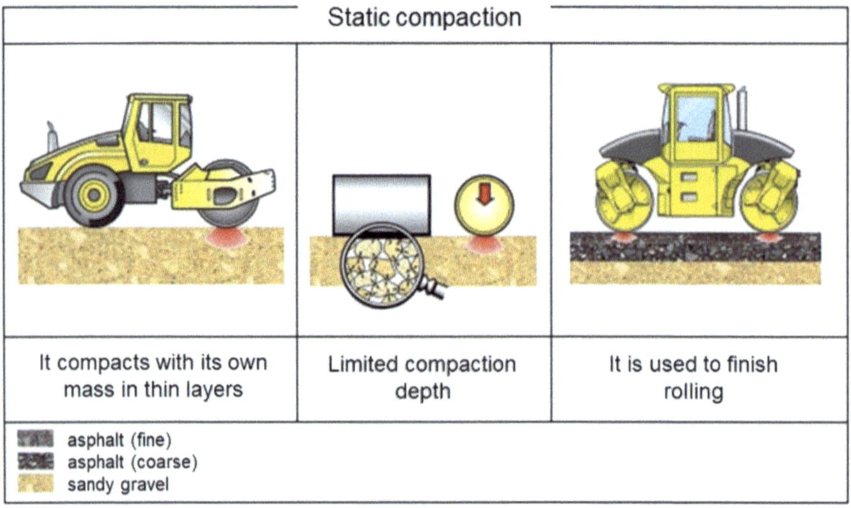

Fig. 8.5 Properties of static rollers with metal compaction cylinders [1]

- heavy, $q = 600\text{–}1000$ N/cm, $N = 6\text{–}10$, asphalt

Roller dimensions:

- roller diameter, $D \approx 55\sqrt{q}$ [mm]
- roller width, $b \geq (1.0\text{–}1.2)D$ [mm]

8.3 Spiked Rollers

Spiked rollers (*Hedgehog roller*) are rollers with spines (spikes). Spiked rollers are made on the basis of smooth articulated rollers, Fig. 8.6. They are used for compacting embankments or layers of fine-grained and binding materials, as the first ones in the technological sequence of use. They successfully replace smooth static and vibrating rollers where the material needs to be crushed beforehand, because their small surface area enables the crushing of larger pieces of stone or pulverization before compaction. They can be divided into light spiked rollers: mass 2–5 t, roller diameter 1 m (without spikes), roller width 1.2 m, spike length 0.18 m, and medium rollers weighing 8–15 t, diameter 1.5 m, width 1.5 m, spike length 0.23 m.

A larger number of roller spikes reduces the number of passes, but the stickiness of the soil between the spikes increases. The spikes are round and square in section. Round-section spikes compact thicker layers better due to deeper penetration. The conical shape prevents accumulation between the spikes. Square profile spikes give good results on thin layers. Instead of spikes, the trailed rollers can have a sheath made of grids, for compacting clay soil.

Fig. 8.6 Spiked roller [2]

Depending on the size of the roller, the number of spikes is 15–25/m^2 of the cover. In order to increase the pressure, the rollers are designed with the possibility of additional loading (water, sand). Spiked rollers can be made with a vibration device, which increases the efficiency of use compared to static action. They are often used in combination as trailed vibratory rollers. The thickness of the filled layer can be up to 1.2 the length of the spikes. At the beginning of the machine operation, the spikes are completely inserted into the embankment, and towards the end there are small penetrations. They can be self-propelled or trailed. With the advent of pneumatic rollers (rubber wheels), with more versatile applications, the requirements for spiked rollers are lower.

8.4 Pneumatic Rollers

Pneumatic rollers statically compact fine-grained binding material, gravel, asphalt, Fig. 8.7. A number of wheels with compaction tires are placed on each axle. These rollers are used in the construction of roads, dams, and airstrips. It is common to use pneumatic rollers directly behind the paver, before smooth rollers. The tires have an advantage when compacting asphalt, because they do not leave a rut behind, making good use of the change in air pressure, from smaller to larger values, in order to achieve stability of the upper layer.

The most commonly used pneumatic rollers weigh up to 25 t. They achieve a pressure per unit of shaft length equal to or greater than the pressure of smooth rollers, with additional load (sand,…). The compaction thickness of coherent soil depends on the wheel load and tire pressure and it is 15–25 cm, and with maximum load even 35–45 cm. For that, 4–5 passes are sufficient, at a speed of up to 10 km/h.

Compaction with tires promotes the convergence of fine-grained binding soil particles (b), without major movements towards each other (acceleration of consolidation). That is why a lower machine speed is important. To increase the compaction,

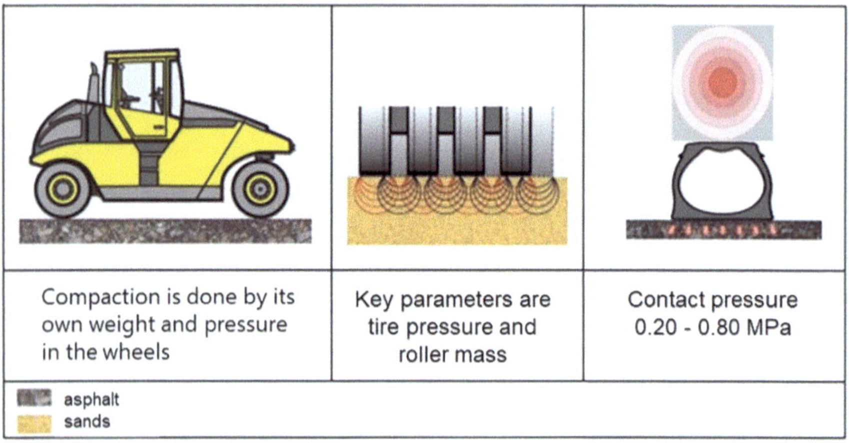

Fig. 8.7 Properties of pneumatic rollers [1]

you can go for a higher load or higher pressure in the tires. With a larger tire diameter, the contact surface increases, and thus also the impact on the depth of compaction.

The tires are set on two or more axles, for example, five wheels at the front, six at the back, in order to always cover the intermediate track during operation. The optimal distance between the tires is: $a_r = (0.4–0.7)B$; B—tire width. Advanced pneumatic rollers are vibrating ones, which provide greater compaction efficiency and productivity.

8.5 Vibrating Rollers

Various types of vibrators have been developed for mechanical soil compaction. According to the principle of vibrating and compacting the soil, these are vibrators for circular vibration and vibrators for oscillatory vibration, Fig. 8.8. Circular vibration implies vertical excitation and oscillatory vibration implies horizontal excitation. Depending on the set compaction of the soil, compaction machines enable automatic switching of compaction, from circular to oscillatory compaction, that is, from aggressive to non-aggressive compaction.

Types of vibratory soil compaction

1. Circular soil compaction (circular vibration, vertical excitation);
2. Oscillatory soil compaction (oscillatory vibration, horizontal excitation).

Rollers with circular vibrations are effective for compacting incoherent soil (sand and gravel), Fig. 8.9a. They act on the ground dynamically (static + vibrations). Circular vibrations increase their efficiency many times (5–10 times) compared to the action of a smooth roller and spiked roller. The vibration frequency is 10–50 (67)

Fig. 8.8 Circular (front roller) and oscillatory principle of compaction (rear roller) [3]

Hz. The forces of vibration in the embankment soil reduce the cohesion forces, the soil particles fill the free spaces, thus creating the highest density of the compacted soil.

The vibrating mass of the roller bounces, the greater the vibrating mass, the greater the amplitude of the vibrations, and therefore, the depth of compaction. When compacting thin layers and fine-grained materials, especially asphalt, low amplitude and high frequency are used, while for thicker layers and coarse-grained materials, high amplitude and low frequency are used. The vibrators can first be divided according to the linear pressure on: light, 70–100 N/cm; medium, 100–300 N/cm; heavy, 350–600 N/cm; very heavy 600–800 N/cm. In terms of mobility, articulated vibratory rollers are known with only a front cylinder for soil compaction.

Therefore, oscillatory compaction changes the direction of impact vibrations around the axis of the drum (variation of the drum around the central axis), Fig. 8.9b. Compared to the classic circular vibrator, the oscillator has slight vertical accelerations, so the cylinder does not bounce.

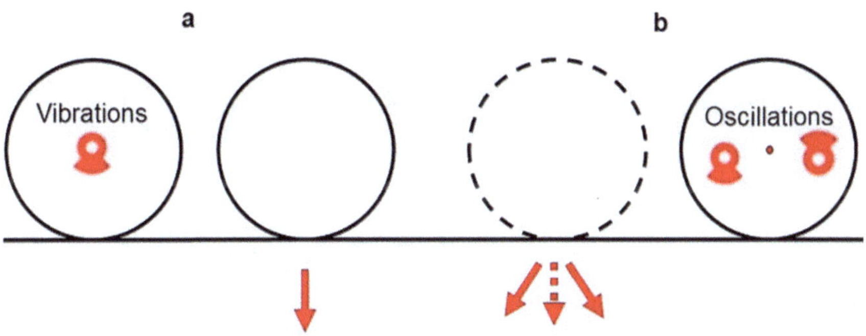

Fig. 8.9 Circular and oscillatory principle of soil compaction

Oscillatory vibration helps the rollers to achieve compaction with less vibrations. Oscillation eliminates about 90% of the stress on materials when compared with a traditional circular vibratory system. This is important when compacting on bridges, over sewers and utility lines and near buildings where a traditional vibratory approach can cause damage. Oscillation uses less force but delivers both vertical and horizontal energy, essentially manipulating the material composition.

Two eccentric vibrator masses—aligned at an angle of 180° from each other—rotate synchronously in the same direction. This allows the drum to rotate rapidly back and forth. As it rotates, the drum does not lift off the base. It remains in constant contact with the substrate, causing static compaction in addition to dynamic shear forces.

8.6 Roller Hydraulic Installation

Wheeled rollers for soil compaction use hydrostatic power transmission to move, vibrate, and control the machine. The scheme of power transmission for a roller with one drum, consists of components of hydrostatic and mechanical power transmitters, Fig. 8.10. The rolling speed is provided by the drive of the rear axle, and the soil compaction is provided by the front drum (cylinder).

The characteristics of hydrostatic rollers with one drum are: mass 6–20 t, pressure 250–450 N/cm, engine power 80–150 kW, roller width 1.6–2.2 m, maximum

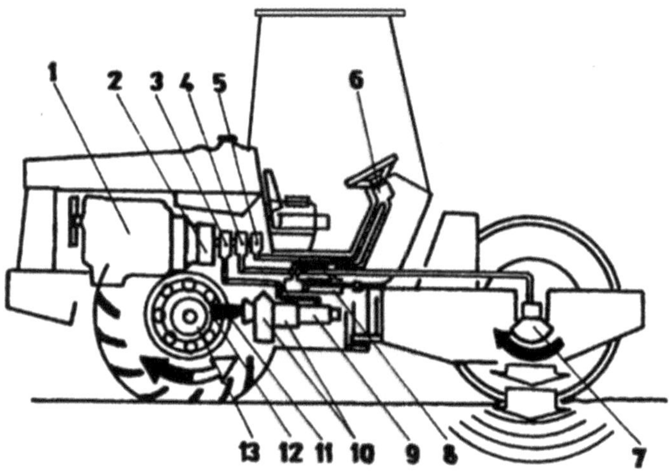

Fig. 8.10 Wheeled vibrating roller with one drum for soil compaction. 1—engine, 2—pump drive, 3—pump for motion, 4—pump for vibration, 5—pump for control, 6—control valve, 7—hydraulic motor for vibration, 8—cylinder for articulated control, 9—hydraulic motor for movement, 10—gearbox (3–4 stages), 11—cardan, 12—differential, 13—reducer in the wheel

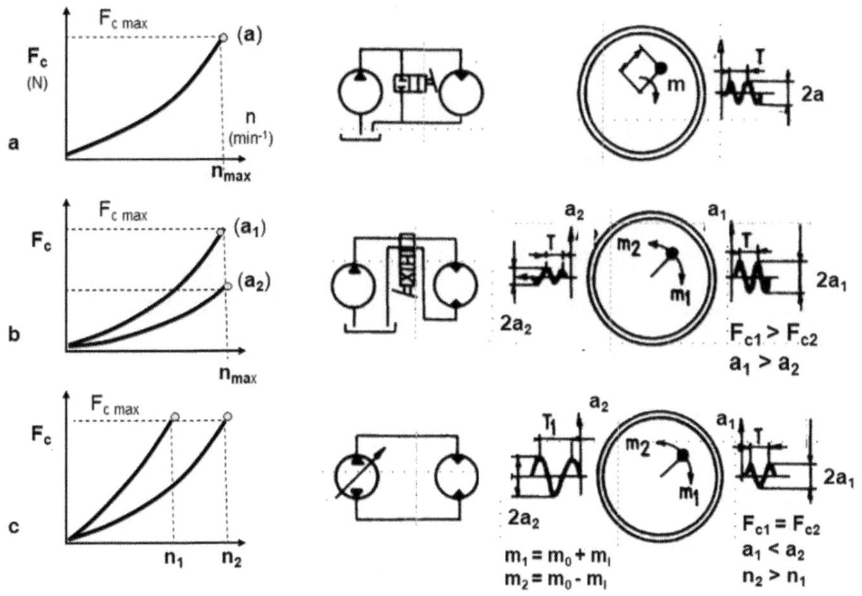

Fig. 8.11 Variants of hydrostatic vibrator drive with circular vibrations [4]

frequency/amplitude (42/0.9 $\div$ 30/1.4) and centrifugal force 10–22 kN. The operating speeds of rollers (v) with three to four speeds are 2–6 km/h and 6–15 km/h, and transport speeds up to 35 km/h. The rollers can overcome slopes of up to 45%.

Basic variants of the hydrostatic drive of the vibrator are shown in Fig. 8.11.

Basic variants of the hydrostatic drive of the vibrator with circular vibrations and the corresponding centrifugal force (F_c):

(a) one frequency and one amplitude (n, a_1),
(b) one frequency and two or more amplitudes (n, a_1, a_2),
(c) two frequencies and one amplitude for each frequency (n_1, n_2, a).

The vibrator drive, Fig. 8.11a, b, uses a hydro-pump and a constant flow hydro-motor. The two-frequency vibrator drive **c** uses a variable-flow piston-axial pump and a constant-flow piston-axial hydro-motor in a closed circulation circuit. A variable pump provides the flow required for one frequency range in one direction (n_1), and in the other direction for another frequency range (n_2).

Hydrostatic drive with tandem rollers, drives both drums, Fig. 8.12. Both roller drums are driving ones, for machine movement and for circular vibration.

For example, the source of the hydraulic power is a piston-axial variable flow hydraulic pump (5). The working pressure of the oil created by the pump in the hydraulic circuit for movement is 250 bar, and the pressure for driving the vibrator and controlling the machine (50–100 bar) is created by gear hydraulic pumps (9, 10, 11). Articulated control of the machine is created by turning the steering wheel to the left or right, using the distributor (8) the oil passage to the hydraulic cylinder is

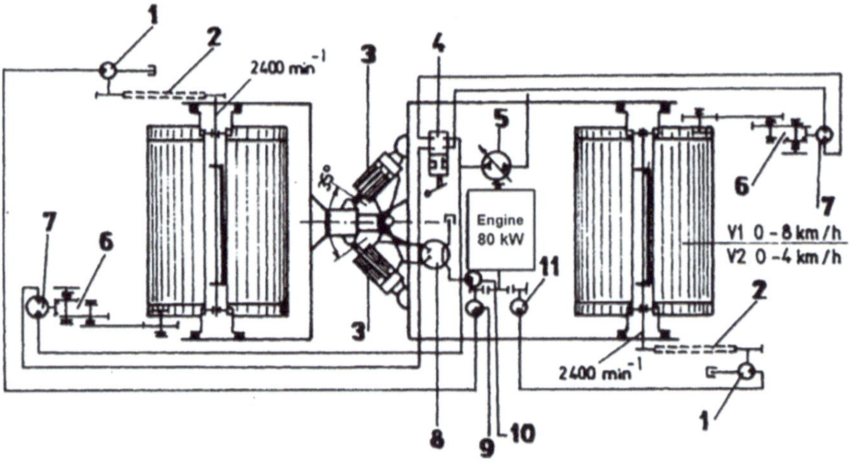

Fig. 8.12 Hydrostatic installation of articulated tandem roller [4]. 1—hydraulic motors for driving the front and rear drum vibrators, 2—gear transmission, 3—hydraulic cylinders for controlling the roller, 4—hydraulic distributor, 5—piston-axial pump, 6—reducers, 7—hydraulic motors for movement, 8—steering distributor, 9 and 10—double gear pumps, 11—single gear pump

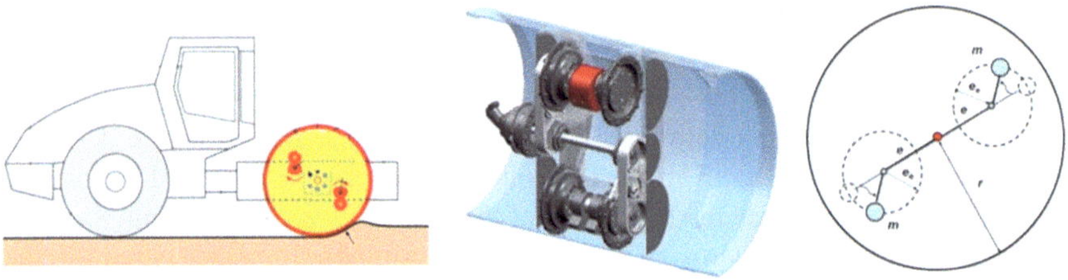

Fig. 8.13 Hydrostatic drive of oscillating drum (adapted [3])

opened to rotate the drum. The gear pump (10) supplies oil to the working cylinder in proportion to the turn of the steering wheel.

In the case of oscillatory rollers, the hydraulic motor enables the vibrator to rotate and the drum to rotate back and forth, Fig. 8.13. Two eccentric (external) axles are driven by a toothed belt, and the steering is electrohydraulic.

8.7 Circular Soil Compaction

Classic vibrating rollers are equipped with a circular vibrator, Fig. 8.14. The circular vibrator model with mass eccentrics produces centrifugal force in the circle of 360°, and in contact with the ground it creates impact forces for soil compaction. The vibrating device is driven mechanically or hydraulically in modern machines.

It consists of a shaft and two eccentric weights, mutually movable to adjust the compaction amplitude. The vibrating device turns on after starting the machine, and turns off before it stops. Modern circular vibrators have a hydraulic drive.

The vibration frequency is adjusted to the compaction technology 10–70 Hz (600–4200 min $^{-1}$), by adjusting the engine speed. The weights of the vibrator (b) are placed so that one of them, "B" is firmly connected to the shaft, and the other, "A" can freely rotate around the shaft up to 180°. Both weights are equal. The eccentricity of the masses from the shaft axis changes.

The excitation centrifugal force ($F_c = mR\Omega^2$) depends on the mass of the weights, the eccentricity of the weights and the number of revolutions of the shaft, Fig. 8.15. The impact force changes according to the harmonic law $F_c\cos\Omega\ t$. It can be compacted with large or small bouncing amplitudes. The amplitude of the excitation force of the weights (AB mass) is followed by the vertical amplitude of the cylinder (*a*):

$$a = \frac{m_{AB} \cdot R}{m_{VC}}\ [\text{mm}]$$

m_{AB}—total mass of vibrator weights A and B (m_1, m_2);
m_{VC}—vibrating cylinder mass; *R*—vibrator eccentricity
Selection of amplitudes with regard to the type and optimal moisture of the soil being compacted, Table 8.1.

Frequency range and vibration amplitude selection

In the frequency range of vibrating rollers 10–50 Hz (Fig. 8.16), the resonant frequency is in a narrower range of 20–30 Hz, when the circular frequency of forced vibrations (Ω) equals the circular frequency of natural vibrations (ω). As soil compaction increases, the frequency range gradually increases. Soil compaction

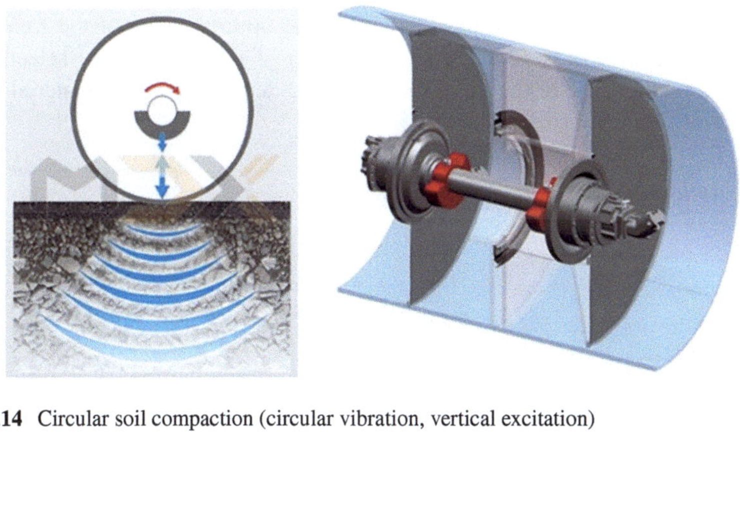

Fig. 8.14 Circular soil compaction (circular vibration, vertical excitation)

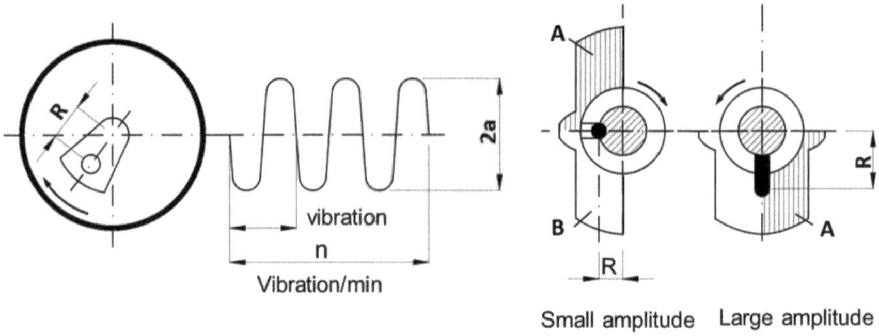

Fig. 8.15 Vibrator parameters. **a** frequency adjustment, **b** amplitude adjustment, (A + B)—unbalanced vibrator mass

Table 8.1 Selecting the vibrator amplitude [5]

Soil, material	Amplitude label	Amplitude selection (mm)
Compaction of protective embankments i supporting layers	4 or 5	0.80–0.90
Compaction of gravel or macadam surfaces	3 or 4	0.65–0.80
Finishing of surfaces from gravel and macadam	2 or 3	0.40–0.65
Compaction of asphalt bonding surfaces	2	0.4
Compaction of asphalt final surface in very thin layers or very hot masses	1	0.25

starts with a low frequency of 10 Hz, then gradually increases to 30 Hz, and increases to 50 Hz when the roller is unloaded.

The compaction (F_c) should be closer to the sub-resonant or super-resonant region, so that the roller does not bounce too much, i.e. that there is control of forced vibrations. Because of this, a continuous change in the frequency and amplitude of vibrations is required. It is considered that the optimal value of the frequency of the vibrating rollers is 40 Hz, and the number of passes 3–6. The selection of the vibration amplitude of the roller is determined depending on the material being

Fig. 8.16 Function of the centrifugal force of the roller (mass 5 t) [5]

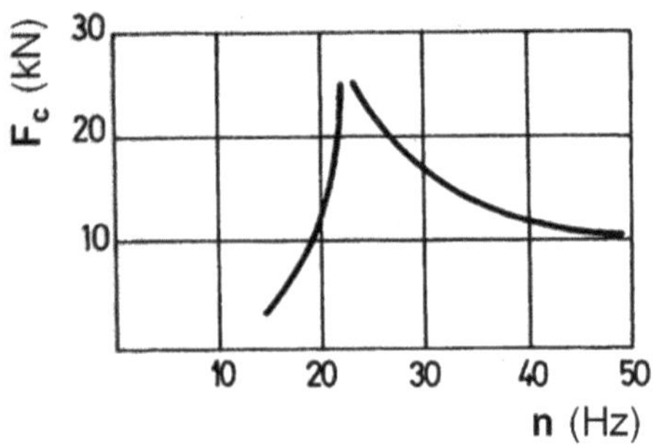

compacted. For medium-class rollers, when compacting the pavement structure, the selection of the amplitude is shown in Table 8.1. Practically, the adjustment of the vibrator amplitude in some machines can be done manually, from the side of the roller (Fig. 8.17) or, in modern machines, automatically from the cabin.

Vibrator torque, Fig. 8.18.

$$M_u = M_e + M_t + M \text{ [Nm]}$$

$$M_u = mR + F_\mu r + J\Omega/t$$

M_e—weight moment of the vibrator eccentric mass
M_t—friction moment in shaft bearings
M_i—moment of inertia of masses in rotation
F_μ—friction force in the bearing, $F_\mu = F_c\,\mu = m_{AB}\,R\,\Omega^2\,\mu$
J—moment of inertia of the shaft
Ω—angular speed of the shaft, μ—friction factor in the bearing

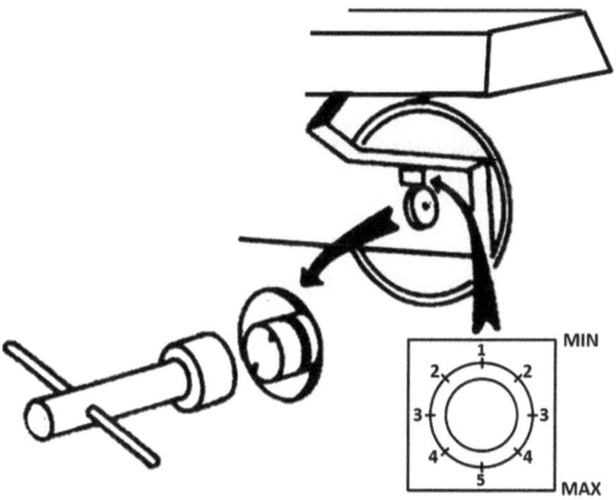

Fig. 8.17 Manual adjustment of the amplitude on the roller drum

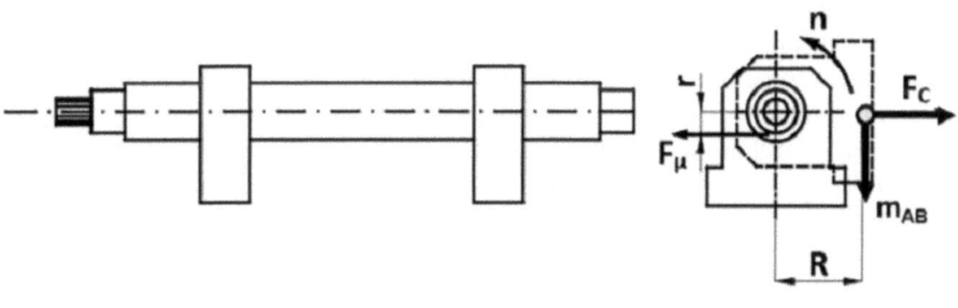

Fig. 8.18 Forces of a circular vibrator

t—time in which the shaft reaches rotation speed.

Vibrator power

$$P_{vib} = M_u \Omega \text{ [W]}$$

Roller movement resistance

$$R_u = R_k + R_i + R_\alpha \text{[N]}$$

$R_k = G f_k$, rolling resistance, f_k—roller rolling factor
$R_i = m_v\, a$, inertia resistance, v—roller speed
$R_\alpha = G \sin\alpha$, slope resistance.

Roller movement condition:

$$F_v = F_o - R_u \leq F_\alpha$$

F_v—traction force, F_o—circumferential force of the roller
F_μ—adhesion force: $F_\mu = G\mu$, μ—adhesion factor
G—total weight of the roller that loads the drive shafts of the roller.

Power required to move the roller:

$$P_{kr} = F_o v \text{[W]}$$

Total power of the roller for movement and vibration compaction of the soil:

$$P_v = (P_{vib} + P_{kr}) / \nu_{tr} \text{ [W]}$$

η_{tr}—efficiency of the power transmission

Tandem vibration rollers

Tandem rollers are suitable for high efficiency of soil compaction, Fig. 8.19. When the front pair of vibrating rollers is raised by a certain amplitude (a), the rear pair of the rollers simultaneously hits the ground (b), and vice versa. Eccentric weights are phase-shifted by 180°. Due to the connection of the vibrator shafts of the front and rear pair of rollers and the phase shift between them, the horizontal forces of the vibrator cancel each other out (c), so the vibrations occur alternately in the vertical direction.

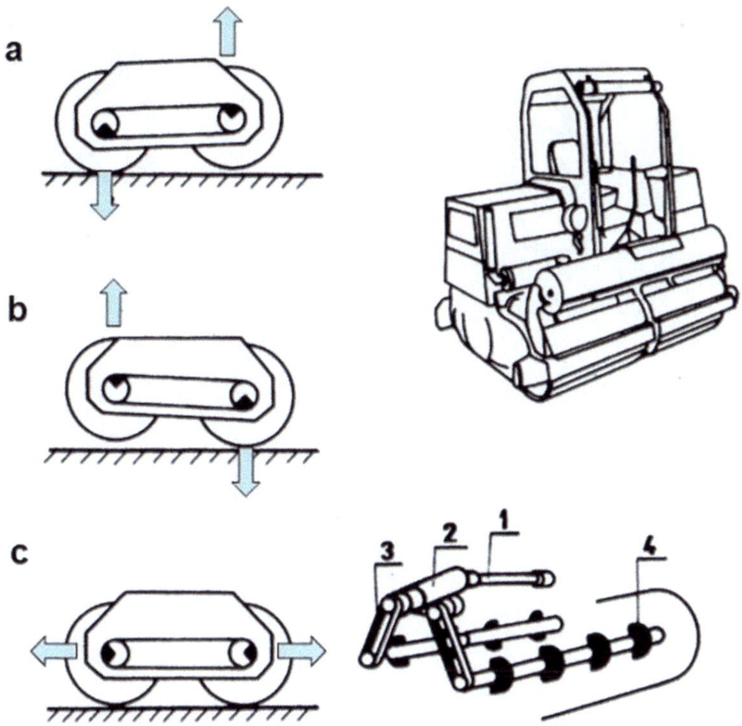

Fig. 8.19 Tandem roller principle. 1—cardan shaft, 2—distribution transmission, 3—toothed belt, 4—eccentrics

Ball and shot vibrators

The vibrator shaft moves the balls in their drum tracks *(ball and track),* Fig. 8.20a. By turning the shaft, the steel balls are set in motion, and they roll on the grooved metal tracks. The steel balls with the mass of the drum cause centrifugal force to compact the soil. A more advanced version of the vibrator is a version with a metal shot inside the torus (b). There are two amplitudes and frequencies, when moving forward (high amplitude and low frequency) or backward (low amplitude and high frequency).

Soil compaction force

Vibrating circular compaction in the contact between the cylinder and the substrate creates vertical compaction forces, which cause the drum to bounce by the height of the amplitude. The compaction force depends on the parameters: the vibrating mass of the cylinder (bandage), the centrifugal force of the vibrator, and the inertia force of the cylinder, Fig. 8.21.

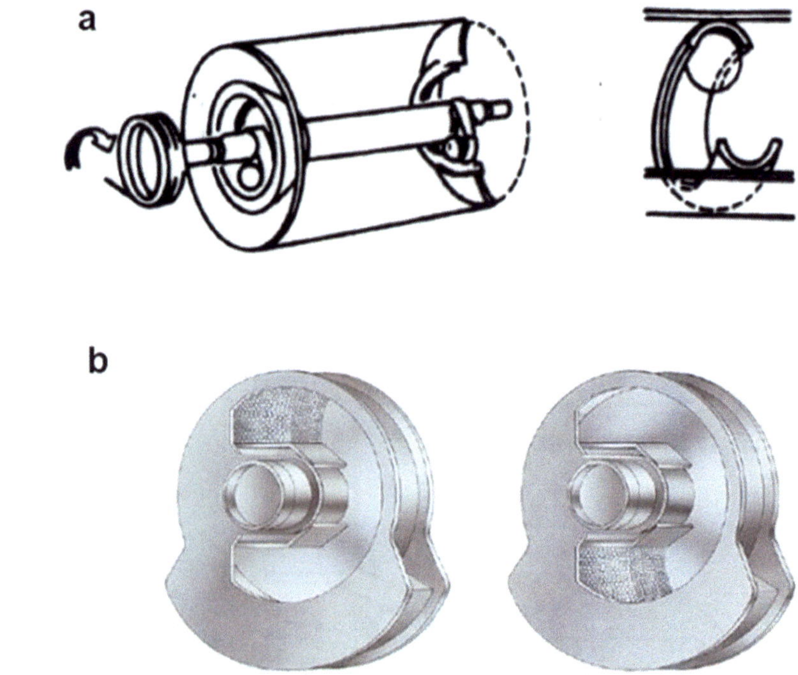

Fig. 8.20 Ball vibrator (**a**) and metal shot vibrator (**b**)

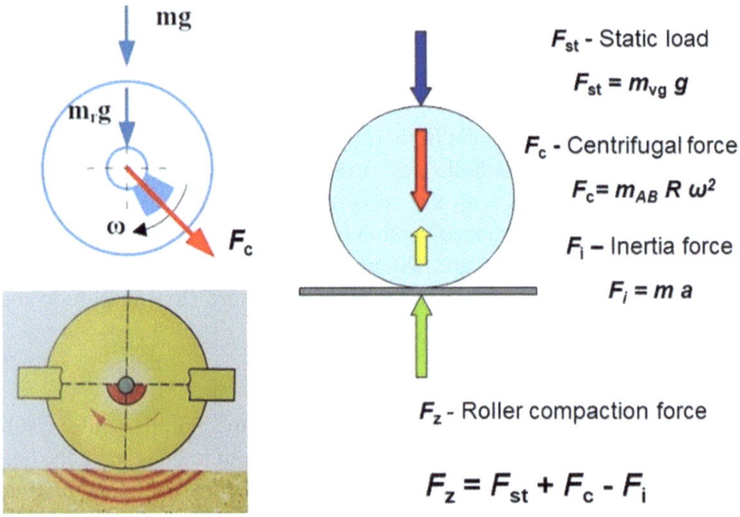

Fig. 8.21 Circular vibratory soil compaction

Table 8.2 Overview of soil compaction parameters with a vibrating roller (adapted [1])

Application	Roller	Static pressure N/cm	Amplitude mm	Frequency Hz	Rolling speed km/h
Stones Clay	 2.5-26 t	≥300	≥1.5	28-35	1-2.5
Gravel Sandy gravel Sands	 	≥100	≥0.4	28-60	2-4
Asphalt Fine asphalt	 2-13 t	100-300	0.35 - 0.9 ≤0.5	30-60 40-60	2-4 2-6

The resulting compaction force:

$$F_z = F_{st} + F_c - F_i$$

- static force: $F_{st} = m_{vg}\, g$
- centrifugal force: $F_c = m_{AB}\, R\, \omega_p{}^2$, according to the harmonic law: $F_c \cos \omega_p\, t$
- inertia force: $F_{st} = m_{vg}\, a$

Overview of soil compaction parameters

The key parameters for compaction with a vibrating roller, in relation to the compaction of a specific soil: amplitude, frequency, and speed of movement (rolling), are shown in an overview in Table 8.2.

8.8 Oscillatory Soil Compaction

The oscillatory vibrator for compaction has a specific design, consisting of a central shaft and two unbalanced masses that rotate in the drum frame, Fig. 8.22. The eccentric masses of the vibrator are shifted by 180^0 around the central axis. Oscillatory compaction changes the direction of action of impact vibrations around the drum axis (variation of the drum around the central axis - drum oscillates torsionally, M_v), which causes shear compaction, Fig. 8.23.

Two eccentric imbalance masses rotate synchronously in the same direction. This imparts a rapid back-and-forth rotational movement on the drum. As it rotates, the

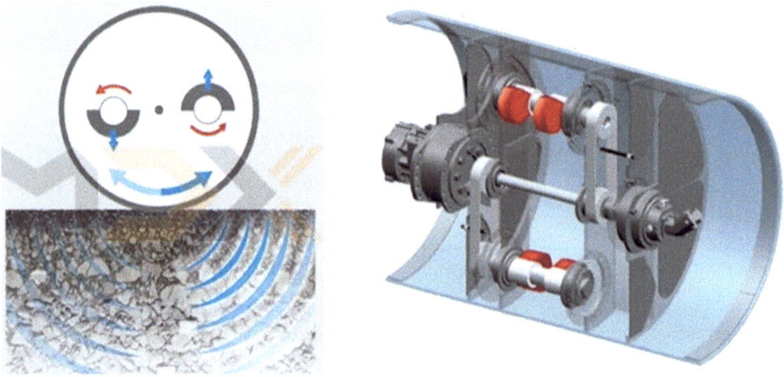

Fig. 8.22 Oscillatory drum for compaction

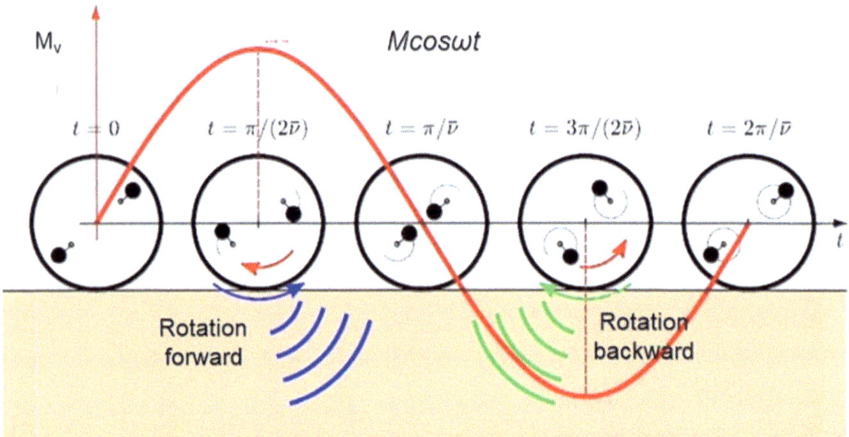

Fig. 8.23 Changing the torque on the roller drum [3]

drum does not lift up from the substrate. It remains in constant contact with the substrate, causing static compaction in addition to the dynamic shearing forces. The oscillation drum is made of extremely wear-resistant steel in order to increase its service life.

Oscillatory compaction is suitable for compaction of any type of asphalt substrate. They are most often used for final layers of asphalt containing larger material fractions. Oscillatory compaction brings the quality of compaction, especially with substrates that are complex for compaction, and especially for construction works on bridges.

There is an option that combines both dynamic compaction systems (vibrations and oscillations) in one drum. This means that it can work as either a vibrating or an oscillating drum. Both systems cannot be used at the same time.

The control unit ensures a continuous phase shift between the rotating masses of the vibrator, which thus enables the quality of compaction, without producing harmful vibrations. Tangential accelerations at the beginning of compaction cause soil creasing, then maximum settlement, and finally plastic deformation of the layer. The modern oscillatory compaction system is an advanced self-adjusting compaction system.

Vario rollers

Compaction that can independently use the advantages of circular or oscillatory vibrators is suitable for all types of soil compaction, Fig. 8.24.

Compaction that combines the advantages of circular and oscillatory vibrators is suitable for all types of soil compaction, Fig. 8.24. Vario rollers (*variomatic*) use directional vibration by means of eccentric masses of two shafts that rotate in opposite directions so as to give directional vibration. By turning one shaft in the opposite direction to the rotation of the other shaft, the direction of the resultant force between the vertical and horizontal axes is automatically changed, according to the changing stiffness of the substrate. There is certain difference between a Vario roller and an oscillatory roller because the shear deformation of the soil is caused by horizontal translation in the VARIO roller, whereas the drum of an oscillatory roller works torsionally.

The roller variant *vibrocontrol* uses a vibrator with three weights; the middle mass of the weights rotates in the opposite direction to the rotation of the other

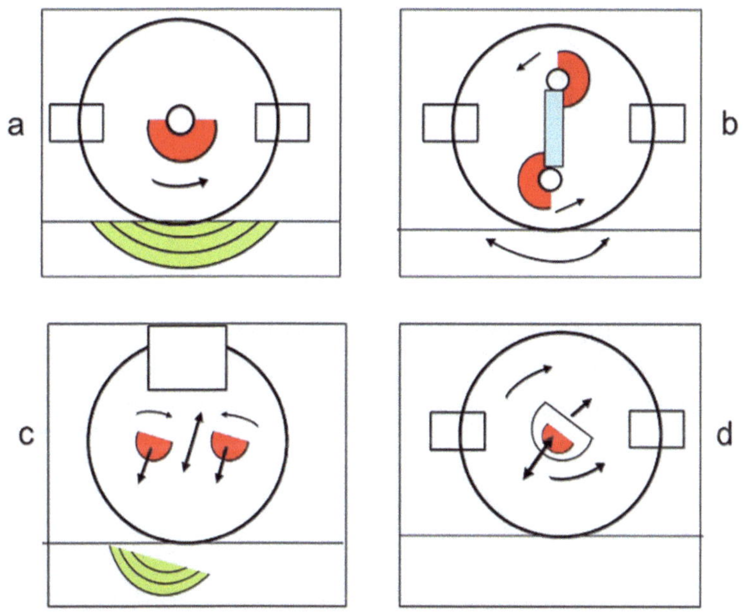

Fig. 8.24 Models of vibration soil compaction [1]. **a** vibrator, **b** oscillator, **c, d** vario vibrator

two masses. Centrifugal forces that arise are added to each other for directional action. Any angle from vertical to horizontal axis can be adjusted. As the compaction increases, the vibration amplitude decreases. The compaction energy is automatically reduced when the stiffness of the substrate becomes too great, which determines the compaction of the soil. The ratio between the force and the path corresponds to the compaction of the substrate. A force and path diagram can be determined.

Vario type oscillating rollers, are practically suitable for compacting different types of soil, Fig. 8.25. The compaction process enables very even compaction with fewer roller passes, which results in less crushing and destruction of grains. The *Vario cylinder* can be gradually or automatically switched from *vertical* to *oblique* or *horizontal* compaction, depending on the soil properties.

Soil compaction control

Modern machines for soil compaction are equipped with a system for soil compaction. The machine operator is left with the option of choosing the frequency, amplitude, and automatic tracking, Fig. 8.26.

Advanced soil compaction machines use continuous compaction control, abbreviated CCC (*Continuous Compaction Control*). Such CCC compaction control systems are *compactometer-CMV* and *terameter-OMEGA*. The systems use a sensor with one or two accelerometers, which are connected to the roller cylinder bearing, then a processing unit and a compaction monitoring screen. In accordance with the increase of soil compaction, the amplitude continuously decreases. The compaction energy is brought to the surface where needed, E_{VIB} (MN/m^2).

A *variomatic* or *vibrocontrol* compaction control system is used for compacting asphalt layers, Fig. 8.27. The Viriomatic system is designed so that the direction of vibration can be adjusted to optimize the compaction effect by directing the compaction force and amplitude.

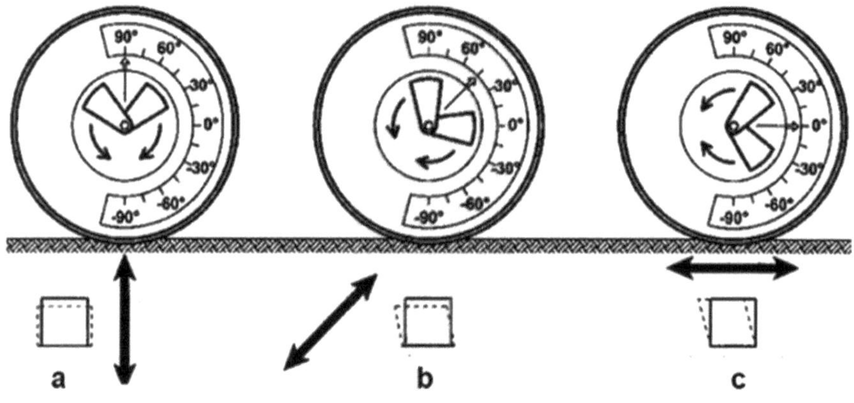

Fig. 8.25 Vario vibrator, adjustment of excitation direction and compaction [6]. Automatic and gradual switching from circular vibration (vertical excitation) to oscillation (horizontal excitation)—depending on local soil properties, **a** vertical, **b** oblique, **c** horizontal

Fig. 8.26 Vibro device and compaction control with a light roller [2]. 1—vibrator hydraulic motor, 2—insulating supports, 3—vibrator housing, 4—vibrator eccentric weights, 5—frequency selection, 6—amplitude selection, 7—automatic vibration control

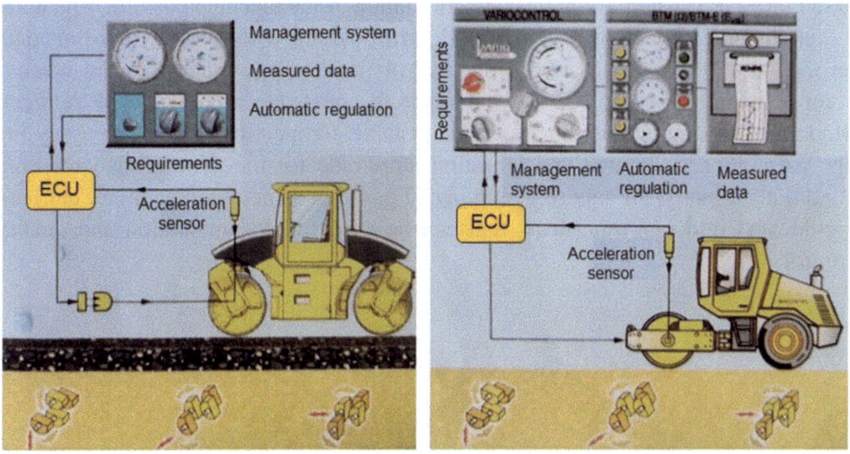

Fig. 8.27 Variomatic and vibrocontrol compaction monitoring system [7]

With vibrocontrol compactors, the vibration direction can be automatically changed from vertical to horizontal using defined control criteria, which enable an optimized compaction process (intelligent compaction) and, consequently, highly uniform compaction. When the maximum compaction force is reached, the excitation direction is changed, so that the applied force does not exceed the maximum force. Two accelerometers, which are placed on the drum bearings, record the continuous acceleration of the drum. The contact force, energy delivered to the ground, and the displacements, are calculated in the process unit, taking into account the parameters such as mass, excitation force, and frequency. The data are transmitted to the integrated system, which automatically controls the compaction parameters.

The vibrocontrol system measures the input energy of compaction (E_{VIB}), using the force on the vibration path. The values of the compaction energy modulus are denoted by CMV and OMEGA. The measurement is based on the acceleration sensor of the compaction cylinder, which determines the compaction force on the compaction surface Fz. The vibration control devices enable automatic adjustment of the compaction amplitude to the required requirements.

Intelligent compaction system

Soil compaction is a complex process that depends on the type of compaction equipment and soil properties. The connection of all these factors leads to the achievement of optimal compaction after a certain number of passes of the roller. The procedures used by conventional compaction machines can result in uneven density, which can be the reason for premature pavement damage.

Intelligent compaction system helps the users to overcome this problem by optimizing the compaction process, and machine control ensures precision and greater working productivity. The concept of intelligent compaction (IC) is based on integrated measuring systems, a built-in computer system, and compaction mapping based on a GPS system for precise determination of the machine position, Fig. 8.28.

Intelligent compaction system provides the operator with valuable real-time data on the state of the quality and efficiency of the works, as they indicate when a compaction job is complete, moving on to the next one at the optimal moment. Modern rollers include automatically adjustable compaction systems to apply the optimal compaction force and vibration amplitude for the material and working conditions. Finally, all data (e.g. number of passes, stiffness, temperature, etc.) are geo-located into real-time maps that can be used for documentation and quality control.

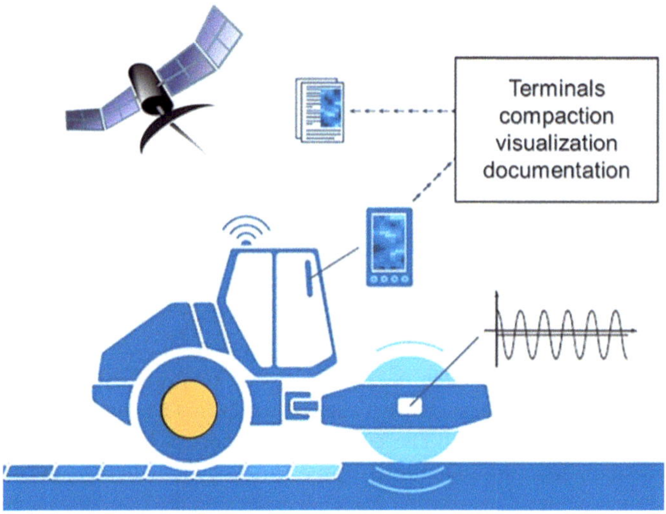

Fig. 8.28 IC system installation

8.9 Vibration Plates and Rammers

Vibration plates are intended for the compaction of incoherent materials (compaction of layers of sand or gravel when building objects in a confined area and hard-to-reach places for rollers). According to the principle of operation, there are two types of vibration plates: vibro plates for circular compaction and vibro plates for directional soil compaction, Fig. 8.29. They are classified according to mass into light up to 500 kg, medium up to 1500 kg, and heavy more than 1500 kg, according to the mobility and control method (manual, remote), and the method of power transmission to the vibrator (mechanical, hydrostatic).

Vibration plates have different base plates, from the lightest 0.15 m^2 to the heaviest 1.2 m^2, often from 0.3 × 0.75 m to 0.7 × 1.1 m. The width of the base plate depends on the depth of compaction: $B = (1\text{–}2)h$. The compaction depth of light vibration plates is 30 cm, and of the heaviest ones up to 180 cm. The vibration plate can be moved only in one direction forward (non-reversible), or in both directions back and forth (reversible). Moving forward or backward, steering left–right or turning in place is adjusted using vibratory masses.

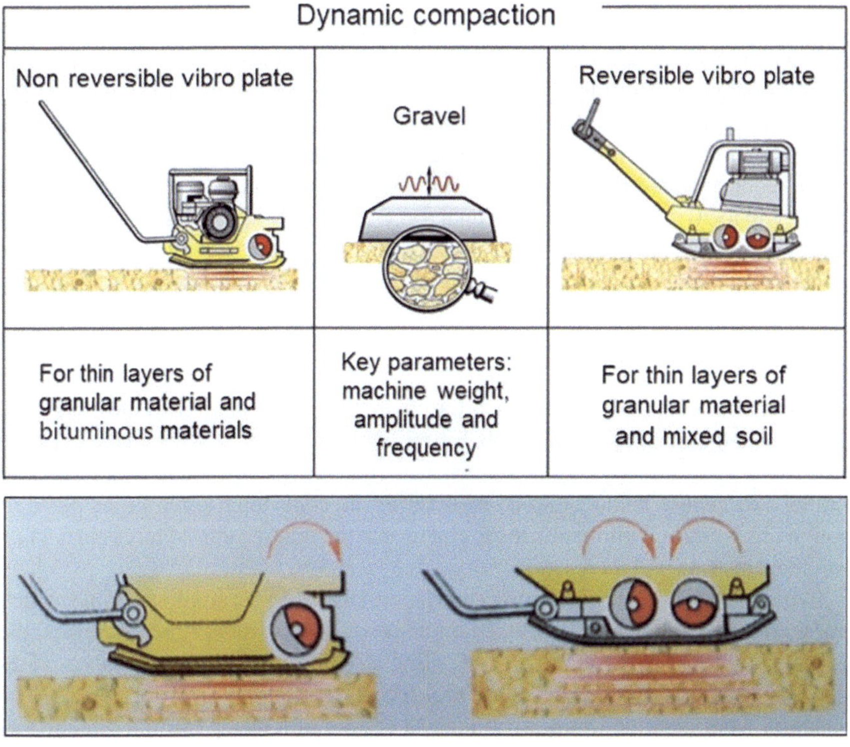

Fig. 8.29 Features of vibro plates for circular and directional compaction of gravel

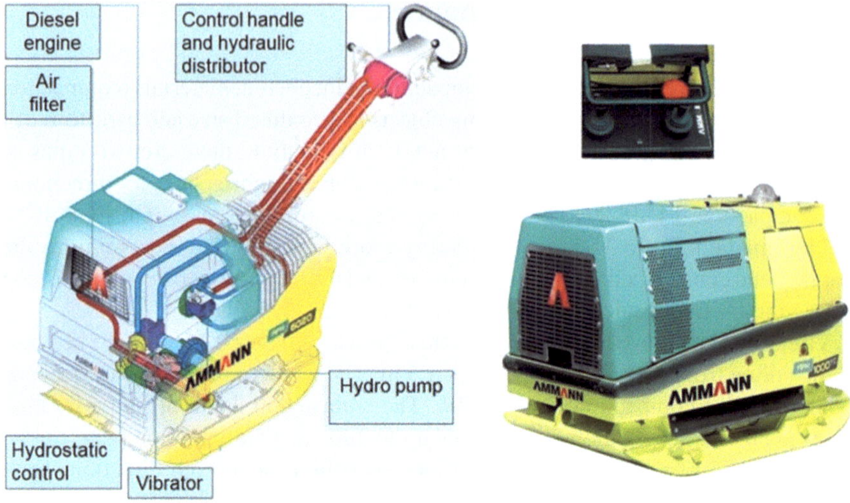

Fig. 8.30 Hydrostatic vibrating plate with manual and remote control (adapted [8])

Advanced designs of vibro plates use a hydrostatic installation for movement and control of compaction, Fig. 8.30. The design of the eccentric and the mode of vibrator operation determine the plate parameters. For example, a vibro plate with a medium weight of 620 kg (*Ammann*) uses a diesel engine (*Hatz, Supra*), with a power of 10 kW/3000/min, which drives the hydraulic pump. The closed hydraulic circuit enables control with circular or directional compaction, which achieves a movement speed of up to 30 m/min and a centrifugal force of up to 70 kN. The soil is compacted to a depth of 90 cm, at a frequency of 46 Hz, the work output is greater than 1000 m^2/h. The Vibro plate can work on gradients of up to 35%. Vibro plate can be manually or remotely controlled.

Vibration parameters of the plate

The movement of the vibration plate is based on the excitation force of the vibrator in contact with the ground, Fig. 8.31a, b. Vibrations of eccentric masses, in the opposite direction of rotation, with equal angular velocities around the same axis or around parallel axes, cause a tangential component of the centrifugal force in the plane of motion ($2F_c \sin \Omega t$). This results in the changes in the relationship of forces in contact with the substrate and its reaction.

Two vibration equal shafts with eccentrics are placed on the auxiliary frame, which can be rotated around the axis by certain angle α. With several vibration plates, the highest speeds are achieved at $\alpha = 45°$, Fig. 8.32. The speed of light vibro plates is $v = 6$–12 m/min and $v = 15$–20 m/min for medium and heavy vibro plates. The transition from positive to negative angle α results in a change in the direction of movement of the vibro plate, back and forth.

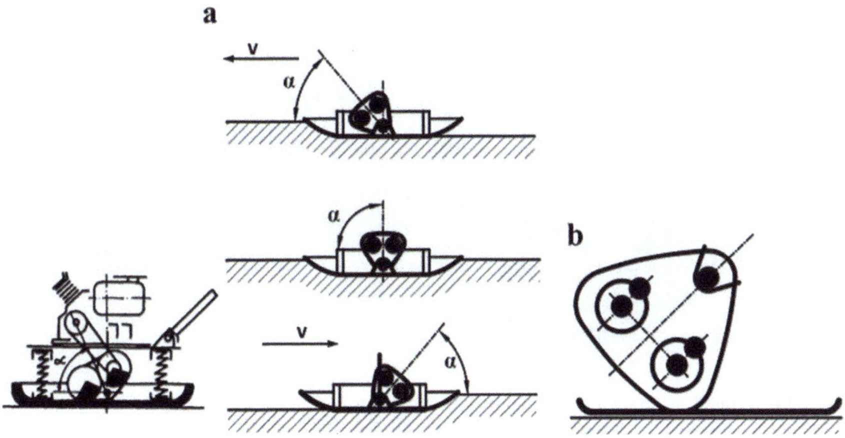

Fig. 8.31 Moving and changing the direction of the plate with two vibrators. **a** turning around the axis below the eccentric, **b** turning around the axis above the eccentric

Fig. 8.32 Speeds of moving the plates according to the angle of the excitation force [7] (at n = 1100 min-1, excitation force: (1) 95 kN, (2) 70 kN, (3) 50 kN, (4) 37 kN)

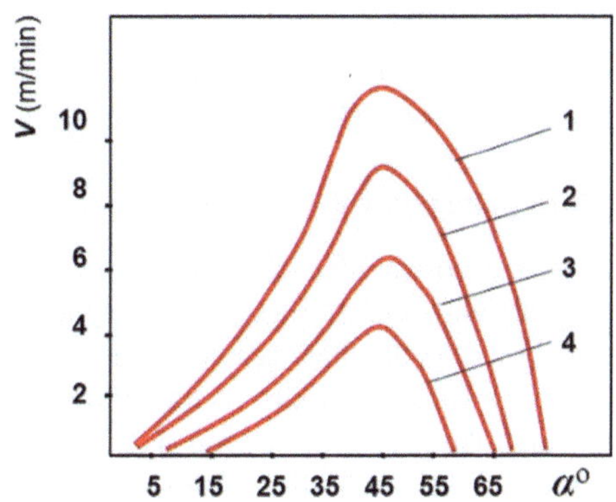

Dimensions of the base plate

- plate width, $B_{\min} = (1\text{–}2)h$, h—layer depth
- plate length, $L = (1.0\text{–}1.5)B$

The number of vibrator revolutions

According to static load

- za $p_{st} = 5\text{–}10$ kPa → $n = 2000\text{–}1200\ \text{min}^{-1}$
- za $p_{st} = 10\text{–}20$ kPa → $n = 1200\text{–}900\ \text{min}^{-1}$

Excitation force

Vibrator with directional action: $F_{p\ max} = 2\ F_c \cos\alpha$.
Medium heavy vibro plates compact layers of 40–80 cm.
600–2000 impacts per minute, impact forces 3.5–20 kN.

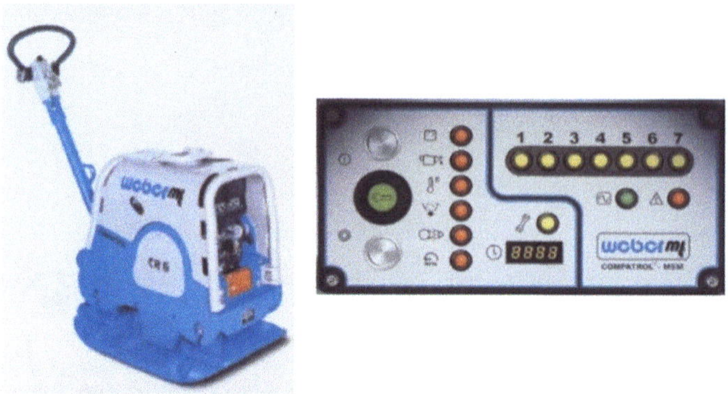

Fig. 8.33 Manual management and control of compaction [9]

Plate vibration amplitude 0.4–1.5 mm, number of passes 2–4

Vibro plate power (P_p)
By frequency

- for n = 20 Hz → 7,5–9 kW/t, for n = 40–50 Hz → 10–13 kW/t
- for n = 50–70 Hz → 13–18 kW/t, for n = 75–90 Hz → 22- 35 kW/t

Vibro plate control

In the case of smaller vibration plates, the control is done manually, i.e. by turning the plate with a lever, Fig. 8.33. Compaction control is monitored by display indicators. based on the plate acceleration sensor. For example, soil compaction has been achieved when the diodes no longer light up.

Control using a remote control makes maneuvering easier and increases the performance of heavy vibration plates, Fig. 8.34. The basic data of the heavy vibro plate are presented in Table 8.3. The remote control system provides safety within the line of sight, the operator is protected from vibrations, noise, and dust.

Vibration compactors

In order to increase the work performance, parallel connection of several vibro plates into one unit is used, which are called compactors, Fig. 8.35. They are connected to mobile machines or special vehicles. The vibrators are hydraulically driven by the base vehicle. Vibroplates with the same technical features are connected. This improves the mobility and control of the compactor. Vibro plates are connected to the front or rear of the machine, and between the axles of the machine.

The deep compaction of the vibro roller is combined with the surface compaction of the plates, increasing productivity and reducing the cost of compaction. This ensures the quality of the compaction and greater performance of the machine. They are often used to stabilize the soil on damaged roads. Vibro plates are raised and

Fig. 8.34 Vibro plate with remote control

Table 8.3 Basic data of the heavy vibration plate [10]

Mass of the plate	1185 kg
Centrifugal force	130 kN
Width of the base	1200 × 1186 mm
Thickness of the base plate	14 mm
Working height	988 mm
Frequency	58 Hz
Movement speed	31 m/min
Effect	2232 m^2/h
Diesel engine (2700/min–3600/min)	Diesel (16 kW/24 kW)
Power transmission	Hydrostatic, 2 circuits
Fuel consumption	4 l/h
Fuel tank	18 l

Fig. 8.35 Combination of deep and surface compaction (vibro roller, vibro plates)

lowered from the vehicle cabin. The width is usually 3.75 m, the speed of movement 0.7–1.0 km/h, and the compaction depth 30–40 cm.

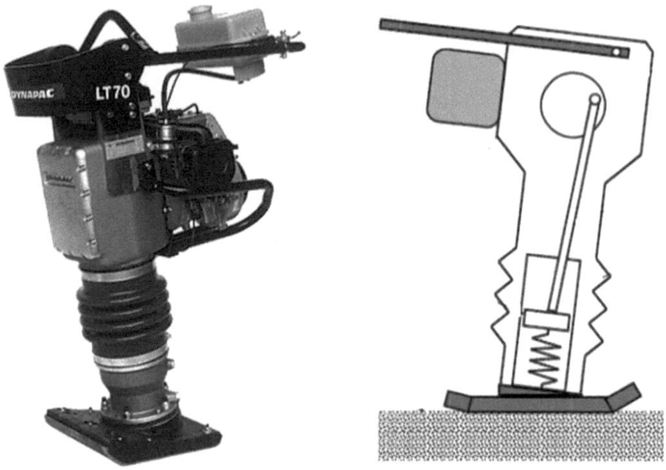

Fig. 8.36 Vibro rammer [11]

Vibro rammers

Vibro soil compactors produce shock vibrations. They are intended for compacting the embankment layer in a limited and difficult-to-access area (sand, gravel, clay and mixed soil, near canals, abutments…). The design of the rammer is shown in Fig. 8.36. The engine drives a crankshaft mechanism with a connecting rod, which is coupled between one or two coil springs in the beveled cylinder of the strike plate. After the rotation of the crankshaft, the plate of the rammer (foot) is raised and after 180° in the expansion cycle, impact soil compaction follows. Practically, rammers achieve about 600 impacts per minute, with amplitudes of 50–100 mm.

The strongest compaction occurs during the expansion stroke with the impact of the mass against the ground. First, the upper part of the rammer is raised to a certain amplitude height, then the lower part, after which the rammer falls to the ground with its full weight and compacts the soil. According to the weight, there are light rammers 50–100 (200) kg and heavy rammers 400–1000 kg, which are called "frogs" due to the slanted foot (cylinder axis) and bouncing.

With a beveled impact plate cylinder on the ground, a horizontal component of the self-moving force is created, which gives the working speed of the machine of around 10 m/min. Optimal compaction is achieved in the frequency range of 10 Hz (540–660 impacts per minute) in the amplitude range of 60–80 mm. Lighter multi-pass compactors (4) can compact a layer 30–40 cm thick on an area of 50 m^2/h.

The basic data of the vibro rammer with a 4-stroke gasoline or diesel engine are shown in Table 8.4.

Table 8.4 Basic data of the vibro rammers [12]

Data	Petrol, BS50-4	Diesel, DS70

Mass	65/72 kg	83 kg
Plate dimensions	280 x 340 mm	330 x 340 mm
No. of blows per min.	656	667
Jump height	61/65	75 mm
Impact power	16/18 kN	
Working speed	7.1/8,8 m/min	8,8 m/min
Surface capacity	119/148 m^2/h	174 m^2/h
Engine	2.7 kW, 3600/min, Honda GXR 120	2.8 kW, 3450/min, Yanmar L48

References

1. Flos R (2002) Tehnika sabijanja u zemljogradnjama i izgradnji prometnih putova, Svezak 1, BOMAG, OK-TEH, Zagreb, https://www.o-k-teh.hr; Brochure. https://www.bomag.com
2. Caterpillar Performance (2020) Handbook, mages available. www.wheeler.com; www.teknox group.com/hr/naslovna/Caterpillar
3. Vibration and oscillation by HAMM (2023) www.mdx.vn/vibration-and-oscillation-by-basic
4. Kelic V. N (1989) Hidroprenosnici. Naucna knjiga, Beograd
5. Mikulic D (2024) Gradevinski strojevi. Teorija i konstrukcija. VVG, Velika Gorica
6. Brandl H (2002) Zbijanje tla i drugih zrnatih materijala za građevine. Građevinar 54, 9, (513–527). https://hrcak.srce.hr/file/17373
7. Briaud L, Seo J (2003) Intelligent compaction: overview and research needs. www.intelligentc onstruction.com/downloads/PapersReports/Texas_BriaudIC%20Report200409.pdf
8. Hydrostatic Vibratory Plate APH 1000 TC (2022) The remote-controlled compaction plate. AMMANN. www.ammann-group.com
9. Weber Compatrol-MSM (2023) www.webermt.com/home
10. Wacker DPU130r (2023) Remote-controlled vibratory plate. www.wackerneuson.com
11. Dynapac AB, Atlas Copco rammer (2023) www.dynapac.com/en/products/compaction
12. Vecker rammers, Petrol, BS50-4/Diesel DS70 (2023) www.wackerneuson.com/products

Chapter 9
Mobile Cranes

Abbreviations

G [N]	Mobile crane weight
Q [N]	Load capacity (cargo mass)
R [m]	Lifting radius
L [m]	Length of the boom
α [−]	Angle of the boom
H [m]	Cargo lifting height
ν [%]	Crane stability coefficient
k [−]	Safety factor
m_{c} [kg]	Counterweight mass
F_{D} [N]	Boom hydraulic cylinder force
S_{k} [m]	Boom hydraulic cylinder extension
F_{T} [N]	Force of hydraulic cylinder for telescoping
S_{t} [m]	Extension of hydraulic cylinders for telescoping
S_{n} [N]	Forces in ropes of pulley assembly
i_{k} [−]	Transmission ratio of pulley assembly
η_{k} [−]	Degree of efficiency of the pulley assembly
M_{d} [Nm]	Torque on the winch drum
M_{hm} [Nm]	Hydraulic motor torque of the winch
v_{t} [m/s]	Load lifting speed
D_{b} [mm]	Drum diameter
d [mm]	Rope diameter
L_{u} [m]	Length of lifting rope
M_{t} [Nm]	Moment resistance in bearing of the superstructure
M_{i} [Nm]	Inertion moment of the superstructure
M_{V} [Nm]	Moment of wind force
M_{HM} [Nm]	Hydraulic motor torque of the superstructure.

D. Mikulić and D. Mück, *Construction Machines*,
https://doi.org/10.1007/978-3-031-67845-5_9

9.1 Purpose and Division of Mobile Cranes

Mobile cranes are used on construction sites to lift and move various cargos, Fig. 9.1. Due to their mobility and passability on construction sites, mobile cranes achieve their work performance in transshipment, assembly, and other works of moving cargos. Modern mobile cranes are hydraulic telescopic cranes.

Mobile cranes can be divided according to several criteria:

a. According to the construction of the boom.

- Telescopic mobile cranes
- Lattice mobile cranes

b. According to the nominal load capacity, classes.

10–20 t, 20–40 t /light and medium
40–60 t, 60–120 t, 120–200 t /heavy

c. According to the mode of operation.

- On the stabilizer feet
- On the wheels of the vehicle

This chapter deals with mobile cranes raised on hydraulic feet.

Fig. 9.1 Telescopic mobile crane on hydraulic feet [1]

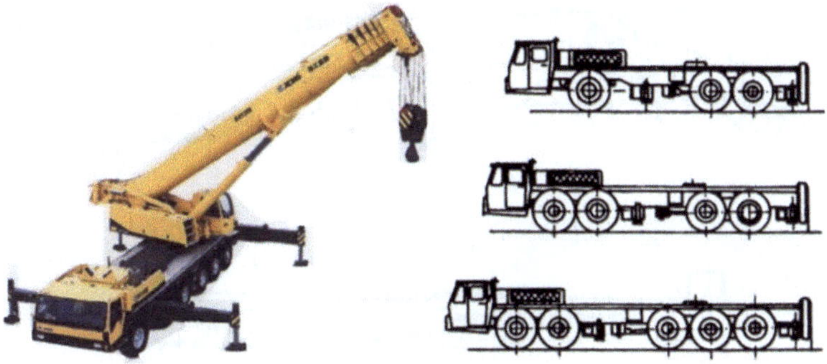

Fig. 9.2 Multi-axle undercarriage of a mobile crane (6 × 6, 8 × 8, 10 × 10)

9.2 Mobile Crane Construction

The mobile crane consists of an upper and a lower plant. The lower structure includes a specially designed undercarriage of truck chassis. The upper structure consists of a rotating superstructure, which includes a working platform and all the devices for working with the telescopic boom and the counterweight. The cargo is lifted vertically using a hook and steel ropes, then rotated and slowly lowered towards unloading.

For moving and working with a crane of smaller capacity, one undercarriage drive engine and one cabin on the vehicle are used from which the vehicle is controlled, and the crane is operated. Truck cranes with larger capacities have two drive engines and two cabins, one on the vehicle for driving the vehicle, and the other lifting cabin for working with the crane on the turntable. The safety of lifting the load is achieved by lifting the truck on hydraulic feet, the so-called stabilizers and the use of safety devices.

Mobile crane undercarriage

The undercarriage (chassis) of the mobile crane is based on a multi-axle vehicle, Fig. 9.2. The transmission of power from the engine to the wheels uses an automatic or semi-automatic transmission. Many gears ensure transport speeds of movement. The wheels have wide tires (445/95 R 25 and 525/80 R 25). The crane is steered with the wheels of one, two or more axles. Due to working in narrow spaces of the construction site, the cranes are made with the possibility of diagonal movement, forward–backward-sideways.

The undercarriage frame is equipped with hydraulic feet to ensure stability of work, Fig. 9.3. Pulling out the feet and levelling the undercarriage as a stabilization base is done from the cabin or from a portable module outside the cabin.

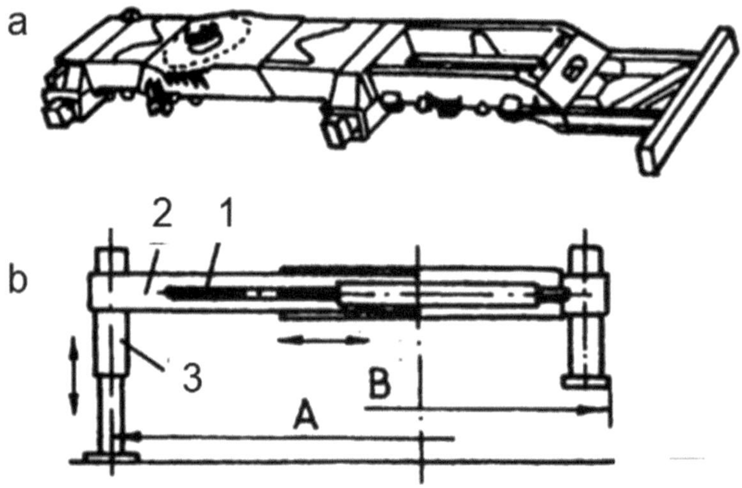

Fig. 9.3 Undercarriage frame (**a**) and telescopic foot (**b**) [2]. 1—Horizontal hydraulic cylinder, 2—Telescopic beam, 3—Vertical hydraulic cylinder

The upper structure of a mobile crane

The main functional units of the upper structure are:

- Telescopic boom,
- Telescoping device (mechanism),
- Device for changing the inclination of the boom,
- Device for lifting cargo with a rope,
- Boom turning device,
- Hydraulic installation,
- Crane counterweight,
- Safety devices.

The telescopic boom of the crane consists of several articles—sections, Fig. 9.4. A telescopic boom of 5–6 articles reaches a extension length of 60 m or more. For special lifting requirements, a light top lattice boom is added. At the back of the boom, a counterweight is attached to the rotating platform, which ensures the stability of the crane. The hydraulic installation enables the management of all functional units of the crane. The cargo is lifted hydraulically with a winch with the help of a pulley and hook coupling. A hydraulic motor with a reducer and a brake rotates the entire superstructure using a small gear on a large gear ring. The crane is controlled by a multi-directional *joystick* operated by the crane operator from the height-adjustable cabin.

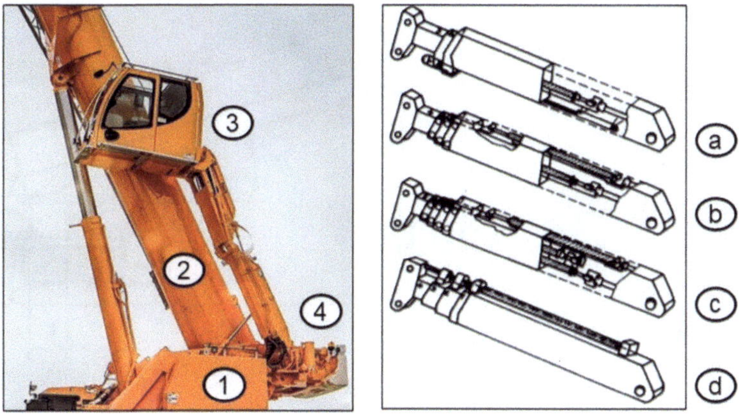

Fig. 9.4 Rotating platform (**1**), telescopic boom (**2**), lifting cabin (**3**), counterweight (**4**). **a**—3 articles, **b**—4 articles, **c**—5 articles, **d**—4 articles and hydraulic cylinder [1, 2]

9.3 Cargo Lifting Parameters

Crane stabilization base

By supporting the crane with the feet on the ground, the stabilization base is set. The maximally extended hydraulic feet, into one and the other side of the crane, form a surface stabilization base with four supports. Hydraulic telescopic feet with four supports are used. The feet are started from the cabin or outside the cabin using a control transfer module, one foot at a time, two at a time, or all at the same time, after which the stabilization base is levelled. The feet are equipped with two hydraulic cylinders. Horizontal hydraulic cylinders are used to pull the beams (articles) out of the undercarriage frame guide. Vertical hydraulic cylinders raise the crane and level the horizontal position, and when the crane is working, they take over the vertical load.

Synchronous extraction of the feet and uniform lifting of the crane, regardless of the asymmetry of the load and the unevenness of the ground, is ensured by the control unit for leveling. If the support force of a foot falls below the permissible value (soft terrain), this foot is independently activated additionally (pressure sensors). The feet can be secured with a mechanical securer.

The basic operating parameters of cargo lifting are:

- Lifting radius (crane reach, crane radius),
- Lifting height,
- Crane lifting capacity.

Lifting radius

Lifting radius R is a function of the length of the boom and the angle at which the boom is located, minus arm K, Fig. 9.5.

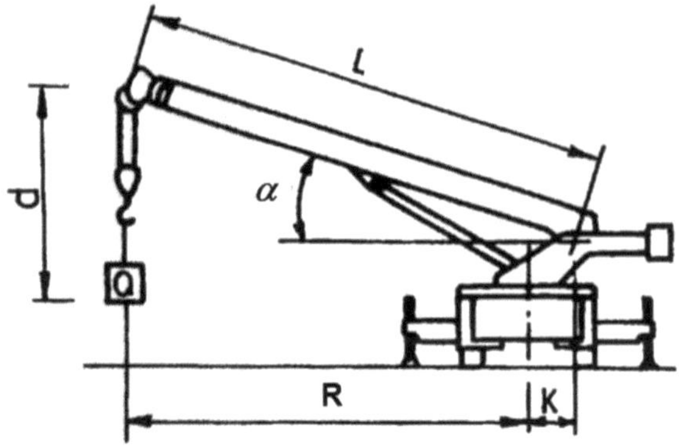

Fig. 9.5 Lifting radius (crane reach, R)

$$R = L\cos\alpha - K[\mathrm{m}]$$

L Length of the boom,
α Angle of the boom with respect to the horizontal position,
K Distance from the pivot point of the boom to the center of rotation of the superstructure (const.).

The lifting radius is a key lifting parameter on which the lifting capacity of the crane depends directly. Due to the variability of the boom length and the boom angle, the crane radius changes, so it must be processed by the processor. The crane operator can set the appropriate lifting radius for a specific cargo on the dashboard and monitor the crane loading percentage. Due to the safety against overturning, during the lifting and transfer of the load, the selected radius of the crane (R) must not be increased (Fig. 9.5).

Cargo lifting height

Lifting of cargo to a certain height depends on the distance of the crane from the ground $\boldsymbol{h}$, the length of the boom $\boldsymbol{L}$, its inclination $\boldsymbol{\alpha}$, and the height of the load suspension $\boldsymbol{d}$.

The cargo lifting height is:

$$H = h + L\sin\alpha - d[\mathrm{m}]$$

h Distance of the crane from the base to the pivot point of the boom,
d Distance from the hook to the cargo.

Crane stability

Loading of hydraulic feet are transferred to the ground. The stability of crane operation is considered first without the load on the hook, then with the load on the hook. Figure 9.6 shows the change in foot loading in the turning circle of the boom, according to the position of the boom and the counterweight, without load and with cargo lifting.

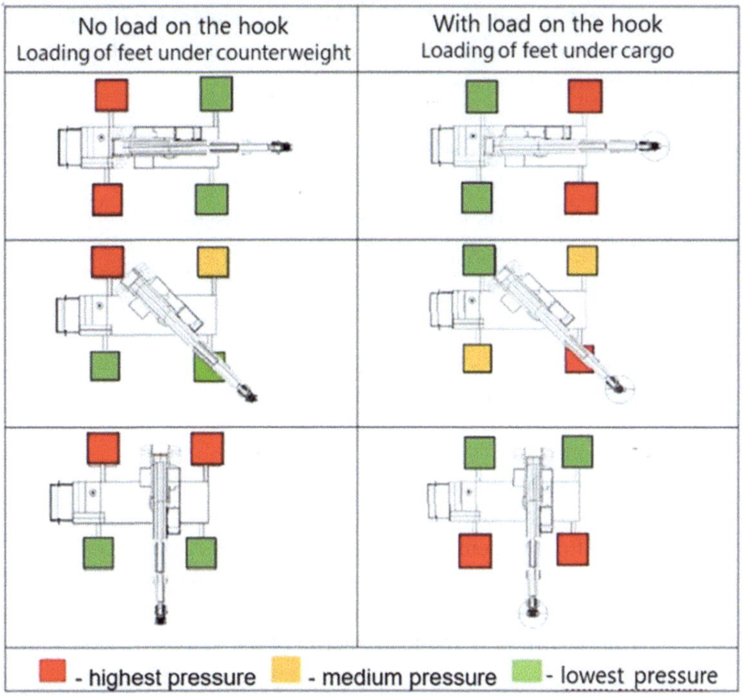

Fig. 9.6 Change in foot loading within the turning circle of the crane boom

When there is no load on the hook, the position of the counterweight determines the position of the loaded feet (rear, diagonal, side). The position of the load on the hook determines the positions of the loaded feet (front, diagonal, side). Based on the pressure measurement in the hydraulic feet, the loading of individual feet is calculated and displayed with load indicators, which is important from the point of view of the crane stability.

For the safe operation of the crane, it is necessary to ensure its stability in relation to overturning. The basic condition of crane stability is that the load and counterweight are in mechanical static balance. Changing the weight of the load changes the length of the boom or the weight of the counterweight. Therefore, the load being lifted must first be properly balanced with the counterweight. Therefore, the cranes are equipped with counterweights of different masses, which are placed at the back of the crane boom. For cranes with smaller capacities, one mass of counterweight is usually used.

The stability of the crane depends on several factors such as static and dynamic forces acting on the crane. The static stability coefficient can be calculated, and it shows the ratio of the value of the moment around the overturning edge. The coefficient must be greater than 1. To calculate the stability coefficient, it is necessary to know the individual masses and loads of the crane, dimensions and other parameters that affect the position of the center of mass.

Stability coefficient of cranes $\boldsymbol{\nu}$ (%) is given as a percentage of safety factor $\boldsymbol{k}$. Lifting standards prescribe the conditions and the procedure for checking stability. The European standards are stricter, for example:

Country	Standard	Stability ν (%)
USA	SAE J765	85
Great Britain	BS 1757	66.6–75
Germany	DIN 15019	75

$$\nu = \frac{1}{k} 100\,[\%]\, k = \frac{\sum Ms}{\sum Mp} \geq 1.2{-}1.5$$

ν (%) Crane stability coefficient,
k Overturning safety factor,
$\sum Ms$, $\sum Mp$ Stability moments, overturning moments.
Stability moments of the crane around the tipping edge (due to weight, in the center of mass of the crane). Overturning moments (due to lifting of the load, inertia of the load, gyroscopic moment).

The limit load capacity of the crane is determined from the overturning conditions, on the edge from the transverse or longitudinal side, Fig. 9.7.

The position of the crane center of gravity (CG) is a function of the position of the telescopic boom, the load and the dimensions of the crane (coordinates of the center of gravity, x_T, x_y, x_z). To determine the center of gravity of the crane, it is necessary to know individual masses (m_i), i.e. forces (F_i), as well as the positions of the centers of gravity of individual masses (x_i).

Transverse coordinate of the center of gravity of the crane:

$$x_T = (\Sigma m_i \cdot x_i)/\Sigma m_i = (\Sigma F_i \cdot x_i)/\Sigma F_i.$$

For example, the limit load (cargo) is calculated for the overturning edge CD, from the condition that the sum of the moments of the overturning forces is equal to the sum of the moments of the non-overturning forces:

$$Q = (\Sigma m_i \cdot x_i)/x_b$$

x_b distance of the feet from the load (R-x), according to the rotation angle of the boom (γ)

The smallest limit load of the crane occurs during lateral lifting, at the smallest rotation angle of the boom ($\gamma = 0°$), and the largest load is at the front ($\gamma = 270°$) or diagonally ($\gamma = 45°$), which determines the stabilization base.

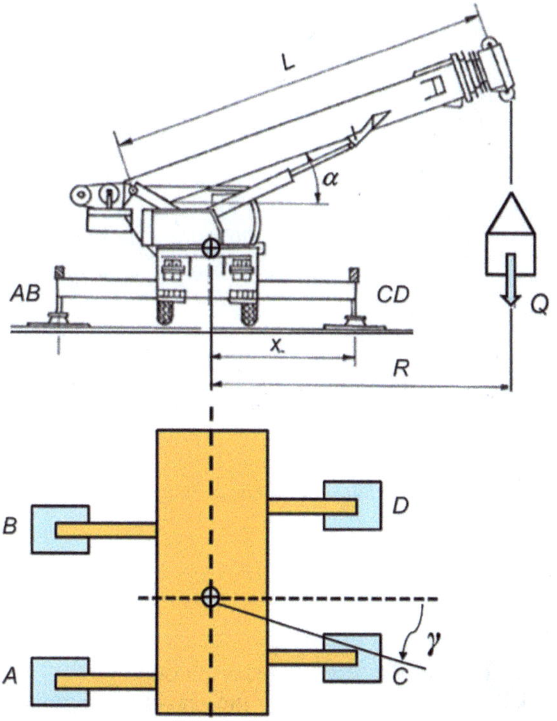

Fig. 9.7 Position of the mobile crane when lifting the cargo

Load capacity table and load diagram

The load capacity of the crane is expressed by mass (kg, t), which is convenient from the point of view of use. In the technical calculation of the construction of the crane, the load capacity is expressed by weight (N), which is suitable for engineering purposes. Practically, the load capacity of the crane means the mass of the cargo (Q), which includes the mass of freely suspended—stationary load, and the mass of auxiliary means for transferring the load (ropes).

The nominal capacity of the crane is the maximum capacity at the smallest radius. The nominal load capacity of the crane means the largest load capacity of the crane at the smallest reach (radius). This figure of maximum capacity is usually not practically feasible, due to the inability of the truck crane to approach the load, which is hindered by its dimensions, limited by the cabin, feet, and the dimensions of the load.

The load capacity of the mobile crane is defined when working on feet and on wheels:

a. For the case of working on feet (at a certain lifting radius) and full rotation of the boom, Table 9.1.
b. For the case of working on wheels and moving the crane on wheels.

The operator determines the lifting parameters for safe lifting of the load. The load to be lifted must be within the permissible load capacity, according to the table or

Table 9.1 Lifting capacity of the crane on hydraulic feet (23 t) [2]

Length of the boom (m)				
Crane reach (radius) (m)	8	8–11	11–13	13–14
	Load capacities in a full turning circle (kg)			
	10	60	70	20
3.00	23.000	19.700	19.200	17.000
3.50	20.000	18.200	17.700	16.100
4.50	16.000	15.500	15.500	14.500
6.00	12.800	12.800	12.800	12.300
7.50		8.600	8.600	8.600
9.00			6.200	6.200
10.50			4.700	4.700
12.00				3.750

10, 60, 70, 20, selection of the operation program

according to the diagram of the crane (radius and lifting height). The lifting capacity of the crane is displayed on the monitor, in the *capacity table* or graphically in the form of a *capacity diagram*. For example, Table 9.1 shows the permitted load lifting capabilities of the medium class of cranes on feet.

The data in the tables are the maximum permissible load capacity of the crane (kg). When the crane boom is placed within the requirements of the load table (radius, boom length), the crane load indicator will be in the safety (green) field. The radius must not change during lifting and carrying. Practically, for the safety of lifting loads, the operators use large reserves in the green field, i.e. lower than the permitted load values in the table.

Lifting capacities above the thick line in the load capacity table on the feet are limited by the strength of the structure, and the capacities below the line are limited by the overturning limit (corresponding to 75% of the allowable overturning load). The numbers in the header of each column of the table (e.g. 10, 60, 70, 20) refer to the lengths of the boom, selected by the switch for the selection of the programmed operations.

If the crane boom is placed according to the requirements of the table, the pointer of the indicator will be in the green load field. The black field indicates the maximum permissible load of 100%. In the event of overload, for example 3–5% of the permissible load capacity (red field), warnings of illegal operations are activated, and after that the lifting function is switched off at the limit load.

Proper positioning of the crane on the ground is very important. The base must be secure, from feet sinking, flat, and solid. If the surface is soft, pads are placed under the stabilizer feet to increase the load capacity. The required area of the pad is calculated based on the load capacity of the pad and the load force of the stabilizer. The crane must not be placed too close to embankments, pits, high voltage, and other specific risks. The crane must be used in accordance with the requirements of safety and protection at work, in accordance with the use of machines with increased hazards. Therefore, the rules of use and safety at work are prescribed for each crane.

The mobile crane capacity diagram combines all three basic lifting parameters in one place, for work on feet or on tires. For example, in the load capacity diagram, Fig. 9.8. shows the parameters of lifting a load 10 t, at the reach of the boom of 7 m, to a lifting height of 12.4 m, with the length of the boom with the second article extended 13.6 m, at an angle of 50°.

Remark In practice, lifting and moving load on tires is avoided, because the risks of overturning are high (condition and slope of the ground, tire pressure, etc.).

Mobile crane specific load capacity

The ratio of the nominal load capacity Q to its own weight G determines the specific load capacity (q), which is an important parameter of the construction. Figure 9.9 shows the diagram of the specific load capacity of telescopic mobile cranes. The

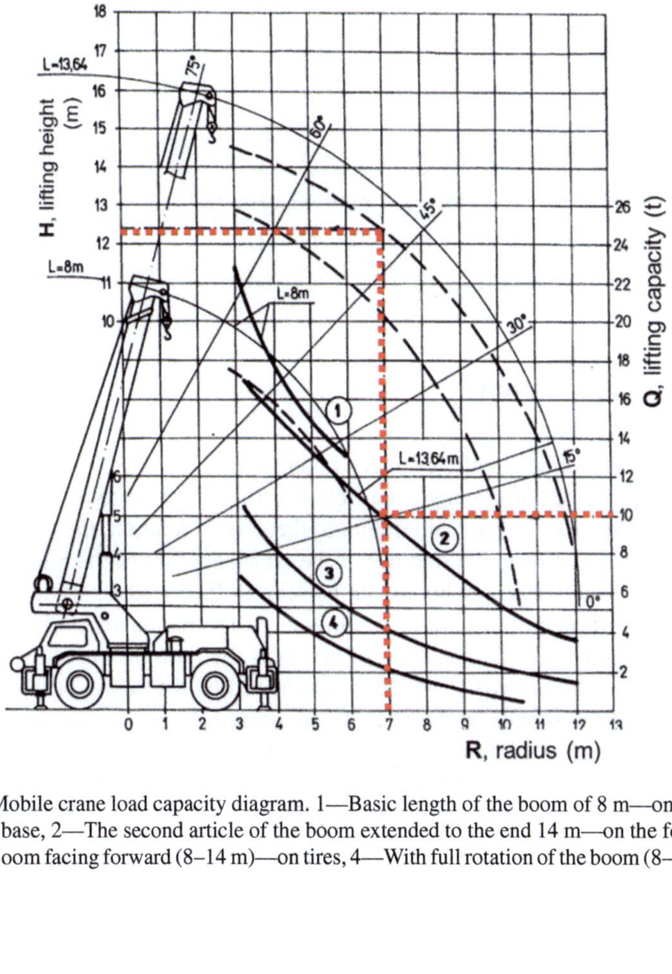

Fig. 9.8 Mobile crane load capacity diagram. 1—Basic length of the boom of 8 m—on the feet of the largest base, 2—The second article of the boom extended to the end 14 m—on the feet of max. base, 3—Boom facing forward (8–14 m)—on tires, 4—With full rotation of the boom (8–14 m)—on tires

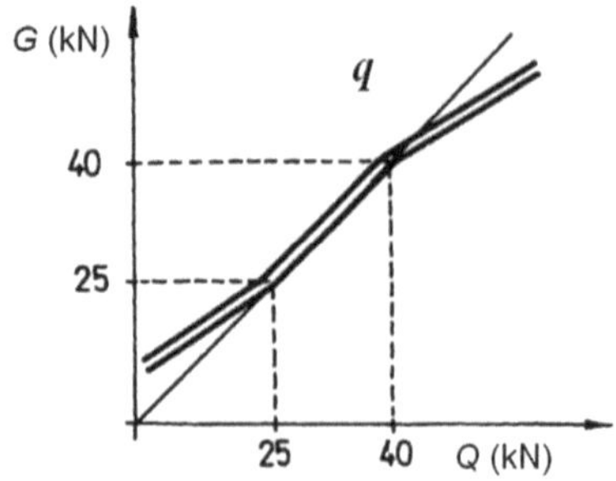

Fig. 9.9 Specific load capacity of mobile cranes

course of the function of the specific capacity shows the perfection of the performance of the cranes according to the criterion of the relationship between the nominal capacity and the total mass (construction of the undercarriage and superstructure of the crane, quality of materials and equipment). Up to a load capacity of 25 t, the mass of the crane is usually greater than the nominal capacity, in the range of 25–40 t, the mass of the crane is equal to the nominal capacity, while with larger load capacities over 40 t, the mass of the crane is less than the nominal capacity, i.e. the mobile crane can lift cargos greater than its weight.

$$q = Q/G$$
$$\text{for } Q < 20, \quad Q < G$$
$$\text{for } 20 < Q < 40, Q \approx G$$
$$\text{for } Q > 40, \quad Q > G$$

9.4 Telescopic Boom

The length of the telescopically extended boom: $L = L_0 + i\, l_k$, (L_0—smallest length of the boom, l_k—step of moving the article, $i = 0.1\ldots\text{n}$). Due to its relatively large mass and its position in space in relation to the axis of rotation, the position of the boom significantly affects the stability of the crane. Profiles of the cross-section of the member of the telescopic boom can be seen in Fig. 9.10, as basic (a, b, c, d), and with reinforcements to increase rigidity (e, f). The parameter of the kinematics of the suspension of the middle class of cranes is the ratio of arms *l/L* (Fig. 9.11). The selection of the ratio (0.16–0.38) is determined by the height of the bogie, the arrangement of assemblies on the base, the stresses of the boom, and the stability of the crane operation. Another important parameter of the telescopic boom is its tilt change, which is most often performed using two hydraulic cylinders. The angle of inclination of the boom in relation to the horizontal position is in the range from −5° to 80°.

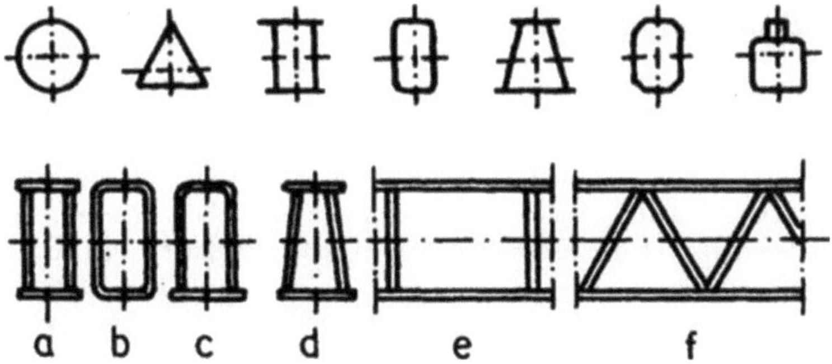

Fig. 9.10 Profiles of the cross-section of the boom

Fig. 9.11 Supports and changes in the inclination of the telescopic boom

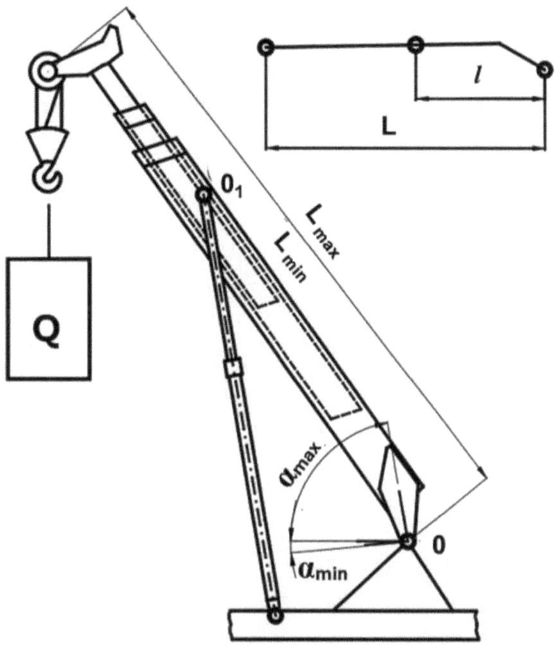

The load on the boom comes as a result of the lifting load, the own weight of the telescopic boom, inertial forces (when raising or lowering, acceleration or deceleration of the load, centrifugal force when turning the superstructure), and wind force on the boom.

9.4.1 Boom Lifting Device

The force of the hydraulic cylinder to change the inclination of the boom $\boldsymbol{F}_{\mathbf{D}}$ is determined from the less favorable load of the boom, i.e. when the loads from the weight of the boom, load, wind force and centrifugal force due to the rotation occur, Fig. 9.12. Otherwise, the change in the inclination of the boom occurs when it is raised or lowered, with one or two hydraulic cylinders, without a load on the hook and without rotation.

Force of the hydraulic cylinder to change the inclination of the boom, $\boldsymbol{F}_{\mathbf{D}}$:

$$F_{\mathrm{D}} = \frac{1}{r_h \sin\gamma}\{[G_0 l/2 + G_1(3l/2 - x) + G_2(5l/2 - 2x) + G_3(7l/2 - 2x) + \psi(Q + G_{\mathrm{k}})(4l - 2x)]\cos\alpha + F_{\mathrm{C}0} l/2 + F_{\mathrm{C}1}(3l/2 - x) + F_{\mathrm{C}2}(5l/2 - 2x) - F_{\mathrm{C}3}(7l/2 - 2x) + (F_{\mathrm{CQ}} + F_{\mathrm{CK}})(4l - 2x) - S\,a_4 + [\psi(Q + G_{\mathrm{k}})a_3 + F_{\mathrm{v}}(4l - 2x)]\sin\alpha\}$$

The extension of the hydraulic cylinder: $S_{\mathrm{k}} = lc_{\mathrm{maks}} - lc_{\mathrm{min}}$.

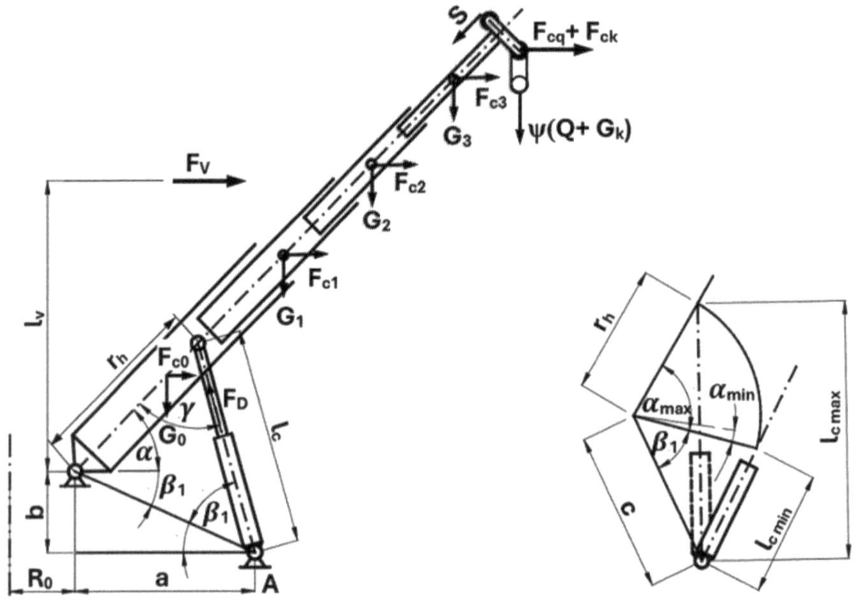

Fig. 9.12 Loading forces of the telescopic boom [2]. G_1, G_2, G_3—weights of articles; $F_{c1,2,..}$—centrifugal forces of articles, G_k—weight of the pulley, Q—weight of the load/carrying capacity, S—force in the rope for lifting the load, ψ—dynamic factor (influence of inertia, centrifugal forces, wind effect on the load)

9.4.2 Device for Boom Telescoping

A model of the telescopic boom is shown in Fig. 9.13. The boom consists of four articles: basic—stationary (to which a hydraulic cylinder for lifting the boom is connected), and three moving articles. The first article is extracted by means of its hydraulic cylinder, while the second article is moved by a chain that is attached to a stationary article at one end. Gal's chain is transferred over the previous sprocket fixed at the end of the second article. The extraction of the third article is carried out before the start of work, and it is closed in the upper or lower position with an electromagnetic lock, or in another way. Around the boom articles there are longitudinal and lateral sliding bearings.

The force of the hydraulic cylinder to change the length of the boom F_T is determined from the unfavorable loading conditions, Fig. 9.14, i.e. when the boom is extracted improperly under load (cargo on the hook).

The force of the hydraulic cylinder to change the length of the boom, F_T:

$$F_T = F'_{t1} + F''_{t2} + F_{l1} + S + [G_1 + G_2 + G_3 + \psi(Q + G_k)]\sin\alpha.$$

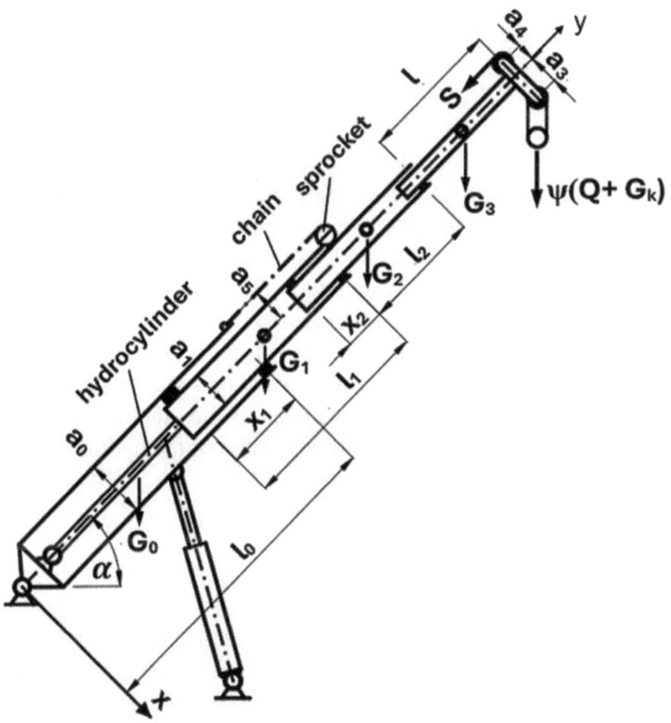

Fig. 9.13 Calculation model of the telescopic boom

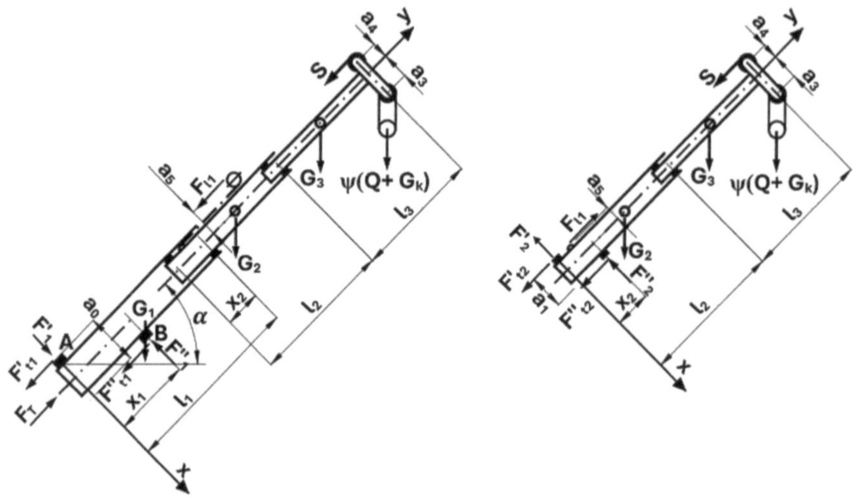

Fig. 9.14 Forces during extension of the boom, and of the second and third article F'_{t1}, F''_{t2}—friction in the article bearings, F_{l1}—force on Gal's chain

Extension of hydraulic cylinders for telescoping:

$$S_t \cong l - x_{min}(l_1 \approx l_2 \approx l_3 = l)$$

Figure 9.15 shows the model of a complex telescopic boom. It is performed using a hydraulic cylinder, a steel rope and a pulley, depending on the design and number of articles (sections). Telescopic article 1 is extended or retracted using a hydraulic cylinder, while articles 2 and 3 are extended or retracted using ropes and pulleys.

An even more complex model uses a pulley system, winding drums and a single piston to cause each article to move in relation to the previous one [3]. The model ensures an even distribution of the load over all parts of the boom. A hydraulic piston that moves inside the base boom pushes the first intermediate article and the tension that this motion develops in the rope causes the other articles to move. The displacement of each article is equal to the extension of the hydraulic piston.

9.4.3 *Load Lifting Device*

Before lifting the load on the hook, the boom of the crane is placed under a selected inclination and a certain length. The load is lifted by a winch rope with a pulley and hook coupling, Fig. 9.16. One end of the winch rope (cable) is tied to the load with

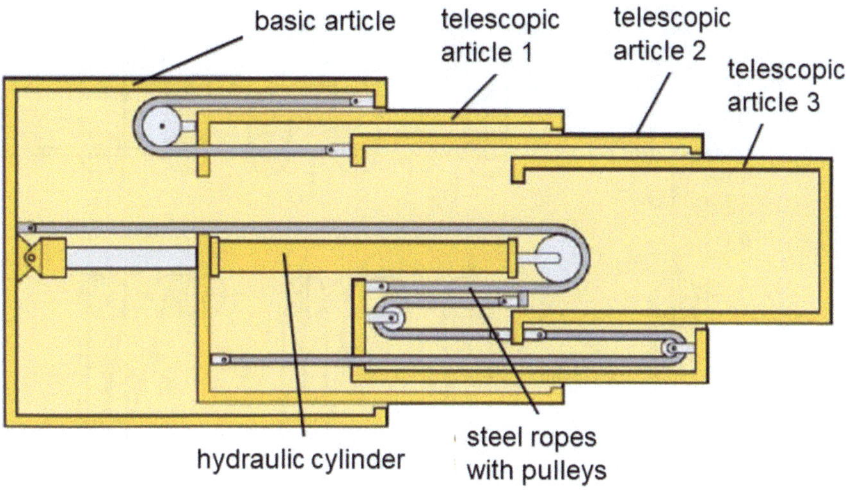

Fig. 9.15 Model of a complex telescopic boom

a hook, and the other end is wound on the winch drum. The lifting device consists of a hydraulic—motor with an automatic brake for holding the load on the hook, a winch reel, one pulley for guiding the rope on the top of the boom, a pulley assembly, and a hook for suspending the load. The upper pulley assembly is stationary with a shaft at the top of the boom, and the lower pulley assembly is driven with a hook. The number of pulleys in each pulley assembly, as well as the number of actively threaded steel ropes, is determined according to the required capacity of the crane, i.e. the weight of the load to be lifted.

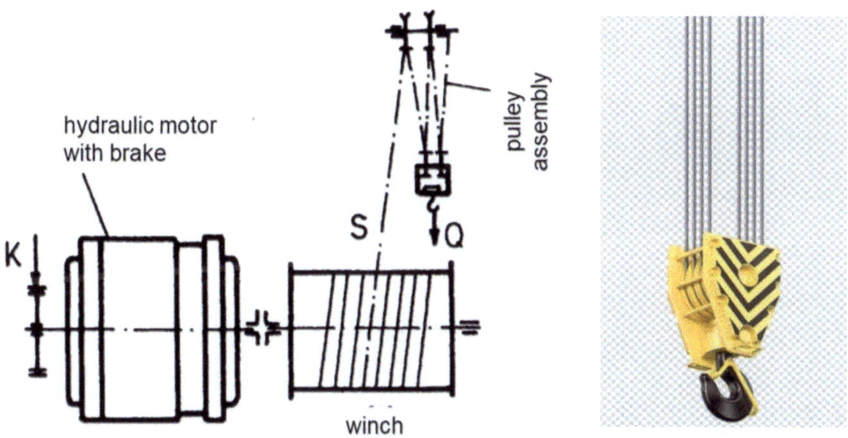

Fig. 9.16 Load lifting device and crane hook

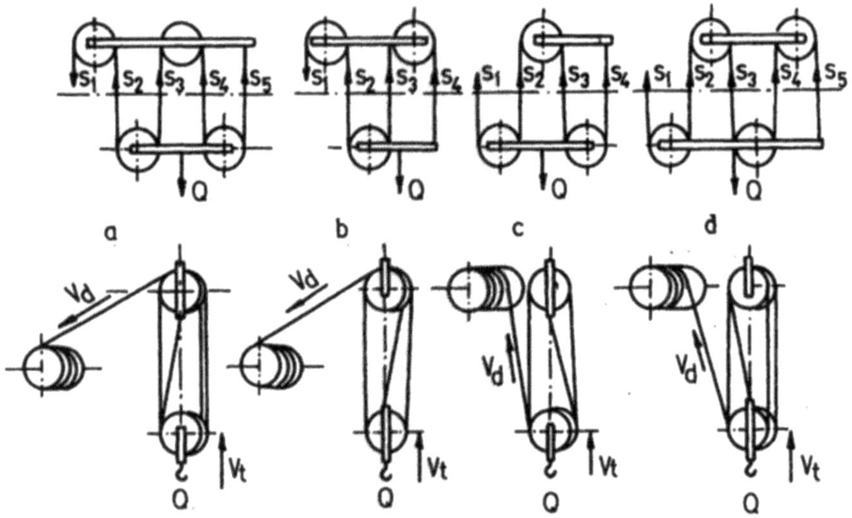

Fig. 9.17 Pulley assembly designs

Pulley assembly

Figure 9.17 shows the formation of the pulley assembly, depending on the load capacity requirements. The number of supporting (active) ropes determines the transmission ratio of such assembly (i_k).

a. Pulley assembly with two movable and two stationary pulleys with transmission ratio $i_k = 4$.
b. Pulley assembly with one movable and two stationary pulleys with transmission ratio $i_k = 3$.
c. Pulley assembly with two movable and one fixed pulley with transmission ratio $i_k = 4$.
d. Pulley assembly with two movable and two stationary pulleys with transmission ratio $i_k = 5$.

Forces in ropes:

$$S_1 = \frac{Q}{i_k \eta_k} [\text{N}]$$

$$S_2 = S_1 \eta_u, S_3 = S_2 \eta_u, S_3 = S_1 \eta_u^2$$

$$S_n = S_1 \eta_u^{n-1}$$

i_k Transmission ratio of pulley assembly.
η_k Degree of efficiency of the pulley assembly (0.94–0.98).
η_u Degree of efficiency of one pulley.

Parameters of a winch for load lifting

Torque on the winch drum:

$$M_{\mathrm{d}} = S_1 D_{\mathrm{d}}/2\,[\mathrm{Nm}]$$

D_{d} Drum diameter.

Torque of the winch hydraulic motor:

$$M_{\mathrm{hm}} = M_{\mathrm{d}}/i_{\mathrm{R}}\eta_{\mathrm{R}}[\mathrm{Nm}]$$
$$M_{\mathrm{d}} = (S_1 + F_{\mathrm{in}})D_{\mathrm{d}}/2[\mathrm{Nm}]$$

i_{R} Transmission ratio of the reducer, in front of the hydraulic motor
η_{R} Degree of efficiency of the reducer

Inertial forces of translational and rotational masses:

$F_{\mathrm{in}} = (1+ \mathrm{a})\, F^{\mathrm{tr}}{}_{\mathrm{in}}$, $a = 0.1$–0.3, inertial forces of the rotating masses of the lifting mechanism
$F^{\mathrm{tr}}{}_{\mathrm{in}} = G\, v_{\mathrm{d}}\, /g\, t_{\mathrm{u}}$, v_{d}—circumferential speed of winding the rope, $\mathrm{t_u}$—drum acceleration time

Speed of lifting and lowering the load:

$$i_{\mathrm{k}} = v_{\mathrm{d}}/v_{\mathrm{t}}$$

v_{d} Speed of winding the ropes on the winch drum
v_{t} Load lifting speed

The speed of winding and unwinding of the steel rope is variable and depends on the winding layer of the rope (first layer, second layer, etc.), pulley assembly, friction, and engine power. With a smaller number of carrying ropes of the pulley, i.e. by turning off the number of pulleys on the movable pulley assembly, the speed of lifting and lowering the load increases, but with a decrease in the load capacity.

Stopping the load during lifting and lowering

Braking torque:

$$M_{\mathrm{k}} \geq M_{\mathrm{d}} + M_{\mathrm{in}}^{\mathrm{tr}} + M_{\mathrm{in}}^{\mathrm{r}};\, M_{\mathrm{k}} \geq M_{\mathrm{hm}}$$

M_{d} Moment on the winch drum from the load,
$M^{\mathrm{tr}}{}_{\mathrm{in}}$ Moment of inertia from translational masses (from the load on the hook),
$M^{\mathrm{r}}{}_{\mathrm{in}}$ Moment of inertia from rotating masses (from the drum …).

Brake safety level: $\beta = M_k / M_d$, $\beta = 1.5$–2.0
Dimensions of the winch and rope

- Ratio of drum diameter to rope diameter, $D_b/d = 15$–25
- Drum diameter, $D_b = (500$–$600)\delta$
- Wire diameter in the rope strand, $\delta = 0.8$–2.5 mm
- Width of drum B depends on the required length of rope L_u and the diameter of rope d.
- length of lifting rope:

$$L_u \approx l\, i_k + l_g + (1 \div 2)d + (0.05 \div 0.1)d[\text{mm}].$$

l	Length of the rope (boom maximally raised, hook touching the ground)
l_g	Length of the rope from the drum to the top of the telescoping boom
(1-2)d	Accessory for fastening the rope to the drum
(0.05-0.1)d	Free distance between ropes.

- Rope diameter:

$$d \approx k_c \sqrt{F_{maks}}[\text{mm}].$$

F_{max}	Maximum permissible rope force
k_c	Factor of the number of working cycles, 6-60 h^{-1}, $k_c = 0.28$–0.38

To lift the load with a winch, steel braided ropes are used, of strength $\sigma_L = 1400$–1800 N/mm^2, with rope diameter $d = 6$–24 mm. The steel rope is dimensioned with a considerable degree of safety (5–10), according to the strain for extending, bending, constant strain due to braiding, or complex strain. The service life of the steel rope is determined based on the fatigue of the material, by bending it over the reel and the drum. Steel rope lasts longer if there are fewer bending changes. Bilateral bending has a less favorable effect on a straight braided rope than on a crossed braided rope. The lower the load on the rope and the larger the diameter of the reel, the greater the number of bending changes that can be achieved. One way to pre-emptively replace the rope is to monitor the repeated loads.

Auxiliary means of carrying load

Various auxiliary means are used to transfer the load (ropes, chains), depending on the type, size, and weight of the load. The basic requirements for these devices are safety against breakage or tearing, as well as against accidental release of the load, quick attachment and release of the load, as little own weight as possible, non-damage of the cargo during handling and suspending. The means for suspension are equipped with a certificate of correctness (atestation) and with a permanent visible mark of the nominal load capacity (stamp). If different loads are transported with the crane according to shape, size, and type, then the crane is equipped with universal means

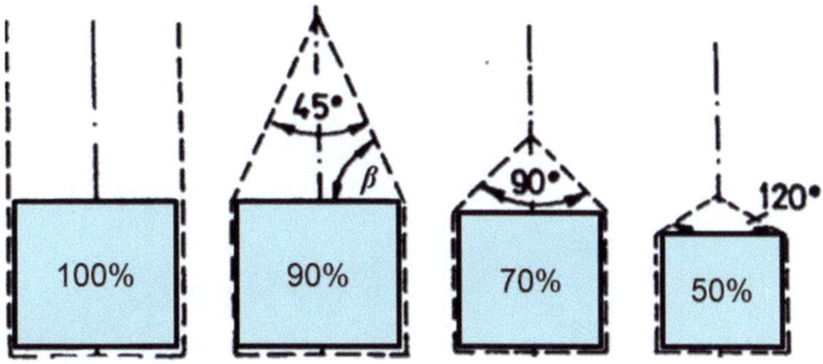

Fig. 9.18 Carrying capacity of the means of load transfer

for transferring the load, and the load is attached to them with auxiliary means. When only certain loads (always the same, or very similar) are transported with the crane, then the crane is equipped with the means that enable simpler, easier, safer, and faster handling of the load. After tying down the load, it should be then lifted from the substrate on a trial basis, at the same time checking the stability of the means of transfer as well as the crane.

It is important to know the permitted carrying capacity of the means of carrying load, which depends on the angle between the arms of the rope or chain, Fig. 9.18. The magnitude of the force in arm S_k is obtained as a component of force Q separated in the direction of the arms. The angle between the arms of the rope or chain, when tying the load as shown in the figure, should not be greater than 120°, and this connection should not be loaded with a weight greater than 50% of the nominal load of the rope. For an angle of 45° between the arms, the permissible load can be maximally 90% of the nominal load of the rope. You should aim for as small an angle as possible between the arms of the rope (45° and less). If the so-called stamp of the rope indicates a nominal load of 12 t capacity, this connection can be loaded with a maximum of 8.4 t at an angle of 90°, and with a load of 6 t at an angle of 120°.

Load on the arms of the steel rope:

$$S_k = Q/i_k \sin\beta [\mathrm{N}]$$

Q Cargo mass
i_k Number of supporting ropes
β Rope arm angle

9.4.4 Device for Turning the Boom

By means of a small gear on a large toothed wreath, the boom turning device with hydraulic motor, reducer and brake turns the entire superstructure, Fig. 9.19. The

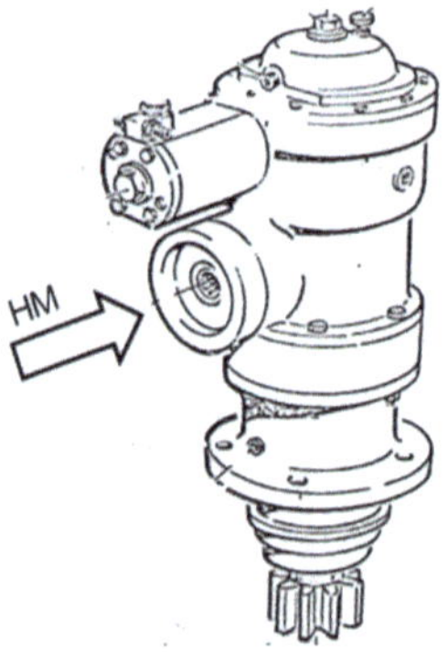

Fig. 9.19 Boom rotation reducer and gear ring

loads of the hydro-motor result from the moment of rolling resistance in the bearing, the moment of the wind force, the moment of the crane superstructure due to the slope of the terrain, and the torque of the resistance of inertial forces.

Parameters of the boom turning device

The moment of rolling resistance in the bearing depends on the load on the bearing:

$$M_t = G\, d_l/d_k \mu(\textit{superstructure weight},\ \textit{bearing/ball diameter},\ \textit{friction}).$$

Tilting moment of the crane

Due to the subsidence of the terrain, the crane is located on a certain slope.

Tilting moment:

$$M_n = M_G \sin\beta$$

M_G Moment of all vertical forces acting on the rotating part of the crane.
β Permissible tilt angle of the crane due to ground subsidence.

Moment of inertia of the boom due to rotation:

$$M_i = J_u \omega_0 / t_u$$

J_u Total moment of inertia of the rotating part, $J_u = \Sigma\, M_i\, r^2{}_i$.
ω_0 Angular velocity ($n_0 = 1$–$10\ \text{min}^{-1}$)
t_u Acceleration time.

Moment of wind force

$$M_{V\max} = F_{VQ} L + F_{VK} l_v$$

F_{VQ}	Maximum wind force acting on the load, $F_{VQ} = C_t\, q\, A_Q$
C_t	Load shape factor, q—unit wind pressure
A_Q	Load surface exposed to wind
L	Arm of the wind force acting on the cargo
F_{VK}	Wind force acting on the surface of the boom
$F_{VK} = C_g\, q\, A_K$	Position of the boom perpendicular to the wind direction
C_g	Boom shape factor
A_K	Surface of the boom exposed to the wind
l_v	Arm of wind force

Moment of wind and inclination of the crane: $M_{Vn} \approx 0.7\,(M_{Vmax} + M_n)$.

9.5 Hydraulic Installation of the Mobile Crane

Figure 9.20 shows the diagram of the hydraulic installation of a medium-class mobile crane. The undercarriage of the vehicle houses the components of the hydraulic installation for the stabilizer feet, the components for controlling the front and rear wheels, and the control of the winch for self-extraction. On the upper structure, all the components of executive devices are located, for lifting the boom, telescoping the boom, lifting the load and turning the boom. The diesel engine drives a double constant flow pump that creates the required oil pressure and flow. The working oil pressure in the device installation is about 250 bar. The amount of oil in the installation is about 100 l, and about twice as much is in the oil tank. The flow of oil between the undercarriage and the superstructure in all operating positions of the cranes is carried out through the rotor coupling assembly.

Hydraulic control circuits:

- Changing the inclination of the telescopic boom,
- Setting the stabilization base,
- Telescoping the article of the boom,
- Lifting the load with a winch rope,
- Turning the crane boom,
- Steering the front (and rear) wheels,
- Management of the extraction winch.

Hydraulic cylinders for changing the inclination of the boom

FD—required force for lifting the boom, p_{max}—oil pressure
Diameter of one hydraulic cylinder: $D_d = (2\, F_D - /\, \pi\, p_{max})^{1/2}$ [mm].

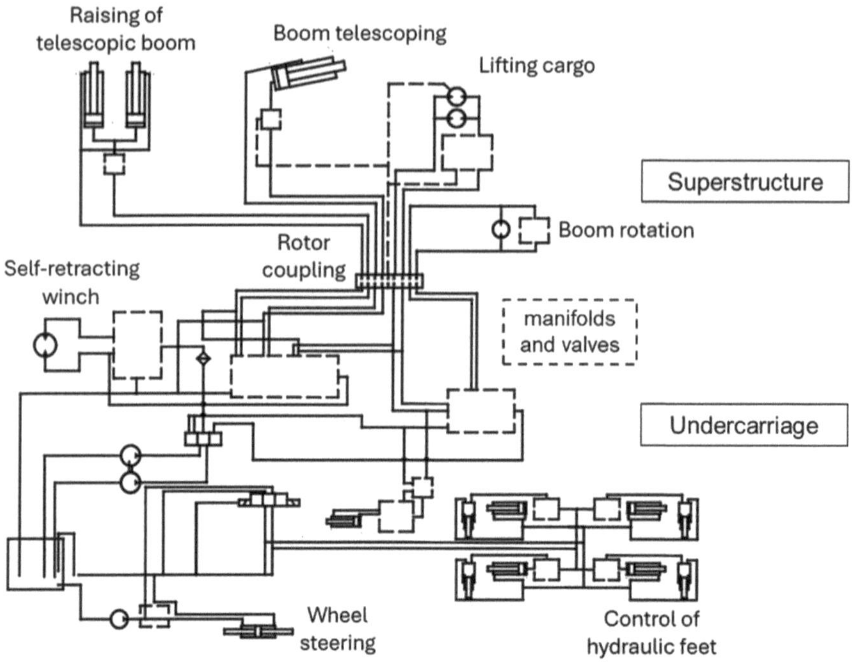

Fig. 9.20 Hydraulic installation of a mobile crane (*23 t*) [2]

Hydraulic cylinder for telescoping

F_T—required force for telescoping the boom, p_{max}—oil pressure
Diameter of hydraulic cylinder for telescoping: $D_t = (2\,F_T\,/\,\pi\,p_{max})^{1/2}$ [mm]

Hydraulic motor for load lifting winch

$$M_{hm} = M_d/i_R\eta_R[\text{Nm}].$$
$$M_d = (S_1 + F_{in})D_d/2[\text{Nm}].$$

M_{hm} Torque of the hydraulic motor

Hydraulic motor for turning the boom

$$M_{HM} = (M_t + M_{V\max} + M_n + M_i)/i_R\eta_R$$
$$M_{HM} = [M_t + 0,7(M_{V\max} + M_n) + M_i]/i_R\eta_R$$

M_{HM} Hydraulic motor torque
M_t Moment of resistance in the bearing

M_{Vmax} Moment of wind force
M_n Tilting moment
M_i Moment of inertia
i_R transmission ratio of the reducer and η_R—degree of efficiency.

9.6 Mobile Crane Safety System

The mobile crane safety system ensures the safety of lifting and transferring load. It allows the crane operator to monitor the loading of the crane, shows the elements of work safety and prevents damage or overturning of the crane. Mobile cranes are equipped with an electronic safety system. Load signaling is visual and audible, with automatic shutdown in case of overload. Before that, the device warns the operator about the deviation from the safe operation.

The electronic safety system controls the parameters of the crane using sensors and calculates the current dynamic loading of the crane. By comparing the measured loading on the hook with the permissible load, which is memorized in the computer for the operation of the crane on feet or wheels, the instruments show the percentage of the loading from the maximum permissible load. When the permissible loading from the table is exceeded, the system warns the operator of improper operation. If the stability of the crane continues to be disturbed, the system automatically stops the irregular operation until the complete shutdown of the engine. The components of the crane safety system are shown in Fig. 9.21.

When using the crane, the operator should know approximately the basic information:

- About the weight of the cargo to be lifted,
- About the boom reach (crane radius),
- About the height to which the load should be lifted.

The computer shows the percentage loading on the meter, as shown in the principal scheme of the safety system, Fig. 9.22.

The loading percentage indicator on the instrument panel is usually divided into several fields (green, yellow, black, red). For example, the hand in the green field shows a loading of 0–75%, the hand in the yellow field shows the transition towards the maximum allowed load (100%, black field). The red field is the condition of extreme overload of the crane, in which the alarm warning devices are automatically activated for the operator, after which, in case of further lifting, the operation is automatically switched off.

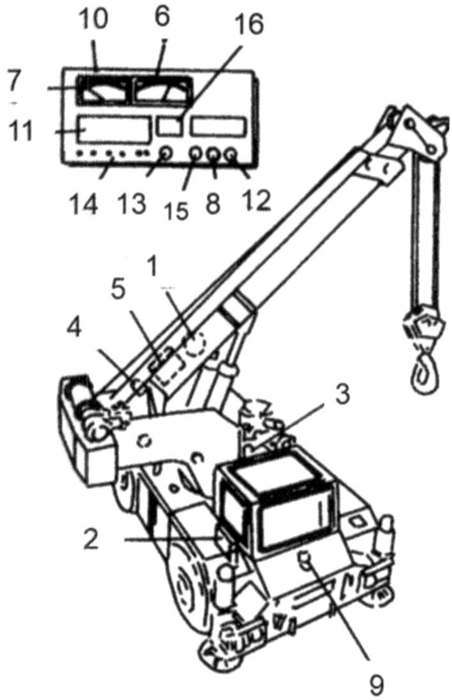

Fig. 9.21 Mobile crane safety system. 1—Boom extension sensor, 2—Computer, 3—Oil pressure sensor, 4 and 5—Boom hydraulic cylinder angle sensors, 6—Loading percentage meter, 7—Boom reach (radius) meter, 8—Warning light, 9—Warning bell, 10—Instrument panel, 11—Crane capacity table, 12—Test switch, 13—Operation selector, 14—Boom extension lamp, 15—Fuse, 16—Operation table

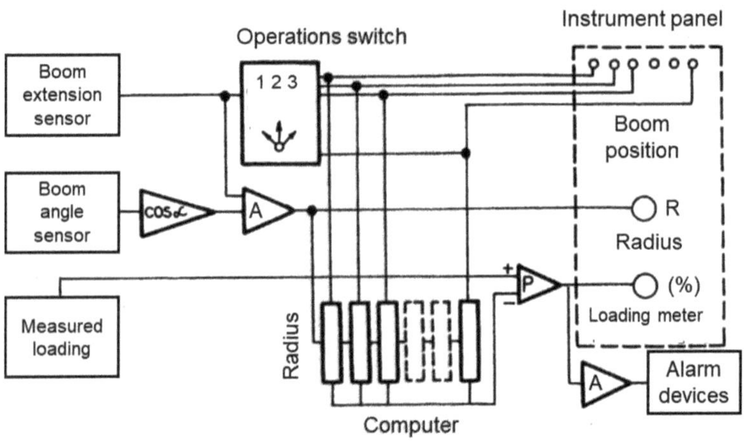

Fig. 9.22 Crane safety system diagram

A practical way to measure the loading on the hook is by measuring the oil pressure in the hydraulic cylinders to change the angle of the boom. Some systems measure the loading on the hook based on the deflection of the boom at specific points or by extending the lifting rope. The length of the boom is measured by the extension sensor, and the tilt angle by the boom angle sensor. The measured pressure in the hydraulic cylinders determines the measured loading of the crane. The accuracy of loading measurement depends mostly on the following factors: inertial forces, wind forces, boom deflection, crane tilt and others.

Service life of the crane

The design calculation is carried out in accordance with SAE J1078—the recommended method of analytical determination of the competence of hydraulic telescopic cantilevered crane booms, which includes cross-sectional dimensions, checking of bending, shearing, stretching, and others [4].

When you want to ensure a long working life of the machine, it is necessary to allocate reliability to the components, ensure their strength, material fatigue, stability, and self-control. The key parts of the metal construction are calculated for a working life of 10,000 working hours [5]. Other assemblies, parts and components with a shorter service life are subject to self-diagnosis and preventive replacement programs, especially the elements of the crane hydraulic and safety system.

Loading forces and moments of resistance are set for cases of the highest complex load of the crane. The calculation procedure requires the application of the finite element method (software *Solidworks* or similar). By simulating the loading according to the load spectrum and the fatigue curve, the durability of the parts can be determined, and the reliability of the parts can be determined based on the probability of working and critical loadings. For the key parts of the mechanism of the working device, materials with increased tensile strength, increased strength, high toughness at low temperatures, and resistance to brittle fracture are used.

9.7 Advanced Load Lifting System

The advanced lifting system is shown on the example of a telescopic mobile crane with nominal capacity of 100 t, Fig. 9.23.

Basic data of the mobile crane *LTM 1100–4.2* [6]:

- Maximum load capacity at a minimum radius of 2.5 m—100 t
- The argest length of the telescopic boom—60 m
- Maximum lifting height with double folding lattice (option)—91 m
- The largest radius with a double folding lattice (option)—68 m
- Counterweight set (options), 2.5–28.2 t
- The largest mass of the mobile crane is 48 t
- Superstructure, Diesel engine 129 kW
- Undercarriage, Diesel engine 350 kW

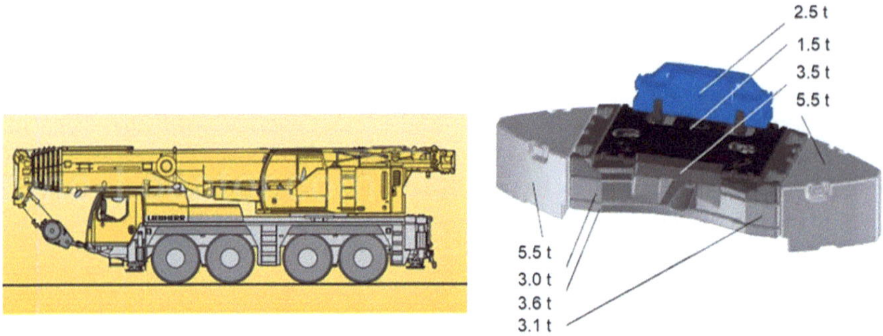

Fig. 9.23 Mobile crane 100 t and counterweight mass (*Liebherr*)

- Automatic transmission AS-TRONIC, ZF 12 and power distributor
- Drive axles 2, 3 and 4, (8 × 6), 8 × 8 option
- 1 and 2 steering axle, 3 and 4 as needed
- Hydropneumatic suspension
- Wheels, 8 wheels 14.00 R 25 (16.00 R 25)
- Maximum movement speed75 km/h
- Modern brakes with ZF intarder.

The telescopic boom is made with 6 articles and a foldable lattice boom. The profile of the cross-sections of the boom has an oval shape of the third generation (G3), which proved to be the most resistant to complex strain.

Counterweight mass (m_c)

The counterweights of the crane are placed on the front part of the undercarriage, usually in front of the cabin. The counterweight mass that can be transported on the undercarriage depends on the permissible axle loading, and the excess is transported by truck. To mount the counterweights, the boom is turned backwards, and the required mass of the counterweight is raised using two hydraulic cylinders. The installed counterweight must be in accordance with the data from the loading table. It is allowed to use only combinations of the counterweights that are prescribed. In the program "equipment" of the control system, it is necessary to enter the exact mass of the mounted counterweights. Each lifting capacity of the crane includes a counterweight weighing 2.5 t. The basic counterweight has a mass of 17.2 t. The additional counterweight package can be of different weights, up to 25.7 t, a total of 28.2 t.

The *Liebherr* mobile crane is equipped with two electronic systems. The LICCON system enables the adjustment of the stabilization base according to the available space and safe lifting of the load. The TELEMATIK system ensures variable length of the telescopic boom based on automatic locking of individual articles.

Program settings of the electronic system of the mobile crane include:

- Initial settings program—"Set up" program. The configuration program appears automatically. A graph of programmed payload values depending on the radius and length of the extended telescope follows.
- Working program—"Crane operation" program. The crane operation program assists the operator by displaying the data on the monitor. An audible signal accompanies critical displays.
- Program for managing the telescopic boom—"Telescoping" program. The screen shows the operator the locking of an individual article, the position and the state of extraction of the telescopic cylinder (a company innovation, the TELEMATIK system).
- Working range limitation program—"Working range limitation" program. Manually setting the working area. The pulley length limit, the radius of the crane, the point of rotation of the right and left sides, the edge limit, etc. are set. After entering the parameter, a graphic display of the permitted working space follows.
- Program for displaying support forces in the feet—"Support force monitoring" program. In this program, it is possible to monitor the permissible loading for each foot, as well as the current loading of the stabilizer, which is displayed on the screen.

Crane control functions using the LICCON system are shown on the hydraulic installation, Fig. 9.24. The diesel engine drives two hydraulic variable pumps and one gear pump. All working devices, for turning the superstructure, the hydraulic cylinder of the telescope, the cylinder for lifting the boom, and the winch for lifting the load—are driven by a hydraulic control block that is directly controlled, including the number of revolutions of the engine. All functions are controlled by the control block of the system.

The operator determines the working area of the crane, the radius, the height of the boom, edge restrictions, so work is enabled only in that area, which is important for the safe work with a standard base or variable stabilization base, in an open or confined space.

Figure 9.25 shows the graphical representation of the standard and variable space of the crane in the 360° turning circle on the stabilizer feet.

Procedures for controlling the variable stabilization base:

- Measuring the current length of each stabilizer foot and comparing it with the maximum length stored in the software and displaying it on the monitor of the safety control system and the Bluetooth remote station.
- Measuring the forces in the supports of the feet of the support cylinders and comparing with the given permissible forces set in the safety control system.
- Calculating the current permissible lifting capacity of the crane and displaying it on the monitor.

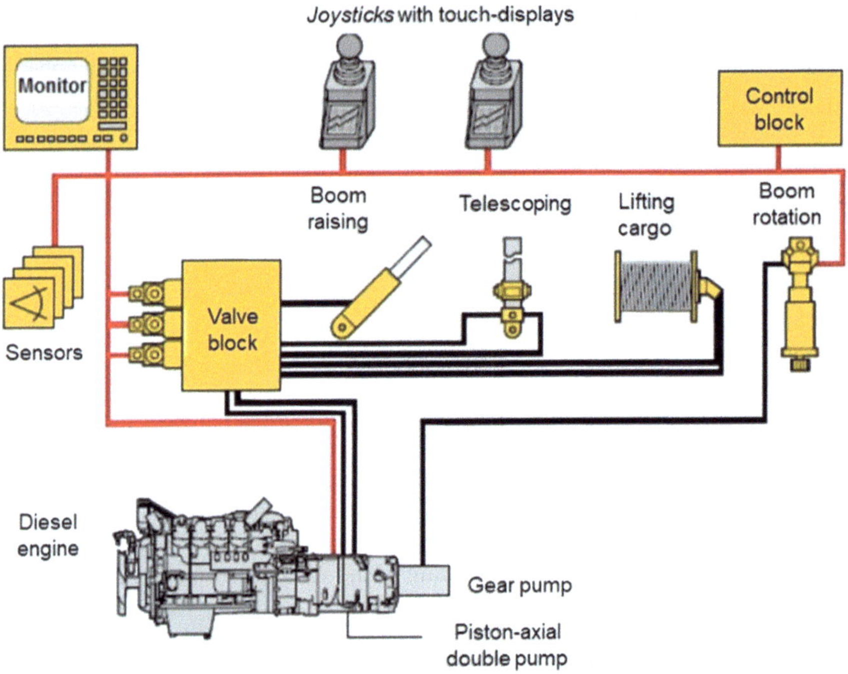

Fig. 9.24 Scheme of the hydraulic installation of the crane [6]

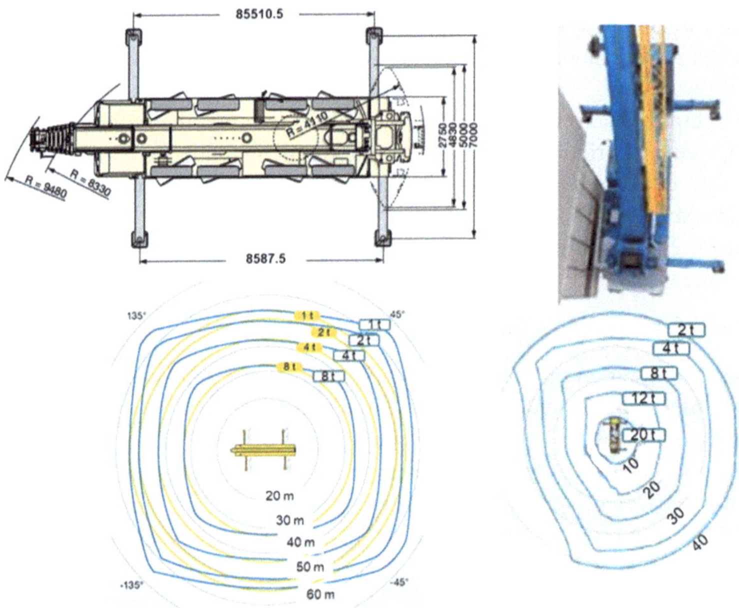

Fig. 9.25 Standard and variable stabilization base surface [6]

The advantages of adjusting the stabilization base is safe work in a confined working space

The LICCON system works on the principle of comparing the actual loading and the maximum permissible loading programmed and stored in the central memory according to the loading capacity table and the number of rope windings on the reel. The loading of the crane consists of the lifting loading moment and the boom moment. The boom moment is calculated from the boom angle information, telescoping boom length information, and boom mass data. The reach is calculated using the boom angle information, telescoping boom length information, and crane geometry data. The bending of the boom due to its own weight and the weight of the load is considered. A large number of sensors provide information on the load management parameters: weight/force sensors, tilt sensors, length sensors, pressure sensors, position sensors, speed sensors, pedal sensors, and limit sensors.

The TELEMATIK system automatically telescopes the boom and ensures the desired length using a special multi-stage hydraulic cylinder, Fig. 9.26.

The multi-stage hydraulic cylinder is of double action with hydraulic locks for locking individual articles in a certain position. The order of telescopic extraction is from larger to smaller cylinder, which gives a large working reach. The operator can combine the extraction of individual articles with the same boom length. For the

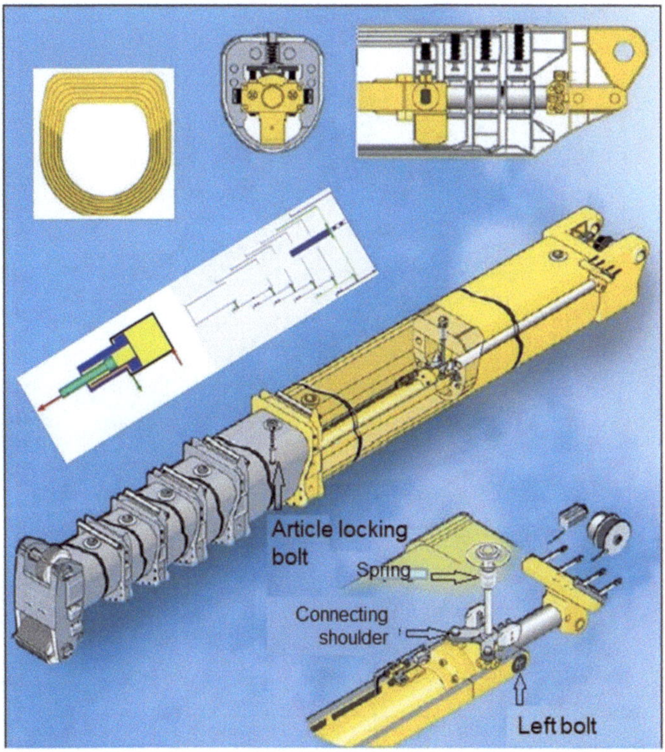

Fig. 9.26 Telematic telescopic boom system [6]

same boom length, there are different load capacities, which depend on the crane current center of gravity. For example, with one combination you get the highest load capacity at the smallest reach, but also the smallest load capacity at the largest reach. With another combination, a higher load capacity is obtained at the largest reach and a lower load capacity at the smallest reach.

Through the automatic boom telescoping process, the crane operator does not have to worry about locking and unlocking individual sections of the telescopic boom. The four-way control stick (*joystick*) hydraulically determines the direction and the speed of the article movement, and the operator is always in control of the crane.

Figure 9.27 shows the loading capacity diagram of the mobile crane.

- The load capacity can be read on the tilt curve of the telescopic boom, in tons,
- The reach can be read on the lower horizontal scale in meters,
- The height can be read on the right vertical scale in meters,
- The length of the boom can be read on the telescopic boom, in the position of the combination for locking the articles.

The *grey field* on the inclination curve shows the minimum required length of the telescopic boom. Only the height from the hook to the pulley head at the top of the boom is added to the desired lifting height. In a real situation, it is necessary to add

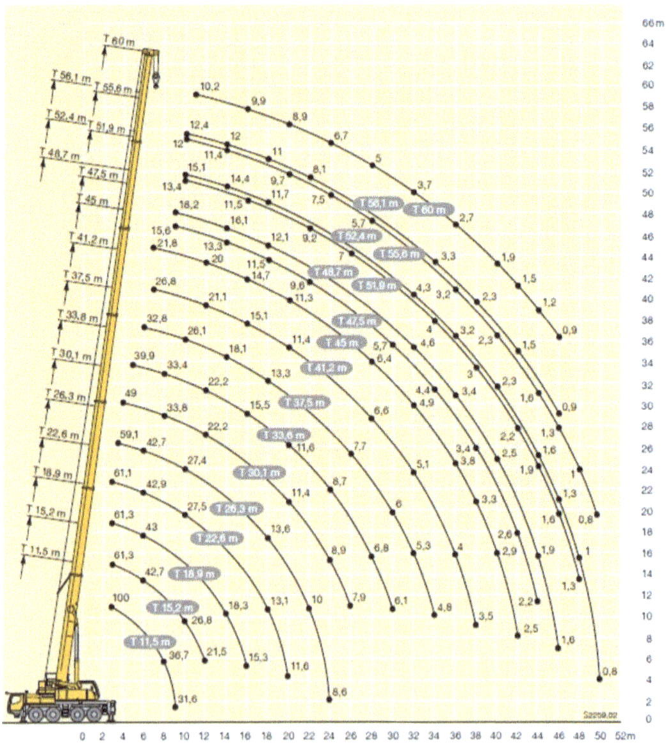

Fig. 9.27 Load capacity diagram of the mobile crane on the stabilizer feet [6]

also the height of the means for transporting the load. The distance between the hook and the head with the pulleys on the top of the boom can be regulated using a limiter that serves to disable lifting.

Black dots indicate the maximum values of the load capacity on the reach that can be read on the lower scale and the height on the right vertical scale. When the mass of the load reaches the value indicated in the black dot, the safety system indicates the loading of 100%, activating the overload protection system.

The load capacity diagram does not replace the load capacity table but serves only as a temporary orientation so that the correct configuration of the telescopic boom can be selected.

An example of lifting a load: At a radius of 7 m, it is necessary to raise a concrete block weighing 15 t to a height of 18 m. On the lower scale, look for a radius of 10 m (from the center of rotation to the end of the chassis + pure reach) and then look vertically for the intersection with the horizontal of the required height of 18 m. Let the first curve be the one on which it is read in the grey field, the minimum length of branch T = 22.6 m and the permitted load 27.5 t. From the required lifting height to the height of the hook, there remains a distance of 1.5 m. Then, according to the dimensions of the block, it is necessary to correctly determine the angle of acceptance of the load using the means for suspension of the load and add their height, for example 3 m. For this height of the transfer means, the height is not sufficient and for this reason, the next higher curve is chosen with a boom length of 26.3 m also of a permitted load capacity of 27.4 t, because there is a distance of 6 m between the load and the hook, which meets the height of 3 m.

Crane capacity table

The maximum permissible carrying capacity of a mobile crane at a certain radius, a certain boom length, and a certain counterweight is shown in Table 9.2. The load capacity includes the weight of the load bearing members, pulley assembly, hook, and the means of suspending the load. The weight of the load to be lifted is reduced by the specified weights.

- Icons with crane settings are displayed above the table
- The far left and right columns show the reach in meters
- The upper horizontal row—scale, shows the length of the boom in meters
- In the columns below the length scale, the carrying capacity in tons is shown.

The crane should never be loaded to the maximum. The crane lifting capacity is located at the intersection of a certain radius (horizontal alignment) and a certain boom length (vertical alignment).

Finally, the lifting capacity of the crane depends on the following settings:

- The length of the telescopic boom,
- Use of folding lattice—boom extension,
- The crane counterweight mass,

Table 9.2 Carrying capacity of the mobile crane on the stabilizer feet (t)

Reach, m	Length of Boom, m													
	11,5		15,2	18,9	22,6	26,3	30,1	37,5	41,2	45,0	47,5	51,9	55,6	60
3	100	61,4	61,3	61,3	61,1									
3,5	72,8	59,6	59,5	59,4	59,4	59,1								
4	67,1	56,8	57	56,80	56,8	56,4	49							
4,5	61,8	52,5	52,8	52,9	52,8	52,8	48,6							
5	57,1	48,5	48,9	49	49,1	48,9	47,9							
6	48,5	42,1	42,7	43	42,9	42,7	42,3	32,8						
7	41,9	37	37,6	38,3	38,4	38,3	38	32,6	26,80	22,1				
8	36,7	32,7	33,6	34,1	34,3	34,2	33,8	31,9	26,5	21,8				
9	31,6	29,1	30,1	30,6	30,7	30,6	30,3	29	26	21,9	18,2			
10			26,8	27,4	27,5	27,4	27	26,1	25	21,7	18	13,4	12	
12			21,5	22,1	22,2	22,1	22,2	21,9	21,1	20	17,5	12,9	11,7	10,2
14				18,3	18,5	18,3	18,7	18,1	17,7	17,1	16,1	12,2	11,4	10
16				15,3	15,5	15,9	15,8	15,5	15,1	14,7	14,1	11,5	10,9	9,9
18					13,1	13,6	13,4	13,3	13,2	12,7	12,1	10,8	10,4	9,5
20					11,6	11,6	11,4	11,3	11,4	11,3	10,7	10	9,7	8,9
24						8,6	8,9	8,8	8,5	8,4	8,4	8	7,5	6,7
28								6,8	6,6	6,4	6,4	6,2	5,7	5
32								5,3	5,1	4,9	5	4,6	4,4	3,7
36									4	3,8	3,8	3,5	3,2	2,7
40										2,9	2,9	2,6	2,3	1,9
44											2,2	1,9	1,6	1,2
48												1,3	1	
50													0,8	

- The surface of the stabilization base,
- Boom turning angle,
- Number of active pulleys.

References

1. Liebherr Mobile Cranes (2020) Product range. Images available, www.liebherr.com
2. Mikulic D (2024) Gradevinski strojevi. Teorija i konstrukcija. Veleuciliste Velika Gorica, Velika Gorica
3. Chandgude A et al (2016) Design and analysis of telescopic boom for mobile cranes. Int J Adv Eng Res Sci (IJAERS) 3(4). India. www.ijaers.com
4. A Recommended Method of Analytically Determining the Competence of Hydraulic Telescopic Cantilevered Crane Booms (2010) SAE J1078-202001 (ISO/TR 19961 Cranes)
5. Kogan J (1976) Crane design. Theory and calculations of reliability. John Wiley & Sons NV, Toronto, Israel Universities Press, Jerusalem
6. Mobile Crane LTM 1100-4.2 (2020) Technical Data Brochure. Liebherr-Werk Ehingen GmbH, www.liebherr.com

Chapter 10
Telehandlers

Abbreviations

G [N]	Telehandler weight
G_z [N]	Fork weight
Q_n [N]	Nominal load capacity
Q_d [N]	Load capacity in full reach
H [m]	Load lifting height
α [^{0}]	Lifting angle of the telescopic boom
F_c [N]	Force of hydraulic cylinder for lifting the boom
F_t [N]	Force of hydraulic cylinder for telescoping the boom
F_z [N]	Force of hydraulic cylinders for turning the fork
σ_d [N/mm^2]	Strain of telescopic boom
δ_d [mm]	Deflection of the telescopic boom
S_{long} [−]	Longitudinal stability factor
S_{tran} [−]	Transverse stability factor
R [m]	Turning radius
v [m/s]	Machine speed
M_c [Nm]	Centrifugal moment
M_z [Nm]	Gyroscopic moment
G_p [N]	Front axle load
G_s [N]	Rear axle load

10.1 Purpose and Division of Telehandlers

For lifting and moving light loads on the construction site, telehandlers, i.e. telescopic forklifts, have been developed, Fig. 10.1. Light loads up to 5 t are lifted and delivered by telescopic forks to a certain position of the construction object.

D. Mikulić and D. Mück, *Construction Machines*,
https://doi.org/10.1007/978-3-031-67845-5_10

Fig. 10.1 Construction telehandler (*telescopic forklift, JCB*) [1]

Telehandlers combine the properties of forklifts and telescopic cranes. The efficiency of telehandlers in construction compared to truck cranes and classic forklifts lies in the speed of moving the cargo. Holding the load on the forks and leveling is ensured automatically by adjusting the forks. Telehandlers can use other types of tools (loading shovel, grapple, lifting hook, etc.). They are often used in agriculture and internal transport as telescopic forklifts. Control of the telehandler using a multi-directional *joystick* is performed by the machine operator.

Regarding the connection between the telescopic boom and the undercarriage, two types of telescopic loaders are distinguished: fixed (classic) and rotary ones, Fig. 10.2. ***The fixed telehandler*** is designed so that the telescopic boom is placed on the right side of the cabin and firmly connected to the frame of the machine undercarriage. Manipulation of the load on the forks is connected to the wheel steering of the two steering axles, which gives the necessary maneuverability. Larger cargos can be lifted by activating the supporting front feet. These machines transfer the load from one part of the construction site to another and set it at different heights of the facility.

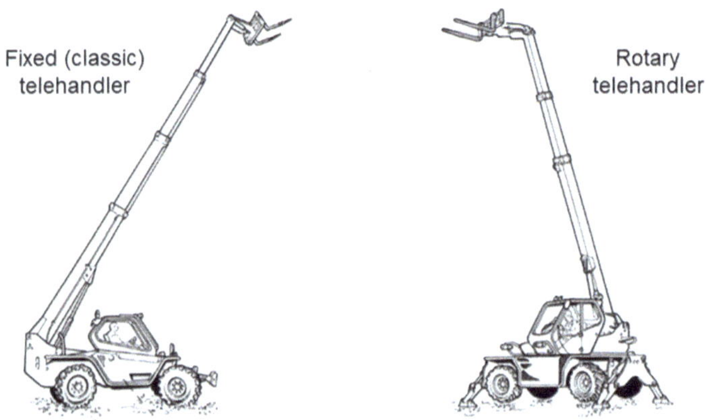

Fig. 10.2 Types of telehandlers [2]

The rotary telehandler is made according to the principle of a mobile crane on hydraulic stabilizer feet, i.e. the boom with the cabin can be turned and it lifts the cargo on the undercarriage. These machines can add cargo only from a stable construction site surface to different object heights. For quick adaptation to uneven terrain, hydraulic feet are most often designed diagonally.

The advantages of telehandlers compared to mobile cranes and forklifts:

- A telehandler can deliver the cargo vertically and laterally, for different reaches and heights on self-levelling forks. The mobile crane can lower the load only to the desired location directly, vertically.
- A forklift requires a flat and solid surface (asphalt, concrete), while the telehandler can move on the terrain, with control using all four wheels. The forklift lifts the load vertically while the telehandler lifts it diagonally and adds the load telescopically to the required height.

In the case of internal transport and transshipment of cargo in closed and open spaces, the application of faster methods of cargo manipulation is required. The solution was found in a ***telescopic forklift*** with a variable reach in space. Such compact machines with high maneuverability are used when sorting palletized and non-palletized cargo because they provide greater efficiency than forklifts. Powerful telescopic forklifts are used for very demanding lifting and maneuvering of port containers.

10.2 Telehandler Construction

Telehandlers use one drive engine and one cabin from which the machine is controlled. A telescopic boom is placed on the side of the cabin, which is in direct connection with the forks and the load. The telescopic boom enables the height reach in confined and hard-to-reach spaces on construction sites. Compared to forklifts, the telehandlers can reach greater lifting height and provide greater visibility when working with the load, owing to the lateral position of the boom.

Telehandlers are designed in accordance with the standard (EN 1459, tab. B), which prescribes the conditions for telescoping the load on the forks, when the machine is stationary and when lifting on a flat and solid surface. To ensure greater stability and greater lifting height, the stabilizer feet are used on the front of the machine, which relieves the front axle. The basic data of several well-known brands of classic telehandlers used in construction are shown in Table 10.1.

The carrying capacity characteristics of classic telehandlers are more or less equal. Their mass, dimensions, drive, and wide off-road tires with a pressure of 4.5 bar ensure the soil trafficability and accessibility to the facilities on the field. The height of the undercarriage with 15.5 R 25 wheels is 40 cm, and the turning radius is about 4 m, Fig. 10.3.

Telehandlers use all-wheel drive. This ensures soil mobility and work stability. The JCB telehandler has a hydrodynamic drive system, uses a torque converter and a

Table 10.1 Characteristics of fixed (classic) telehandlers

Manufacturer	JCB (VB)	CAT (SAD)	Merlo (IT)	Manitou (FR)
Model	540–170	TH 83	P40.7	MT 1740
Max carrying capacity (t)	4.0	4.0	4.0	4.0
Machine weight (t)	12.0	10	10.3	11.2
Engine power (kW)	90.0	81	74.5	74.5
Length (m)	6.39 (v)	5.72	5.80	5.98
Width (m)	2.44	2.45	2.40	2.43
Height (m)	2.69	2.45	2.51	2.61
Wheelbase (m)	2.75	2.97	2.87	2.77
Undercarriage height (m)	0.40	0.45	0.48	0.48
Turning radius (m)	4.10	4.20	4.05	4.17
Max. speed (km/h)	29.0	32.0	40.0	25.0

Fig. 10.3 Classic telehandlers (*JCB 540–170; 4 t / 12 t*)

powershift gearbox, with 4 forward and reverse gears, Fig. 10.4. MERLO telehandler uses a hydrostatic drive, hydraulic pump—hydraulic motor, Fig. 10.5.

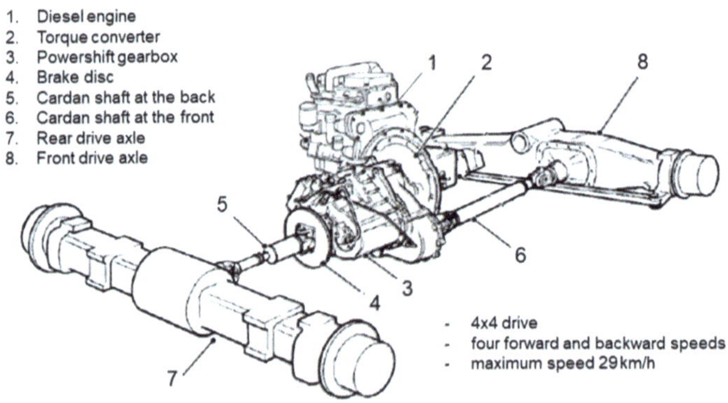

Fig. 10.4 Hydrodynamic transmission (*JCB*) [1]

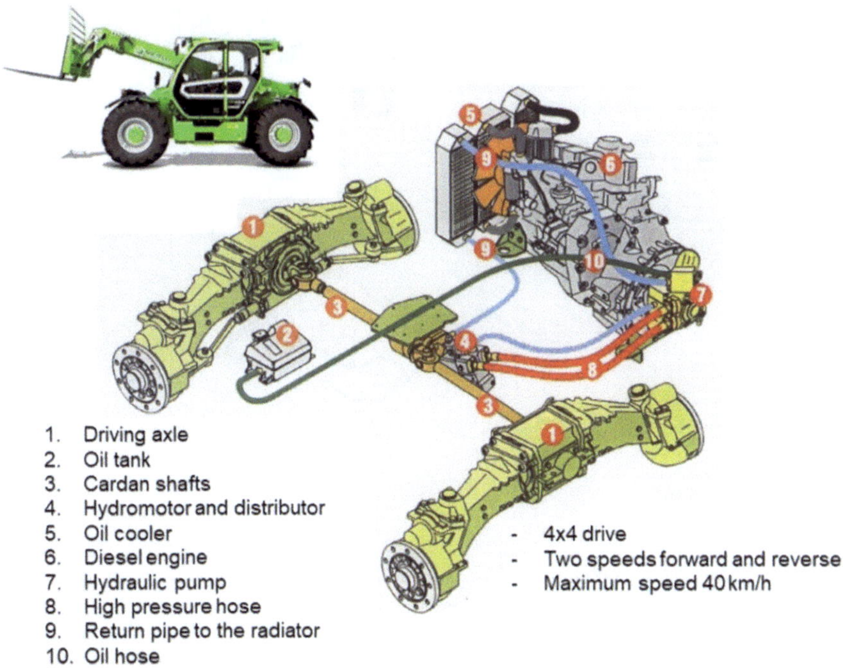

Fig. 10.5 Hydrostatic transmission (*MERLO*) [3]

The drive bridges of the front and rear axles are of rigid design, with the possibility of turning the wheels and steering. In addition to the axle transmission, differential, shaft, and planetary reducer in the wheels, a set of lamellas for braking on all four wheels are installed on the wheel shafts, because during the operation it is necessary to ensure the stability and that the machine does not move uncontrollably (when working with a raised boom).

Possibilities of selecting the steering mode, Fig. 10.6.

- "normal" (B) when the rear wheels are stiff and the front wheels turn,
- "all four" (A) when all four wheels turn (with the rear wheels turning opposite to the front wheels)
- "dog walk" diagonal steering (C), when all four wheels are turned but in the same direction.

Telescopic boom

The structure of the telescopic boom is shown in Fig. 10.7. One hydraulic cylinder on the bottom side serves to raise the boom. The other two hydraulic cylinders are used to extend the boom, one of which is located on the boom and the other inside the boom section. The boom is also equipped with a compensation cylinder for leveling the forks. The compensating cylinder starts automatically with the boom being raised,

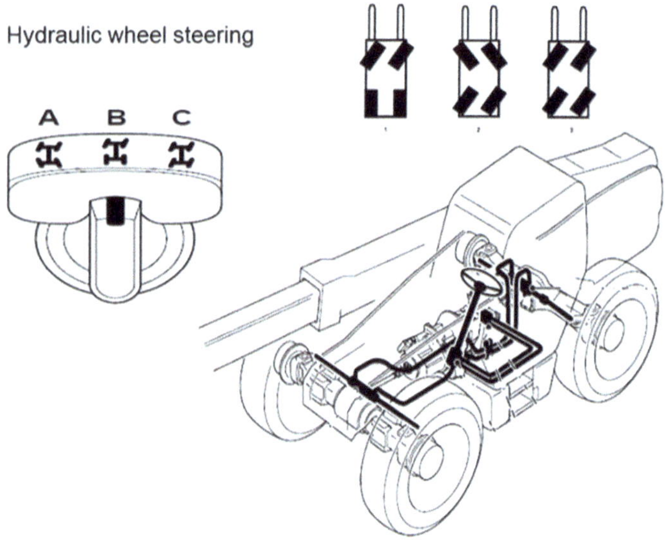

Fig. 10.6 Wheel steering options (*JCB*)

and it is connected hydraulically to two fork cylinders, which ensures a constant position of the forks and the load when the boom is being raised.

The telescopic boom consists of three articles: outer (1), central (2) and inner (3). When extending or retracting, the inner and outer articles move simultaneously, based on the mutual connection of the inner and outer hydraulic cylinders (4, 5). Hydraulic hoses that pass through the boom (6) supply oil to the inner cylinder (5), the cylinder for tilting the forks (10), and the connection for various tools. The hoses can move freely through the inner article using a pulley (7) on the lower part of the inner cylinder. Sliding plates (8, 9) are installed between individual articles of the

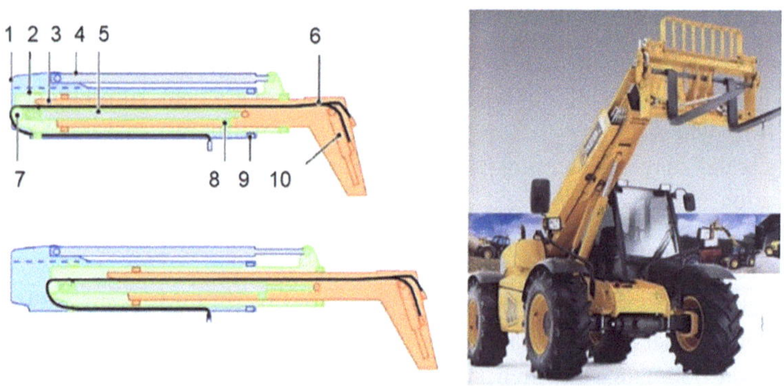

Fig. 10.7 Telescopic boom (*JCB*). 1–2–3 outer—central—inner joint, 4–5 inner and outer hydraulic cylinder, 6—hydraulic hoses, 7 pulley, 8–9 sliding plates, 10 fork hydraulic cylinder

telescopic boom, with the possibility of adjusting the clearance. Sliding plates are installed above, below and on the sides of the articles.

The hydraulic gear pump is the power source of the telescopic boom (maximum pressure $\Delta p_{max} = 350$ bar, flow rate $q = 130$ cm^3/min (2500 min^{-1}). The auxiliary gear pump (175 bar) enables turning the wheels and steering the machine. The total amount of oil VG 46, is 185 l.

10.3 Load Capacity of the Telehandler

Figure 10.8 shows the load capacity diagram of the telehandler on the front feet and the diagram of load capacity on the wheels is shown in Fig. 10.9. The load capacity is defined with 75% safety. The center of gravity of the cargo on the forks is assumed to be in the middle of the forks.

The most important features are:

- Maximum load capacity at a given reach,
- Maximum reach height.

According to the capacity diagram on the feet, the machine has a maximum capacity of 4 t with the telescopic boom fully retracted. It is also possible to extract the boom with the same cargo at an angle of 73° up to a height of 10 m (darker field). The extension of the boom is possible because of the fact that the force arm is smaller, the more the angle of the boom increases. This field of action is not fully usable

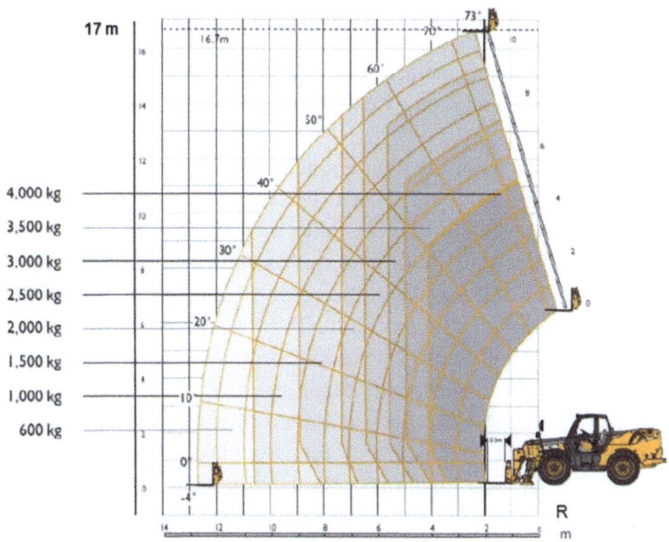

Fig. 10.8 Loading capacity diagram of the classic telehandler on feet (*JCB*) [1]

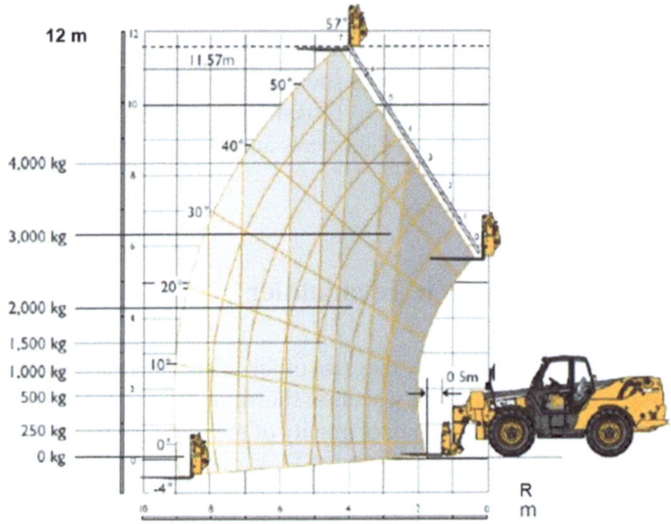

Fig. 10.9 Loading capacity diagram of the classic telehandler on wheels (*JCB*)

because the machine cannot add cargo above itself. Two hydraulic feet serve as lifting supports, resulting in greater load capacity and greater lifting height, up to 17 m. In addition, the hydraulic feet relieve the wheels of the front axle, and a larger machine stabilization base is obtained, thus achieving greater safety of lifting the cargo.

With the horizontal boom fully extended, the machine can only carry 600 kg. It is evident from the diagram that it is possible to completely pull out the telescope with a cargo of 2.5 t, but only in the case when the load is lifted almost vertically to the possible lifting height. This is achieved by setting the machine on the feet, placing the boom at a large angle, and then by extracting the boom, the cargo is raised to the maximum height.

The working capabilities of the machine when lifting on wheels are not as good as when the machine is on its feet. When the feet are raised, the machine support moves 0.8 m backwards. Due to the elasticity of the tires, the wheeled machine is much more unstable. The maximum load with a fully extended boom is 2.5 times smaller and amounts to 250 kg. Also, the reach of the machine is smaller, so the maximum height of adding the load is 12 m. To avoid the possibility of overturning, an electronic system for automatic load monitoring is installed on the machine.

Figure 10.10 shows the load capacity diagram of rotary telehandlers on stabilizer feet. As with mobile cranes, the key parameter is the reach (radius). The rotary telehandler ensures a higher load capacity than the classic telehandler at every load lifting radius. Radius (R) is a function of the length of the boom (L) and angle (α) under which the boom is located, minus arm (K)—the distance from the pivot point of the boom to the center of rotation of the superstructure, $R = L \cos \alpha - K$.

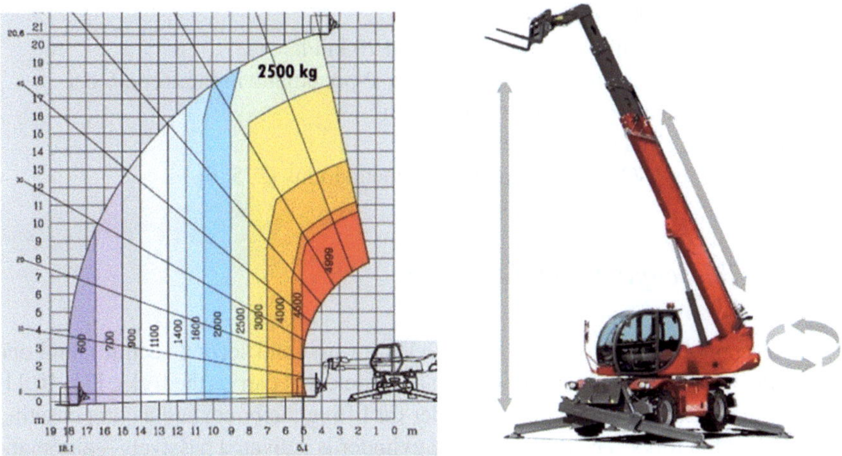

Fig. 10.10 Loading capacity diagram of the rotary telehandler (*MAGNI*) [4]

10.4 Calculation of the Machine Construction

This chapter presents a simple example of the construction of a telescopic handler with a nominal load capacity of 3 t. The machine is intended for internal transport of loading and transshipment of goods. Its features must be better than those of a classic forklift for loading and unloading trucks. These advantages are truck loading from only one side, smaller dimensions, maneuverability, load capacity, high movement speeds, with the main task of unloading and sorting palletized and non-palletized goods, which provides better use of storage and workspace. The basic load capacity requirement is 1500 kg with telescoping the cargo in full reach.

Telescopic handler (*telescopic forklift*)[1] requirements

- Empty telehandler mass m = 5000 kg
- Nominal load capacity $Q_n = 3000$ kg
- Nominal load capacity in full reach $Q_d = 1500$ kg
- Maximum load lifting height 4000 mm
- Lifting angle of the telescopic boom −100 to + 650
- Extension of the telescopic boom, reach 1500 mm
- Forklift equipment with standard forks T180, A-type, 1200 mm
- Loading the truck from one side, side or rear
- Strength and safety of boom structures in accordance with the standards
- Static and dynamic stability in accordance with norms, ascent, lateral inclination
- Longitudinal and transverse stability during movement and braking
- Safety control against overturning, using the loading indicator
- Diesel engine, 45 kW/2500 min^{-1}

[1] Technological project STIRP TP-06/0192–01, MZOS, Brodarski institut d.o.o. Zagreb.

- Lifting and transferring loads on wheels
- Automatic powershift transmission, speed 20 km/h
- Wheels 14", choice: tires SE-superelastic, P-pneumatic in the front, 6 bar
- Rear wheels steering, turning radius 2.3 m
- Operation on climbs and slopes, min. 10°
- Application of ISO standards and recommendations [2, 5, 6].

10.4.1 Telescopic Boom Structure

The telescopic boom consists of two articles (of first and second section), and one hydraulic cylinder is used for telescoping the article that carries the forks, Fig. 10.11. The boom is also equipped with a cylinder for lifting the boom and a compensating cylinder for the forks. The compensating cylinder is activated passively—automatically when the boom is raised, and it is hydraulically connected to two fork cylinders, which ensures a certain leveled position of the forks and the load when the boom is raised. T180 type forks are mounted on a standard support (ISO 2328), and can rotate by 110° and move sideways 1.1 m.

Compensating cylinder

The compensating cylinder is connected in a hydraulic circuit with the cylinders for turning the forks. The diameter of the compensating cylinder is equal to the diameter of the tool turning cylinder. The stroke of the compensating cylinder (s_k) is twice the total stroke of two cylinders for turning the forks (s_v). From the equality of volumes when moving the pistons of hydraulic cylinders, it follows:

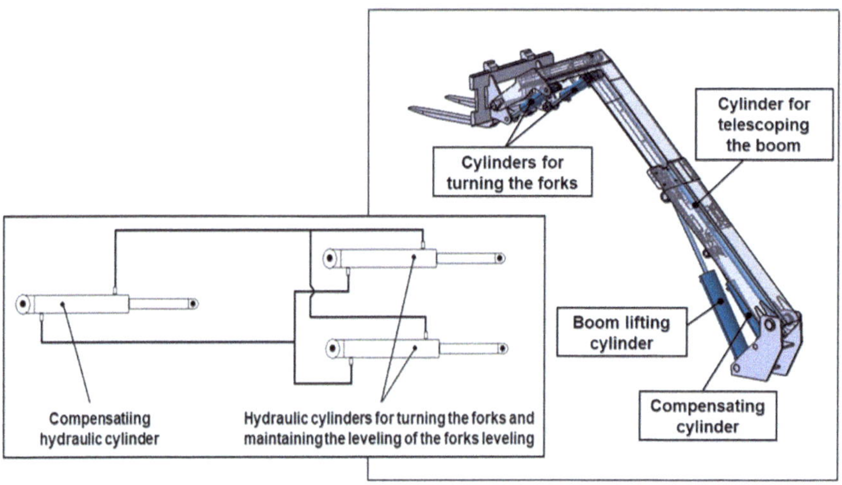

Fig. 10.11 Telescopic boom structure [7]

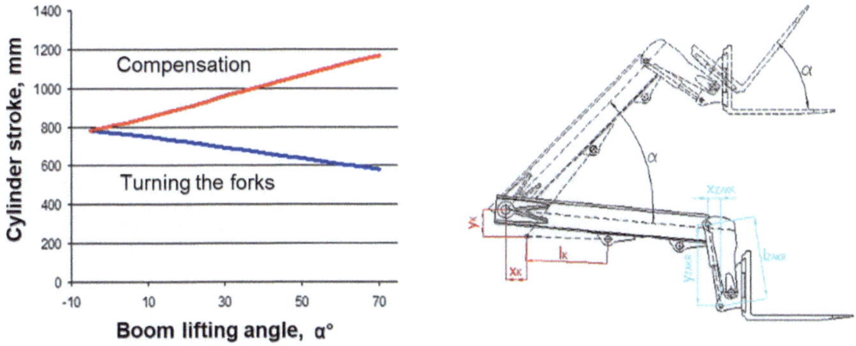

Fig. 10.12 Diagram of fork leveling maintenance

$$V_k = 2V_v$$

Stroke of the fork hydraulic cylinder piston $s_v = s_k/2$

V_k, V_v oil volume of compensating cylinder/fork cylinder

s_k, s_v piston stroke of the compensating cylinder/fork cylinder

The stroke of the compensating cylinder and cylinders for turning the forks, in relation to the angle of the boom rotation during its lifting, is shown in the diagram of maintaining a horizontal or some other working position, Fig. 10.12. To keep the cargo securely on the forks during transport, the machine operator can correct the inclination of the forks and the load by means of a joystick.

10.4.2 Hydraulic Installation of the Telehandler

The diagram of the telehandler hydraulic installation is shown in Fig. 10.13. The power source is a gear pump, with a maximum working pressure of $p_{max} = 200$ bar and a flow $q = 36$ cm^3/rev. An auxiliary gear pump (150 bar) enables the turning of the rear axle wheels. The total amount of oil, specification ISO VG 46, is 70 l. This ensures the operation of all hydraulic cylinders for lifting and extending the boom and turning the forks. When the machine is overloaded, the electromagnetic safety valve cuts off the supply of hydraulic oil to the telescopic cylinder to prevent the machine stability disruption and overturning. An auxiliary gear pump provides operating pressure to turn the rear wheels and steer the machine.

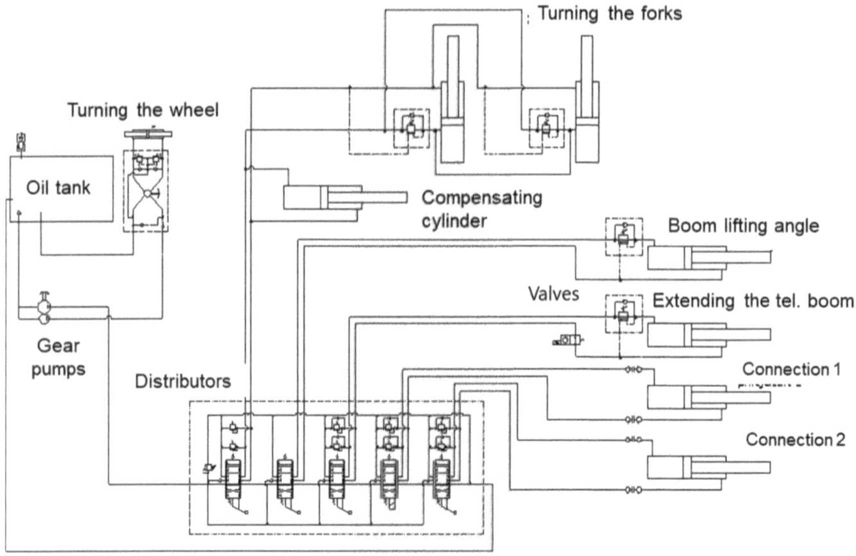

Fig. 10.13 Scheme of the hydraulic installation of the telehandler

Determination of hydraulic components

(a) Hydraulic cylinder for lifting the boom

Equation of balance moments, Fig. 10.14.

$$G_0 \cdot g \cdot \frac{l_0}{2} \cdot \cos\alpha + G_1 \cdot g \cdot (l_0 - x_1 + \frac{l_1}{2}) \cdot \cos\alpha$$
$$+ \psi \cdot (Q + G_v) \cdot g \cdot \left[(l_0 - x_1 + l_1) \cdot \cos\alpha + c_Q\right]$$
$$+ \psi \cdot (Q + G_v) \cdot g \cdot \sin\alpha \cdot (d + c_Q \cdot \sin\alpha)$$
$$= F_C \cdot \sin\gamma \cdot r_h + F_C \cdot \cos\gamma \cdot r_v$$

Based on the maximum force of the hydraulic cylinder for lifting the boom (F_c) at the horizontal position of the boom and the highest pressure, there follows the selection of the hydraulic cylinder for lifting the boom, with a diameter of 140 mm, and a pump flow of 70 l/min for a lifting speed of 0.1 m/s.

(b) Hydraulic cylinder for load telescoping, Figure 10.15.

$$G_1 \cdot g \cdot \sin\alpha \cdot \frac{a_1}{2} + G_1 \cdot g \cdot \cos\alpha \cdot \frac{l_1}{2} + \psi(Q + G_v) \cdot g \cdot \cos\alpha \cdot (l_1 + c_Q \cdot \cos\alpha)$$
$$+ \psi \cdot (Q + G_v) \cdot g \cdot \sin\alpha \cdot (\frac{a_1}{2} + d + c_Q \cdot \sin\alpha)$$
$$+ \mu \cdot F_2 \cdot a_1 - F_2 \cdot x_1 = F_T \cdot \frac{a_1}{2}$$

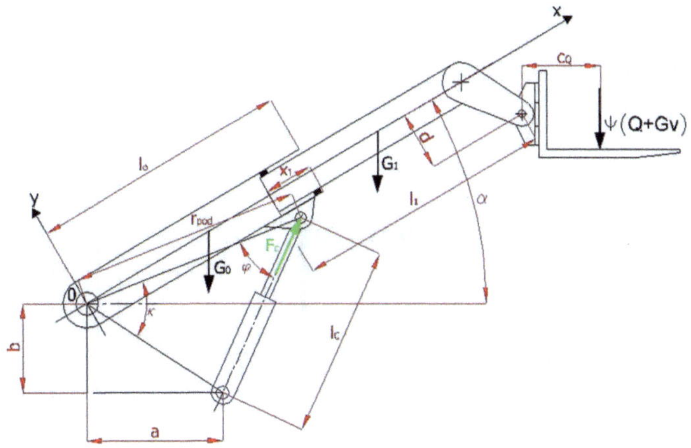

Fig. 10.14 Calculation scheme of the hydraulic cylinder for lifting the boom

Based on the maximum force of the hydraulic cylinder for telescoping (F_T) at the highest inclination of the boom and the highest pressure, there follows the selection of the hydraulic cylinder for the boom extension, with a diameter of 80 mm, and a pump flow of 70 l/min for an extension speed of 0.2 m/s.

(c) Hydraulic cylinders for turning the forks, Figure 10.16.

$$\frac{1}{2}\psi \cdot (Q + G_V) \cdot g \cdot l_Q = F_Z \cdot l_Z$$

Based on the maximum force of the hydraulic cylinders of the forks (F_z) in the horizontal position of the forks ($Q = 3$ t, $Gv = 0.2$ t, $\psi = 1.15$ *inertia factor*), there

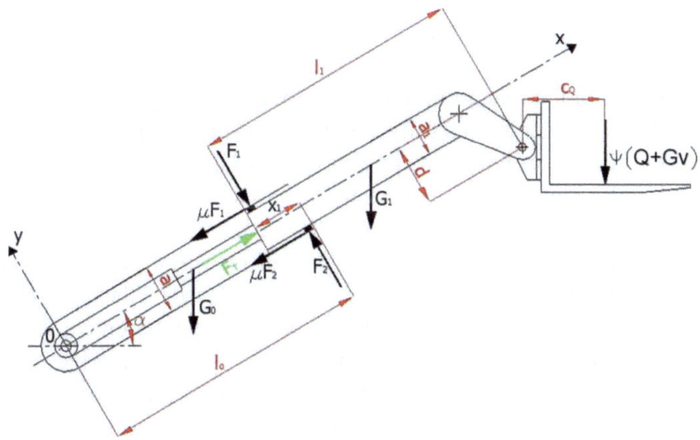

Fig. 10.15 Calculation scheme of hydraulic cylinder for boom telescoping

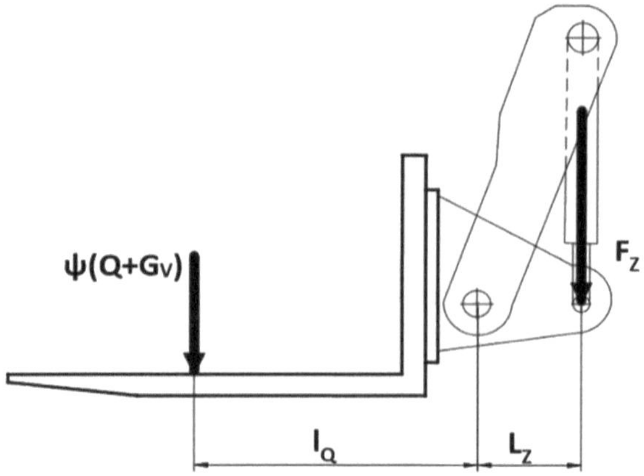

Fig. 10.16 Calculation scheme of a hydraulic cylinder for turning the forks

follows the selection of two hydraulic cylinders for turning the forks, with a diameter of 80 mm, required pump flow of 17 l/min at piston speed of 0.1 m/s.

Selected hydraulic components and their dimensions:

- Boom lifting cylinder: $\Phi 140 \times 950$
- Telescoping cylinder: $\Phi 80 \times 1500$
- Compensating cylinder: $\Phi 80 \times 450$
- Cylinders for turning the forks: $\Phi 80 \times 300$
- Gear pump: $p_{\max} = 200$ bar; $q = 36$ cm^3/rev.

10.4.3 Telescopic Boom Strength

The construction of the telescopic boom is carried out in accordance with FEM and DIN norms for load manipulation, for the first calculated loading (main loadings: load by cargo, own weight, inertia forces), where the safety factor for stress is $S = 1.5$.

The highest boom loading around the rotation axis (Fig. 10.14), with the highest cargo and self-weight factor, on a fully extended telescopic arm, of one meter reach, determines the maximum boom loading moment $M_{\max} \approx 111$ kNm. For the boom geometry, it is sufficient to generate 10 loading cases of the supporting structure for the range of the boom lifting angles from -10 to $65°$.

The finite element method (*CATIA/Mechanical Design + Analysis and Simulation*) was used to analyze the strength of the structure. Based on the highest equivalent strains, the dimensions of the telescopic boom profile are determined. A box-shaped boom profile is adopted, with a wall thickness of 8 mm for the basic article and 10 mm for the extending article, Fig. 10.17. Calculated stresses and deflections of the telescopic boom are shown in Figs. 10.18, 10.19, and Table 10.2.

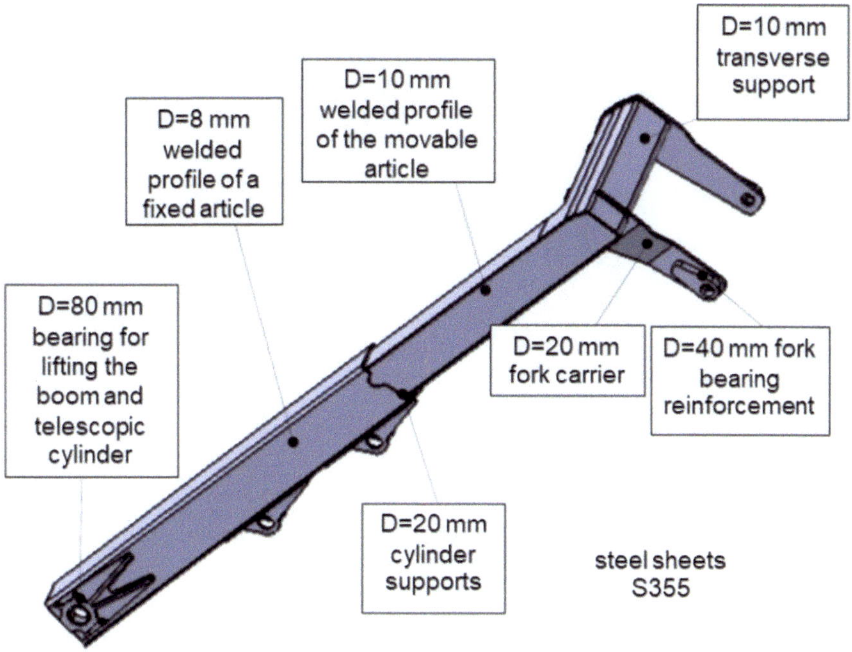

Fig. 10.17 Determining the telescopic boom profile

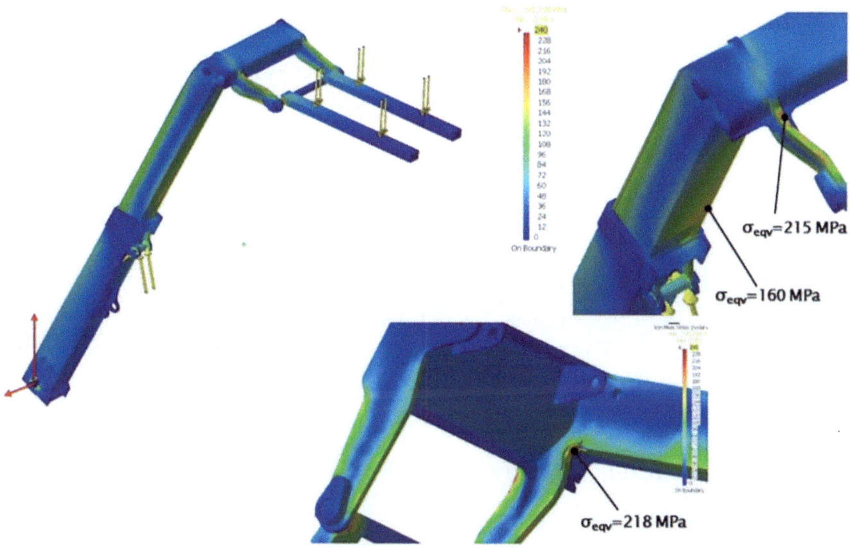

Fig. 10.18 Equivalent strains of the telescopic boom parts, boom position 65° [7]

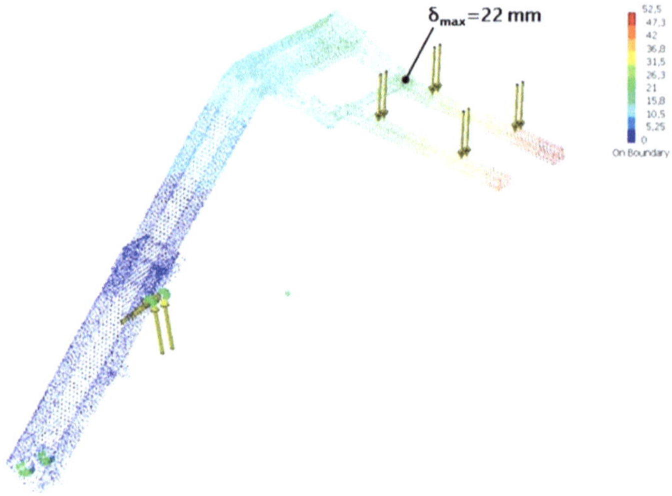

Fig. 10.19 Equivalent deflection of the telescopic boom (65°, $\delta_{max} < \delta_d$)

Table 10.2 Strains and deflection of the telescopic boom

Boom material, EN 10027		Permissible strain, FEM 1.001
Telescopic boom, S355		
Thickness, mm	R_{p02}, N/mm^2	σ_d, N/mm^2
t ≤ 16	355	237
16 < t ≤ 40	345	230
40 < t ≤ 63	335	223
Bolts, 42CrMo4		
Thickness, mm	R_{p02}, N/mm^2	
40 < t ≤ 100	650	433
Permissible deflection of the telescopic boom "Das grosse Buch der Farhrzeugkrane", Rudolf Becker		
$L/\delta > 42$		
Length of the telescopic article, mm		δ_d, mm
≈2000		47

10.4.4 Telehandler Stability

The longitudinal and transverse stability is analyzed. The telehandler stability can be presented by the stability triangle, Fig. 10.20, which is imagined between the front wheels and the pivot point of the rear axle (the rear axle is not rigid, but a

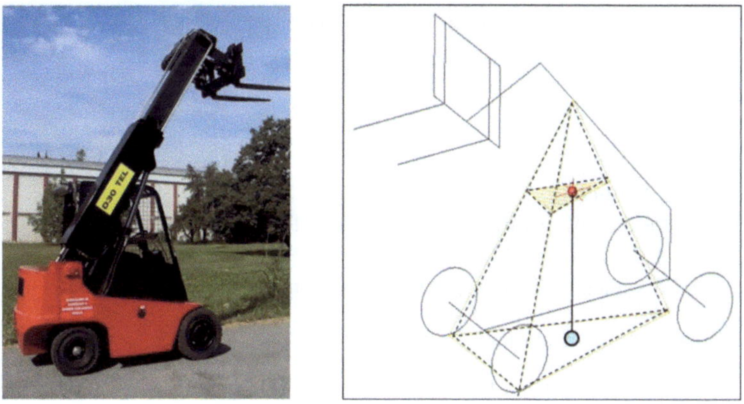

Fig. 10.20 Changing the center of gravity of the machine and the stability triangle

freely oscillating shaft, which is why the rear wheels are excluded from the stability triangle). The load should be lifted vertically, either on level ground or by using the frame leveling function.

As long as the machine center of gravity falls within the stability triangle, the telehandler is stable. By lifting the cargo, the center of gravity shifts forward, closer to the limit of the stability triangle (the theoretical limit of stability from tipping forward is when the center of gravity coincides with the axis of the front wheels), which increases the risk of the machine tipping forward due to the action of dynamic forces generated during sudden braking or in case of backward movement.

By lifting the telescopic boom the center of gravity (CG) moves upwards and backwards, which reduces the area of the stability triangle. Changing the position of the telehandler center of gravity is a function of the extension and inclination of the telescopic boom. A smaller area of the stability triangle does not leave much room for moving the center of gravity, which is a problem when working on uneven terrain, on a side slope, and on uphill and downhill slopes, where the risk of overturning is the greatest. By reducing the stability triangle and the relatively small shift of the center of gravity, caused by the slope of the terrain or dynamic forces when driving, accelerating, or braking, the center of gravity is shifted beyond the limits of the stability zone, which can cause tipping of the machine forward or to the side.

(a) Longitudinal stability

The change of the coordinates of the machine center of gravity (x_T, z_T) under the action of the inertial force during movement and braking is considered. Longitudinal stability and displacements of the center of gravity are determined from the point of view of sliding and overturning criteria.

Longitudinal stability factor:

$$S_{\text{long}} = (M_{\text{stab}}/M_{\text{in}}) \geq 75\%$$

$$M_{stab} = (Q + G) \cdot g \cdot X_T$$
$$M_{in} = F_{in} \cdot Z_T = (Q + G) \cdot a \cdot Z_T;$$

Q Cargo, kg
g Gravity, 9.81 m/s^2
G Machine weight, kg
X_T Horizontal distance of the center of gravity from the front axle, mm
a Deceleration, m/s^2
Z_T Vertical distance of the center of gravity from the base, mm

Longitudinal stability diagram

The nominal load capacity of the telehandler is 3000 kg (fully retracted boom). The load capacity decreases with the horizontal extension of the boom from the front axle axis (L_Q), so the minimum load capacity (*SWL—safe working load*) in full outreach is 1500 kg, Fig. 10.21. Load capacity is a defined machine safety of 75%, with the distance of the center of gravity of the load from the front of the forks of 500 mm:

- Longitudinal stability of the telehandler in the function of telescoping the boom, Fig. 10.21.
- Practically, graphically in the form of a load capacity diagram, Fig. 10.22.
- Longitudinal stability of the telehandler as a function of deceleration, Fig. 10.23.

The telehandler load capacity areas marked with A, B, C, D and E can be achieved by moving the boom, within the framework of safety against overturning (S_{dop} = 1.33). The machine maximum load capacity of 3 t is achieved on the smallest arm from the front axle axis. It is also possible to extend the boom with the same cargo at an angle of 65° up to a height of 4 m (darker field). The extension of the boom is possible because the force distance is smaller as the angle of the boom increases. With the boom extended in the E area, the machine can carry a load of 1.5 t at full

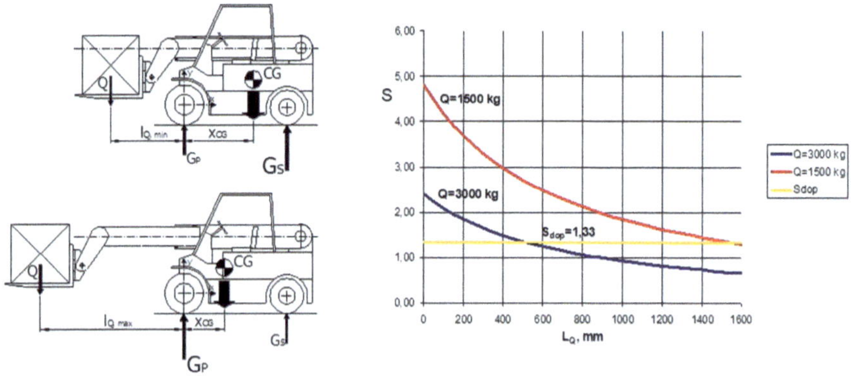

Fig. 10.21 The telehandler stability depending on the horizontal extension of the boom

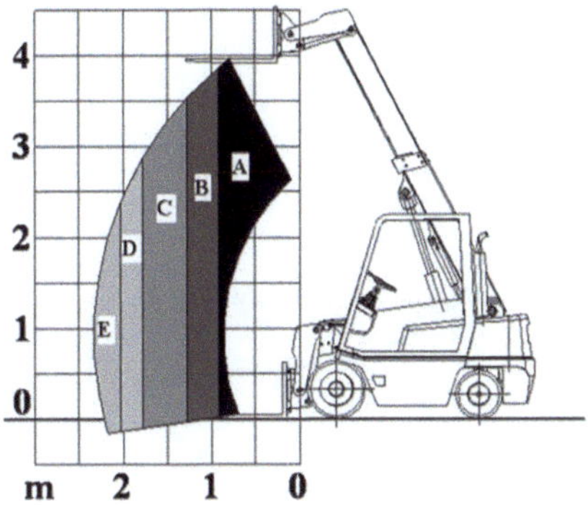

Field	Load capacity
A	3000 kg
B	2500 kg
C	2000 kg
D	1750 kg
E	1500 kg

Fig. 10.22 Telehandler loading capacity diagram

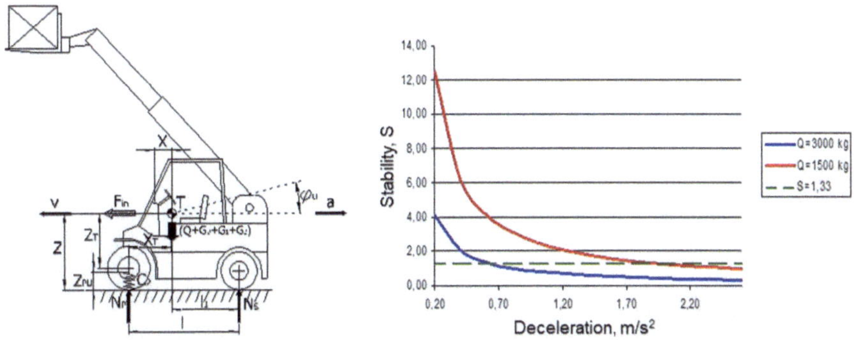

Fig. 10.23 Longitudinal stability as a function of telehandler deceleration

outreach. In areas B and C, it is possible to fully extend the boom with a cargo of 2–2.5 t and add the cargo to the maximum height.

The telehandler longitudinal stability during braking and stopping indicates caution when transporting the maximum load of 3 t, because the permissible deceleration is only 0.7 m/s^2, Fig. 10.23. When transporting a load of 1.5 t at full outreach, the permissible deceleration is 1.7 m/s^2.

(b) Transverse stability

For the telehandler transverse stability, the changes in the coordinates of the machine center of gravity (y_T, z_T) under the action of the centrifugal force when turning in a curve are considered, and the stability of the forklift on the transverse slope is calculated, Fig. 10.24. Safe machine operation with a smaller (1.5 t) and the largest cargo (3 t), on a side slope of up to 12° can be estimated.

$$\text{Transverse stability factor}: S_{\text{tran}} = (M_{\text{stab}}/M_c + M_z \geq 1.33$$

$$M_{\text{stab}} = (Q + G)p_v/2$$

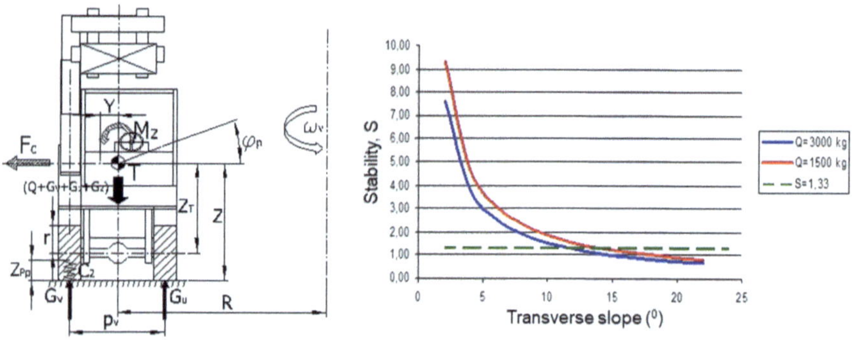

Fig. 10.24 Lateral stability depending on the telehandler tilt

$$\text{Centrifugal moment:} M_c = F_c \cdot Z_T = \frac{(Q+G)}{g} \cdot \frac{v^2}{R} \cdot Z_T$$

$$\text{Gyroscopic moment:} M_z = n \cdot I_p \cdot \frac{v^2}{r \cdot R}.$$

Q Cargo, kg
p_v Wheel track, m
G Machine weight, kg
g Gravity, m/s^2
v Machine speed, m/s
R Turning radius, m
I_p Wheel inertia moment
r Wheel radius, m
n Number of wheels

10.4.5 Telehandler Stability Control

When the machine is working, the load on the front axle increases and the rear axle gradually decreases. Monitoring of the telehandler stability is performed by measuring the relief of the rear axle, to prevent the machine from overturning around the front axle. The built-in sensor measures the load relief on the rear axle/the deformation of the strain gauge and, via a microcontroller and an electronic indicator, shows the machine operator the state of safety against overturning. The tensiometer (force sensor) is attached to the rear bridge with two bolts, in the middle in the place of greatest strain, to monitor and prevent the machine from overturning about the front axle, Fig. 10.25. The machine loading indicator, i.e. the module with lamps in the cabin, shows the operator the information on the safety of the machine. The axle loading sensor and the machine cargo indicator are shown in Fig. 10.26. The loading on the machine is observed as a percentage of the permissible longitudinal stability loading. Overloading of the machine can occur when the force moment of the load is excessive, so that there is a risk of the machine overturning around the axle of the front wheels.

The analysis of the telehandler longitudinal stability can be explained through the load on the axles, Fig. 10.27. The diagram clearly shows the course of increasing the load on the front axle and unloading of the rear axle with the horizontal telescoping of the boom from the axis of the front axle (distance L_Q). The built-in stability factor ($S = 1.33$) determines the permissible lifting and relief of the rear axle loading (G_s, S_{dop}) and the activation of the loading indicator.

Depending on the state of the loading on the rear axle, certain LED diodes light up on the indicator and warn the operator about approaching the limit of the permissible loading, Fig. 10.28. The green light shows the permissible working area. When the loading is close to the limit, the yellow light comes on. At the maximum permissible loading, the red lights light up with a flashing light, with the activation of the audible

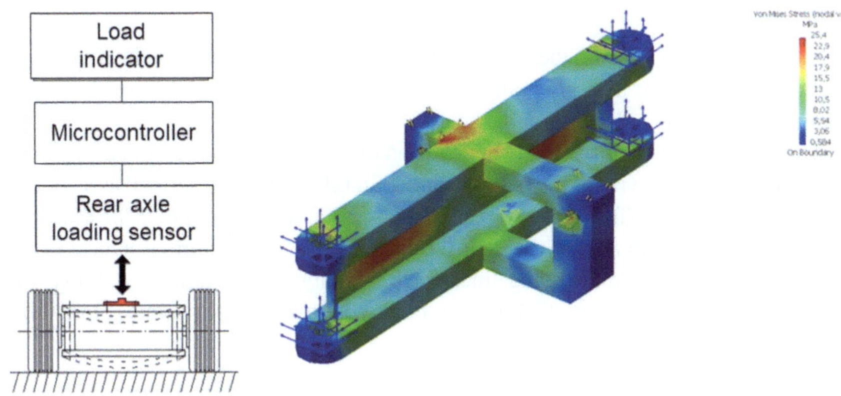

Fig. 10.25 Placement of the sensor in the middle of the rear axle

Fig. 10.26 Rear axle sensor and machine loading indicator

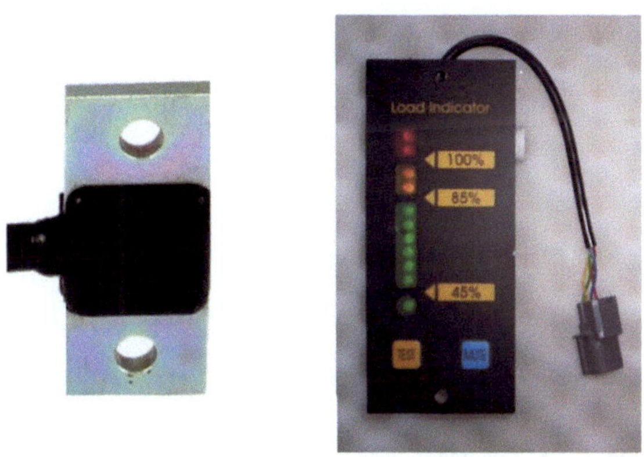

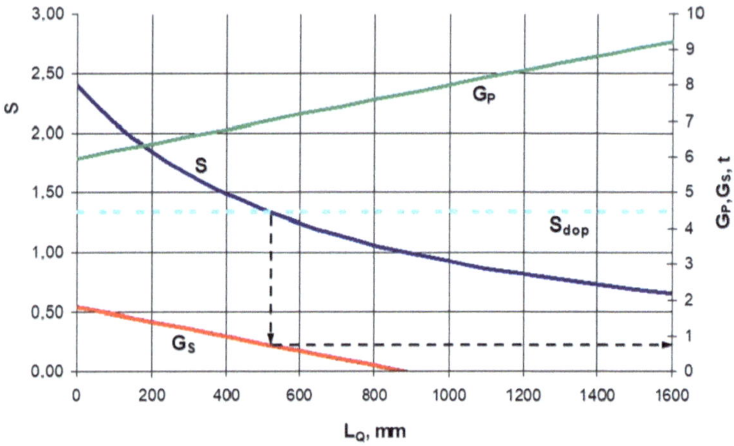

Fig. 10.27 Axle load diagram depending on the load distance (L_Q) G_P—front axle load, G_S—rear axle load

alarm and the blocking of the boom extension. As the loading on the front axle increases and the rear axle gradually decreases, the following lamps turn on: green, orange, red, and an audible signal is heard.

The telehandler stability mostly depends on the weight of the load and its distance from the center of gravity of the base machine. Stability can be described as a zone—or a stability triangle (front axle to rear axle) or a stability rectangle if the rear axle lock is activated (front axle to rear axle). Because telehandlers normally have freely oscillating rear axles, the tipping lines form a triangle. To maintain stability, the telehandler center of gravity (CG) must remain within the stability zone, Fig. 10.29.

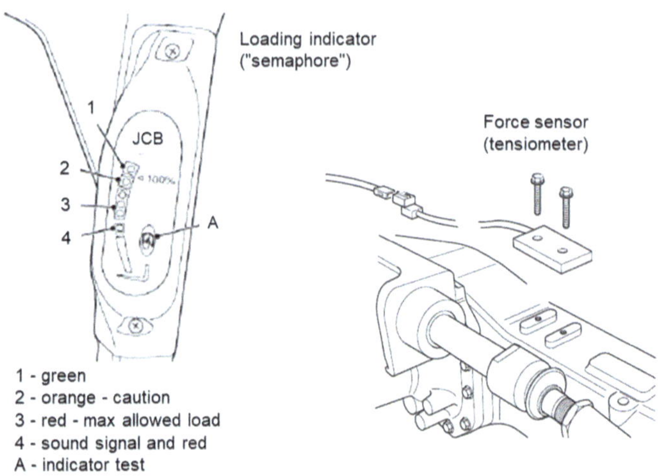

Fig. 10.28 Loading indicator and sensor mounting

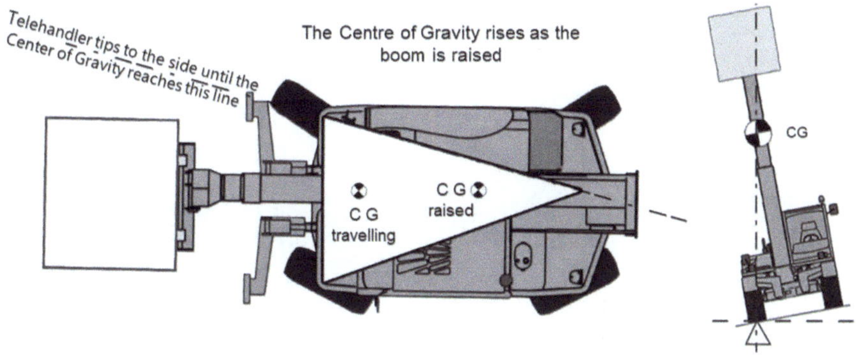

Fig. 10.29 Telehandler stability triangle [2]

Telehandler testing

According to the telehandler load capacity diagram, it is necessary to test and calibrate the safety indicator with the safe working load, which proves the nominal load capacity of the telehandler of 3 t, with a load capacity in full outreach of 1.5 t, Fig. 10.30. Standard load testing is performed by using standard concrete cube modules.

Remark To certify and use a telescopic forklift, it is necessary to meet the prescribed requirements of the national regulations on occupational safety:

- Regulations on occupational safety measures and norms
- Regulations on testing the environment and machines with increased hazards
- Regulations on the list of machines and devices with increased dangers
- Regulations on general measures and safety standards when working with cranes
- Regulations on safety at work when loading and unloading cargo.

Fig. 10.30 Testing the telehandler load capacity (*D30 TEL*) [7]

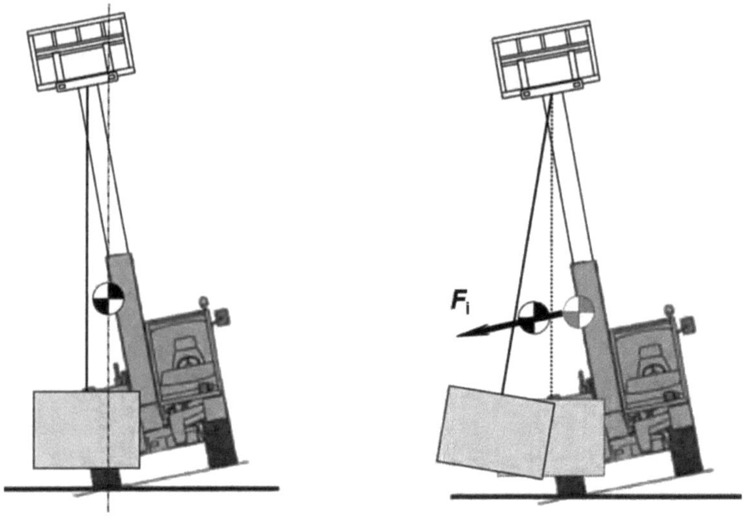

Fig. 10.31 Illegal operations with the telehandler (lifting cargo on a rope and transferring cargo on a rope, especially on a slope) [2]

Maneuvering cargo at high altitudes is considered very dangerous. When lifting and moving cargo to the maximum heights, a machine zone of critical instability of the machine may arise. With an incorrect or careless position of the cargo outside the forks, it is very easy for the machine to tip over, Fig. 10.31. The inertial force (F_i) of the load mostly causes the machine to sway and overturn.

References

1. JCB Product range (2022). www.jcb.hae.org.uk
2. Safe Use of Telehandlers in Construction (2015) Construction Plant-hire Association 27/28 Newbury St London. www.highwayssafetyhub.com/uploads/5/1/2/9/51294565/2d-guidance_on_safe_use_of_telehandlers.pdf
3. MERLO, Stabilized telehandlers (2023) Product range, brochure. Images available. www.merlo.com/aus/en/p/telehandlers/stabilized-telehandlers
4. MAGNI, Rotating telehandlers (2023) www.liftandaccess.com
5. Rough-terrain trucks—Safety requirements and verification (2020) Variable-reach trucks. www.iso.org/obp/ui/en/#iso:std:iso:ISO10896:-1:ed-2:v1:en
6. ISO 13284(2003) Fork-lift trucks—Fork-arm extensions and telescopic fork arms—Technical characteristics and strength requirements (Revised 2022)
7. Muck D, Mikulic D, Koroman V, Klak S (2008) Design of a forklift with telescopic boom. In: 16th Proceedings of the 16th ISTVS international conference of the international society for terrain-vehicle systems, Turin

Chapter 11
Construction Trucks

Abbreviations

G_v [N]	Weight of an empty tipper truck
Q [N]	Load capacity of tipper trucks
F_A [N]	Load on the front axle
F_B [N]	Rear axle load
a [m]	Position of the center of gravity of the empty vehicle from the rear axle
x [m]	Position of the center of gravity of the load from the rear axle
m_A [kg]	Truck load limit
m_B [kg]	Front axle load limitation
m_C [kg]	Rear axle load limit
m_D [kg]	Front load according to the total weight of the vehicle
l_{CG} [m]	Resultant center of gravity
m_i [kg]	Mass of a specific load
l_i [m]	Distance of the center of gravity of the load from the front side
σ_i [MPa]	Truck frame stresses
G_s [N]	Weight of truck box
G_t [N]	Max cargo weight
F_{TC} [N]	Force of the hydraulic cylinder of the tipper truck
L_{TC} [m]	Length of hydraulic cylinder pull
β [^{0}]	Tipper truck hydraulic cylinder angle
V [m^3]	Loading box capacity
N_{dn} [-]	Optimal number of tipper trucks
T_c [min]	Driving cycle time
T_u [min]	Truck loading time
v [km/h]	Truck speed.

D. Mikulić and D. Mück, *Construction Machines*,
https://doi.org/10.1007/978-3-031-67845-5_11

11.1 Purpose and Division of Construction Trucks

Construction trucks belong to the category of cargo vehicles. The construction operative mainly uses trucks for the transport of bulk cargo and other construction materials, concrete mixer trucks, concrete pumps, tanks, and others. Construction trucks are used in combination with a trailer or in combination with a towing vehicle and a semi-trailer (tractor), forming thus a set of freight vehicles.

This chapter considers the most numerous construction trucks used for transporting bulk cargo and other construction materials, Fig. 11.1. These are self-unloading trucks, i.e., tipper (dump) trucks equipped with an open or closed box, with a rectangular or oval tipper box for bulk cargo.

Tipper trucks are common machines on most construction sites, which are loaded with excavators, backhoes, or loaders. Tipper trucks are also used in other areas of application, such as forestry, agriculture, etc. They can also be used to transport palletized goods (with a crane) or to transport containers with a hydraulic arm (*abroll tippers*—roll tipper trucks with lifting hooks).

Division of construction tipper trucks:

- Tipper trucks,
- Tipper trucks with tipper trailer,
- Trucks with tipper semi-trailer (tractors with tipper semi-trailer).

A tipper truck carries a loading box for transportation of cargo. For rear or side unloading, the loading box is raised and lowered hydraulically. The vehicle base consists of two-axle, three-axle, four-axle, or five-axle trucks. Representatives of the trucks most used in construction are four-axle trucks, called standard construction tipper trucks.

A tipper truck with a tipper trailer is a combination of a tipper truck and a tipper trailer, which has its own hydraulic box lifting and the possibility of unloading. The basis of the set of vehicles is the tipper truck (with 2 or 3 axles) with the attached tipper trailer (with 2 or 3 axles), Fig. 11.2.

A tractor with a tipper semi-trailer is a combination of a towing vehicle and a semi-trailer. The box is mounted on the undercarriage of the semi-trailer and it is

Fig. 11.1 Standard construction tipper truck 8 × 4

Fig. 11.2 Tipper truck with tipper trailer and tractor with tipper semi-trailer

hydraulically lifted towards the rear of the vehicle for unloading, Fig. 11.2. The base consists of a tractor (with 2 or 3 axles) to which a tipper semi-trailer (with 1, 2 or 3 axles) is connected. A special semi-trailer (*low-loading trailer*) is a low-floor semi-trailer for the transport of cargo that is not intended for road traffic or machines and equipment that exceed the prescribed dimensions.

In accordance with the technical categorization of vehicles, cargo cars are marked by a category label and description (*Directive 2007/46/EC* and *Regulations 167/2013 and 168/2013*). Cargo vehicles—trucks are classified in the N-category of road motor vehicles, and their trailers and semi-trailers in the O-category of road vehicles[1] [1].

Standard tipper trucks

The most used trucks in construction, especially in the construction of roads and communal infrastructure, are four-axle trucks. These are tipper trucks with four axles (8 × 4, drive on two rear axles), which have off-road capabilities (8 × 6 and 8 × 8), Fig. 11.3.

Four-axle tipper trucks are considered representative trucks in construction operations. This results from the fact that two 8 × 4 tipper trucks can transport more cargo than three classic 6 × 4 or 4 × 2 trucks. Compared to three trucks, this means reduced maintenance costs due to one less vehicle and one less driver. The advantage is that one less vehicle on the road leaves more space for other road users.

[1] In accordance with the regulations on the technical conditions of vehicles in traffic (*Rulebook, NN 85/2016-1864*), the N-category of vehicles is classified and described as follows:

N1: Motor vehicles for the transport of cargo with the maximum permissible mass ≤ 3500 kg
N2: Motor vehicles for the transport of cargo with the maximum permissible weight > 3500 kg but ≤12,000 kg

N3: Motor vehicles for the transport of cargo with the maximum permissible mass >12,000 kg

Also, the O-category of road vehicles, described trailers including semi-trailers, is classified:

O1: Attachment vehicles with the maximum permissible mass ≤ 750 kg

02: Attached vehicles with the maximum permissible mass > 750 kg but ≤ 3500 kg, i.e. semi-trailers with the maximum permissible axle load > 750 kg but ≤ 3500 kg

03: Attachment vehicles with the maximum permissible mass > 3500 kg but ≤ 10,000 kg, i.e. semi-trailers with the maximum permissible axle load > 3500 kg but ≤ 10,000 kg

04: Attachment vehicles with the maximum permissible mass > 10,000 kg i.e., semi-trailers with the maximum permissible axle load > 10,000 kg

Fig. 11.3 Tipper trucks 6 × 4 and 8 × 4 [2]

In addition to higher productivity, the big advantage of 8 × 4 tipper trucks is their cross-country ability and maneuverability and parking, as the two steering axles improve the vehicle maneuverability. Therefore, the standard multi-axle undercarriage provides a variety of superstructure options for the construction equipment, from concrete mixers and pumps, truck cranes, etc., to military superstructures.

The total mass of the truck is the mass of the base vehicle (the mass of the undercarriage-chassis with the cab) and the mass of the superstructure. The permissible load capacity results from the difference between the maximum permissible mass and the mass of the vehicle in the driving condition (the mass of the superstructured vehicle including the mass of the driver 75 kg and the mass of fuel when the tank is filled to at least 90%). The maximum permissible mass of an 8 × 4 tipper truck is 32 tons, of which, for example, at least 30–50% refers to the empty vehicle and the rest to the permissible carrying capacity. This means that the carrying capacity of the vehicle is at least as much as the mass of the basic empty vehicle. For example, the tipper truck *Iveco Trakker 440 8 × 4* has a declared curb weight of 14.8 t, a load capacity of 17.45 t, so the total technically permissible weight is 32.25 t. By using lighter materials of higher strength in the construction, the weight of the truck is reduced, and higher load capacity is achieved. Thus, the load capacity of modern 8 × 4 tipper trucks has been increased to 20 t.

Examples of construction tipper trucks 8 × 4 with the maximum permissible mass of up to 32 tons, *Mercedes-Benz Arocs* 4145 K, *MAN TGS 35.440, IVECO Trakker 360, VOLVO FMX 460, RENAULT Kerax 460. DAF FAD CF 460*, Fig. 11.4.

Fig. 11.4 Standard tipper trucks 8 × 4 (*Mercedes, DAF, Renault, MAN*, …)

11.2 Load Capacity of Construction Tipper Trucks

The load capacity of trucks is their basic characteristic. The load capacity of the vehicle is the difference between the maximum permissible mass of the laden vehicle and the mass of the vehicle in driving condition. The load capacity depends on the permissible axle load and the number of axles. The load of the vehicle at rest on a horizontal surface must not exceed the following amounts:

- Two-axle vehicle 18 t,
- Three-axle vehicle 25 t (26 t),
- Four-axle vehicle 31 t (32 t).

Exceptionally, three-axle and four-axle motor vehicles may have a maximum permissible mass of 26 t or 32 t, respectively, if the drive axle of the motor vehicle is equipped with dual tires and an air suspension that is considered equivalent to the air suspension within the EU or if the drive axle is equipped with dual tires, and the maximum permissible mass of each axle does not exceed 9.5 t.

The maximum permissible mass of a set of vehicles with 5 or 6 axles is 40 t (this includes a two- or three-axle tractor and a two- or three-axle semi-trailer, but it also includes a two- or three-axle towing vehicle and a two- or three-axle trailer). In the case of a set with 4 axles, the maximum permissible weight of the set of vehicles is 36 t (two-axle tractor and two-axle semi-trailer, i.e. two-axle towing vehicle and two-axle trailer).

In the case of a statically loaded vehicle, the axle load at rest on a horizontal surface for single (individual) axles must not exceed 10 t for free axles and 11.5 t for driven axles.

The limitation of axle load capacity is related to the protection of roads from overloading. The following is the permissible load of multiple axles, while respecting the loading of one axle, Fig. 11.5.

11.5 t for axles with inter-distance a mutual distance of less than 1 m

16.0 t for axles at a distance greater than or equal to 1 m and less than 1.3 m

18.0 t for axles at a distance greater than or equal to 1.3 m and less than 1.8 m

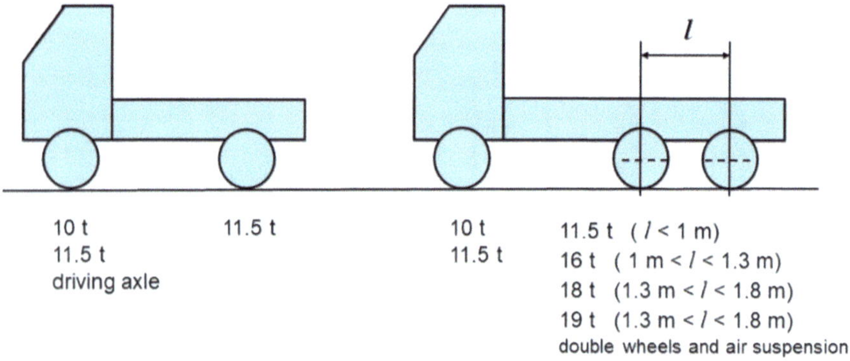

Fig. 11.5 The maximum permissible axle load of the truck [3]

> 19.0 t for axles at a distance greater than or equal to 1.3 m and less than 1.8 m if the drive axle is equipped with double wheels and air suspension or if the drive axle is equipped with double wheels and the maximum permissible mass of each axle does not exceed 9.5 t.

A single axle is defined as any individual axle that is at least 1.80 m away from the adjacent axle. The load capacity is expressed by mass (kg or t) which is appropriate from the point of view of use. When analyzing the vehicle structure, the load capacity is expressed by weight (N or kN). The maxium loading of a two-and four-axle truck is presented in Fig. 11.6.

Axle loads determine the permissible total weight of the vehicle and thus the maximum load mass. In order to ensure sufficient power to accelerate and move the vehicle, the relationship between the gross engine power and the maximum permissible weight is prescribed. For trucks, the minimum required engine power is 7 kW/t. Tipper trucks are powerful machines, whose power of the drive diesel engine is in the range of 250–400 kW. Thus, for a four-axle vehicle (32 t), the power-to-weight ratio is from 8 to 12.5 kW/t.

In addition to the drive axles, a non-drive axle that can be raised can also be installed. By the way, raising the axles is used when driving an empty vehicle in order to reduce the rolling resistance, thereby saving fuel, reducing the wear of tires, bearings, and all other parts that are excluded in that case.

When there is also a requirement for an attachment vehicle, the position of the mechanism for connecting the attachment vehicle (mechanical coupling) is selected depending on the trailer design itself. To reduce the risk of rear-end and side-end collisions in traffic, there are various protections that can be used on trucks. There are different versions of the rear protection mount (most often it is bolted, but sometimes there are also welded versions). In most cases, they are attached to the main frame, and sometimes to the main and auxiliary frame or they form a part of the vehicle structure. The side protection is usually bolted to the main and/or auxiliary frame, but it can also be welded.

Trucks with a built-in payload control indicator can use sensors to monitor the weight of the loaded cargo and thus control the remaining permissible payload

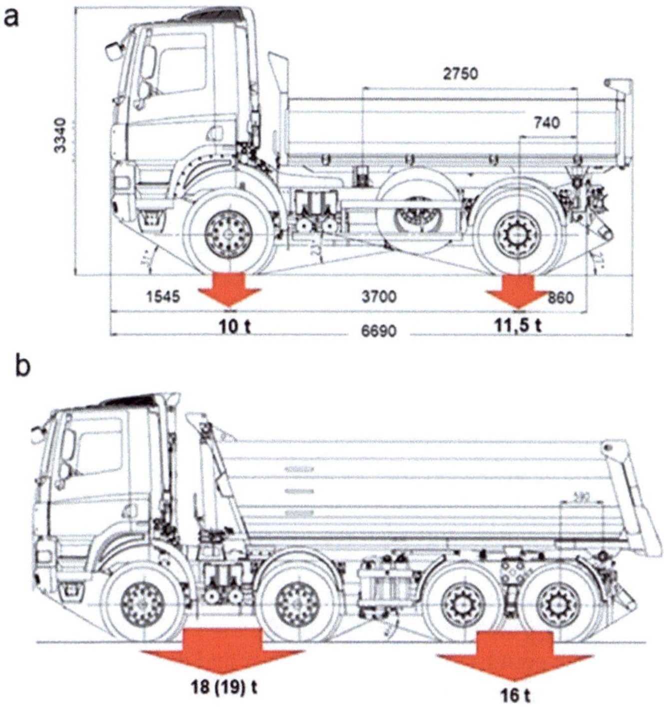

Fig. 11.6 **a** Maximum permissible axle loading of a two-axle truck. **b** Maximum permissible axle loading of a four-axle truck

(suspension displacement, or other). In addition to the load capacity, tipper trucks must meet overall dimensions. The prescribed dimensions of the vehicle refer to the permissible length, width, and height of the vehicle together with the load.

- The maximum permitted height of the truck is 4.0 m.
- The maximum permitted vehicle width is 2.55 m.
- The length of individual vehicles must not exceed 12 m.
- The maximum length of a tractor with a semi-trailer is 16.5 m, and of a towing vehicle with a trailer is 18.75 m. The distance between the rear axle of the motor vehicle and the front axle of the trailer must not be less than 3 m.
- When driving in a circle, motor vehicles and trailers may have a maximum outer circle diameter of 25.0 m, and a minimum inner circle diameter of 10.6 m.

Truck load distribution

The distribution of the load inside the cargo space of the truck is of great importance because it directly affects the axle load, the safety of moving and stopping the vehicle. The cargo can be distributed in such a way that the permissible axle loads are exceeded, but also in such a way that they are not exceeded. It is not only the load that affects the axle loading, but also the mass distribution of the base vehicle. Therefore, it is necessary to know the loading distribution on the cargo space platform. Modern software superstructure packages offer truck and trailer technical calculations (e.g., *TrailerWin)* [4].

Construction tipper trucks can carry different types of cargo boxes for bulk cargo (10–30 m^3) according to the work in a certain soil category or material density (larger box volume for lighter material 1000–1800 kg/m^3, smaller volume for heavier material 1800–3000 kg/m^3). It is important to use an adequate box according to the purpose of transport because an underutilized truck is not economical. For example, a truck with a box of 10 m^3 for transporting gravel with a density of 2000 kg/m^3 should have a load capacity of 20 t. A truck with a box of 20 m^3 should have the same carrying capacity of 20 t for transporting humus with a density of 1000 kg/m^3.

The mass of the vehicle and the mass of the load when moving and stopping affects the loading on the axles, that is, the longitudinal and transverse stability of the truck. In accordance with the superstructure regulations, the manufacturer recommends that the center of gravity be in front of the rear axle of the vehicle. Rarely, but sometimes for certain upgrade configurations (e.g. a crane behind the cab and a smaller box on the rear of the vehicle with a short wheelbase of the base truck) the center of gravity of the cargo area is on the center of the rear axle or slightly behind it. It is important that the permissible axle loadings are not exceeded and that the load on the front axle is sufficient to maintain the vehicle controllability.

Depending on the permissible carrying capacity and the position of the center of gravity of the empty vehicle, each vehicle has its own *load distribution diagram*, so that it is practically used to avoid exceeding the axle loading and unsafe driving. The center of gravity of the load should be within the permissible load curves, Fig. 11.7.

Determining the center of gravity of the vehicle and the center of gravity of the cargo

It is practical to include the mass of the tipper superstructure with the box in the mass of the vehicle (by weighing and calculating, the position of the center of gravity is determined) because the superstructure is permanently attached to the vehicle. Then a loading distribution diagram is created and the permissible position of the center of gravity of the load and the permissible mass of the load in relation to the position inside the cargo area are obtained.

(a) Position of the center of gravity of the empty vehicle from the rear axle ($\Sigma M_B = 0$).

$$a = F_{A,v} l / G_v [\text{mm}]$$

$F_{A,v}$ Loading on the front axle of the vehicle—without load (manufacturer's data).

G_v Weight of the empty vehicle, which includes the tipper superstructure (manufacturer's data).

(b) Position of the center of gravity of the load from the rear axle ($\Sigma M_B = 0$).

$$x = \left(F_{A,max} l - G_v a\right) / Q [\text{mm}]$$

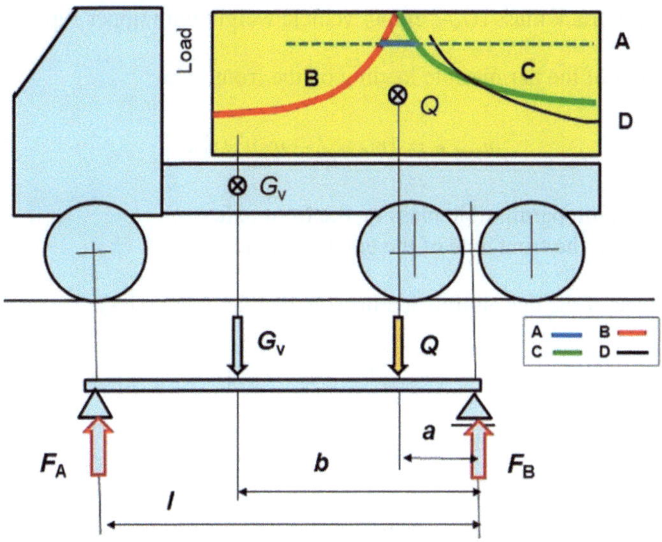

Fig. 11.7 Loading diagram of a three-axle truck. **a** Limitation of the maximum permissible load capacity of the truck. **b** Limitation of the permissible loading on the front axle. **c** Limitation of permissible rear axle (axles) loading. **d** Ratio of the front loading to the total weight of the vehicle

$F_{A,max}$ Vehicle front axle load (calculation, N)
Q Load capacity—max vehicle load (N)

The position of the center of gravity of the load (x) should be in front of the axis of the rear axle. At the same time, the front axle must be loaded with at least 20% of the total load (regulation) or, according to some manufacturers' recommendations, 25% or more.

Axle load calculation

$$\text{Front axle loading with maximum cargo}\left(\sum \mathrm{M_B} = 0\right)$$
$$F_{A,max} = (G_v a + Qx)/l[\mathrm{N}]$$

$$\text{Rear axle loading with maximum cargo}\ (\sum \mathrm{M_A} = 0)$$
$$F_{B,max} = \left[G_v(l - a) + Q(l - x)\right]/l[\mathrm{N}]$$

Formation of loading curves, A, B, C, D [5]

A Limitation of the maximum permissible mass of the truck [kg, tons].

$$m_{max,A} = m_{v,dop} - m_v[\mathrm{kg}]$$

$m_{v,dop}$ Maximum permissible vehicle mass

m_v Empty truck mass (G_v—empty vehicle weight with tipper superstructure)

B Limitation of the permissible loading on the front axle

$$m_{\max,B} = \left(m_{p,dop}l - m_p a\right)/x[\text{kg}]$$

$m_{p,dop}$ Maximum permissible mass on the front axle
m_p Mass on the front axle of the empty vehicle

C Limitation of the permissible loading on the rear axle (axles).

$$m_{\max,C} = \left[m_{s,dop}l - m_v(l-a)\right]/l - x[\text{kg}]$$

$m_{s,dop}$ Maximum permissible weight on the rear axle
m_v Empty vehicle mass

D Minimum front axle loading ratio, as the ratio of the front axle loading to the total vehicle weight (to ensure vehicle controllability).

Front axle loading in percentages, for certain distances from the front axle

$$\left(F_{A,\max}/G_v + Q\right)100 \geq 25\%$$

The relationship between the load on the front axle and the distance of the center of gravity of the superstructure and the load from the rear axle (x), in the formula below %m = 25.

$$m_{\max,D} = m_v(\text{a}/l - \%m/100)/(\%m/100 - x/l)[\text{kg}]$$

Construction tipper trucks carry different types of concentrated and piece loads (load blocks). Figure 11.8 shows an example of a loading distribution diagram, as well as the correct and incorrect position of a piece load [6].

The curves show the permissible axle loadings and the position of the load. As can be seen in the figure, the permissible load capacity of 11.5 t can only be used if the center of gravity of the load is in a small range of 3.1–3.7 m distance from the front side of the loading area. For example, let us transport a load weighing 7 t (*line A*). The *vertical lines B and C* limit the position within which the load can be placed and secured, which satisfies the axle loading limits of the front and rear axles, and the requirement for the minimum loading on the front axle. For the safety of driving the vehicle, the front axle must not be overloaded nor unloaded; it must be loaded at least 25% of the total loading (*D*).

The following is an example of three ways of positioning a concentrated load with the aim of finding the optimal position, Fig. 11.9a. Let the mass of the load be 9 t, its length equal to 6 m, while the distance of the center of gravity of the load from the beginning of the cargo space is 1.9 m. When its position of the center of gravity

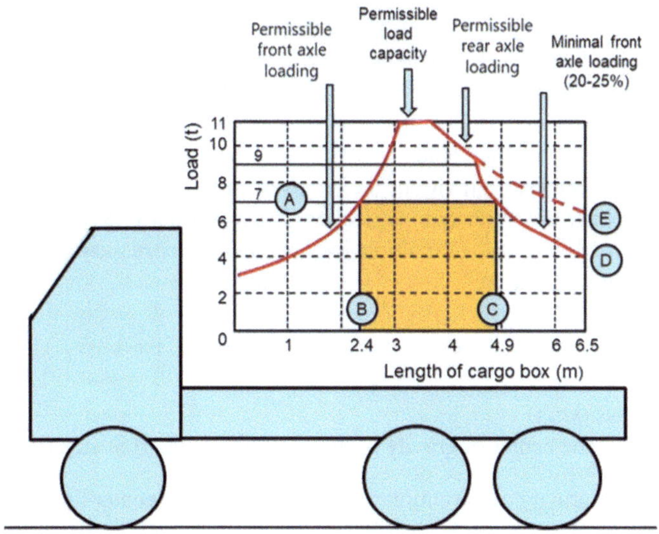

Fig. 11.8 Truck load distribution diagram (Adapted [6])

is compared with the diagram in Fig. 11.8, the conclusion follows that the load of mass 9 t must not be so placed, because for that position of the center of gravity of the load, the maximum load capacity is approximately 5.8 t.

A good way is achieved by moving the load in relation to the first position by, for example, 0.8 m backwards, so that the center of gravity of the load is within the given limits (between 2.6 and 4.5 m). The optimal solution is to rotate the load by 180°, thus positioning the center of gravity of the load within the axle limitation curves (b).

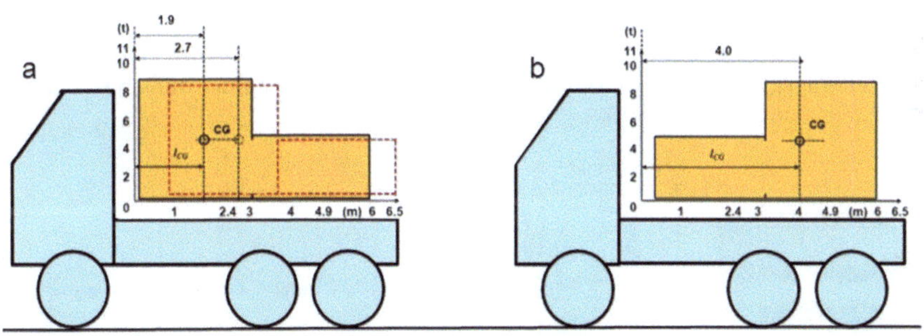

Fig. 11.9 Positioning the load within the axle load limits

To estimate the front axle loading, it is necessary to know the position of the center of gravity of the load. The load can be moved, which can reduce the distance of the center of gravity from the axis of the rear axle, which at the same time relieves the front axle of the vehicle. For the sake of driving safety, the front axle must be loaded with at least 25% of the total load.

Knowing the center of gravity of the cargo is important to place the load in the right place in the truck. With symmetrical bodies, it is usually easy to determine the center of gravity. In the case of multiple load bodies, it is necessary to determine the "resultant" center of gravity:

$$l_{CG} = (m_1 l_1 + m_2 l_2 + m_3 l_3 + -)/(m_1 + m_2 + m_3 + \ldots)$$

m_i Cargo mass (kg, t)
l_i Distance of the center of gravity of the cargo from the front side (m).

The center of gravity of the entire cargo should be on the center line of the vehicle and the cargo should be evenly distributed on each side and properly secured with tie-down straps.

11.3 Features of the Undercarriages of a Tipper Truck

The undercarriage of the tipper truck (often the *chassis*) takes over the entire loading of the vehicle. It consists of the basic frame on which all circuits and power transmission systems from the engine to the wheels, steering and stopping devices, and electronic devices are located. The overall dimensions of the platform for different superstructures (undercarriage with the cabin) are shown in Fig. 11.10.

Traction properties of tipper trucks

Truck users use the traction diagram (*Rimpull-speed*) to evaluate its efficiency. The traction diagram is the result of the joint operation of the engine, transmission, and

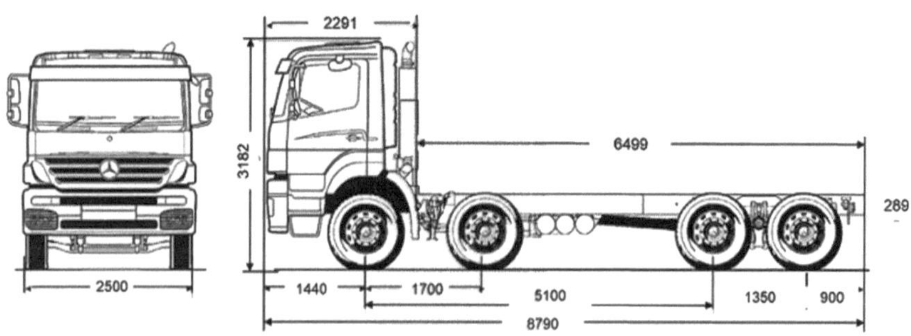

Fig. 11.10 Basic dimensions of a truck with a cabin [2]

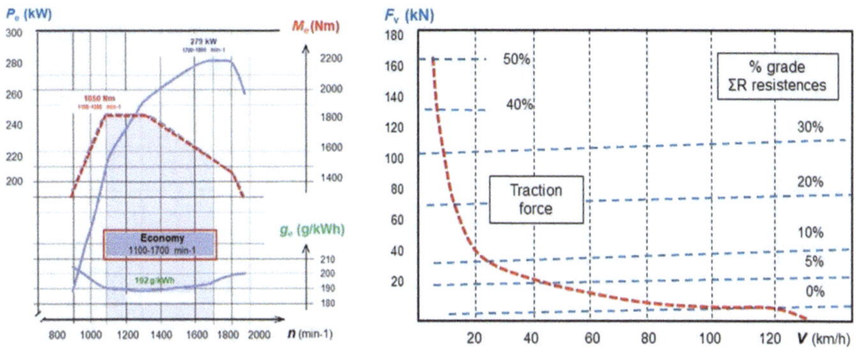

Fig. 11.11 Engine speed characteristic of the engine and traction diagram of the tipper truck (Adapted [7])

the wheels, and it is a graphic representation (*F-v*) of the traction forces on the drive wheels on the one hand and the movement resistance on the other, as a function of the movement speed, Fig. 11.11.

Using the traction diagram of the 36 ton dump truck *Volvo FH12/380* 8 × 4 (*engine D12D380, gearbox VT221 4B*), the performance of the vehicle under full loading conditions can be estimated, which allows the construction operative to use the truck optimally. The truck can achieve a high traction force (160 kN on a 50% grade) on the 8 × 4 drive wheels. The economic speed of the vehicle on the road is 70–105 km/h, at engine speed of 1200–1700 min^{-1}. The maximum speed is 125 km/h on a straight road, 65 km/h on a 3% ascent, and on a 10% ascent the speed is 25 km/h.

Diesel engines allow trucks to move in the most difficult conditions of use. Six or eight cylinder engines, in-line or V-shaped, with *Common Rail* injection, turbocharger, and 4-valve technology are used. The basic indicators of engine performance are the speed characteristics, i.e., power, torque, and fuel consumption as a function of the number of revolutions at full engine loading. The engines of standard trucks develop a power of about 300 kW and torque of about 2000 Nm, whereby the torque is constant in a wide range of engine rotation, and the road fuel consumption is 20–40 l /100 km.

The truck transmission is mostly semi-automatic or automatic, which includes a gearbox, a power distributor between the front and rear axles, axle reducers and differentials, and planetary reducers in the wheels, Fig. 11.12. The driver has at their disposal the choice of manual or automatic gear changes. Automatic gear shifting processes depend on the driving conditions and the position of the gas pedal.

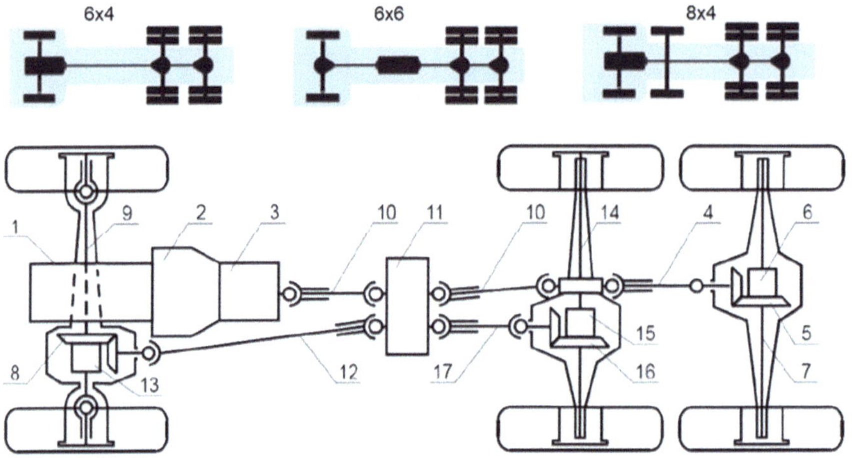

Fig. 11.12 Tipper truck transmission scheme. 1—engine, 2—clutch, 3—gearbox, 4,10, 12, 17—cardan shafts, 5, 8, 16—reducers, 6, 13, 15—differentials, 7, 9, 14—wheel drive shafts, 11—power distributor

Gearboxes offer a large number of gears (9–18), which also includes off-road driving. For example, the 32-ton *Mercedes Arocs* has an automated G 280-16 power-shift gearbox (16 gears), which changes gears quickly and precisely; in off-road mode, it reduces the number of changes to a quarter, which ensures easier driving with a load and on uneven ascents. The trucks have dual wheels on the rear drive axles. The markings on the tires show the dimensions, type of tire and its performance. The designation 315/80 R 22.5 154 M determines the width of 315 mm, the height of the tire profile is 80% of the tire width, R—radial design of the tire, rim diameter 22.5 inches, loading index of double wheels 154/154 at pressure of 8.5 bar and M vehicle speed index (130 km/h).

In modern trucks and tractors, the front axle of the truck is equipped with a *HydroDrive* (MAN) hydrostatic drive for the safe towing of the truck on difficult terrain. It is activated after using the automatic blocking of the other differentials. The mass of the hydro drive is half that of the mechanical axle, and the truck consumes less fuel. The hydro pump gets its drive from the PTO, with a power of about 100 kW, which works under a pressure of about 450 bar and with a flow of 350 l/min of oil. The radial hydro motors in the wheels have 10 radial pistons each, develop a power of 40–50 kW and a torque greater about 6000 Nm.

About 30 L of oil circulate in the hydro system. In cold weather, the oil is automatically heated to the required minimum temperature of around +15 °C, when system activation is possible. The hydro drive is available up to a speed of 25 km/h, after which it automatically switches to passive mode, and when the speed drops below 25 km/h, it automatically turns on again. This drive is also active when driving backwards in the first and second gear. At truck speeds above 60 km/h, it is automatically deactivated, thus protecting the main components and reducing fuel consumption.

Fig. 11.13 Mechanical suspension or air suspension of tipper trucks

Truck suspension can be made with parabolic leaf springs (or semi-elliptical) or air springs, Fig. 11.13. The air suspension system provides automatic regulation of the ride box height and ground clearance, as well as greater comfort.

Steering of an 8 × 4 truck is performed with the wheels of the two front axles. Servo control devices enable easy control. In curves, the steering wheels have different turning angles, so the steering mechanism is complex. The turning radius of an 8 × 4 truck is about 21 m.

Truck braking is pneumatic. The braking force is developed with the help of compressed air. The air is pressed into the tanks with the help of a compressor that is powered by the engine. Each circuit of the two-circuit system has its own air tank, Fig. 11.14. The driver acts on the brake pedal, i.e., opens the brake valve, and the air pressure (8–10 bar) allows the braking of the wheels. Trailer braking is also pneumatic. The trailer starts braking before the towing vehicle. In case of separation of the trailer from the towing vehicle, the trailer is automatically braked by its own pneumatic device.

The pressure regulator (2) ensures the working air pressure. The air dryer (3) has the role of removing moisture from the compressed air. The regeneration tank (4) enables the dryer to regenerate during the time when the air does not pass through it but is released into the environment by the regulator. The four-circuit safety valve (7) ensures working pressure in both circuits of the service brakes, the parking brake circuit, and the trailer braking circuit. Compressed air tanks (8) provide the necessary amount of compressed air for the operation of all brakes. The compressed air from the two tanks is led to the service brake valve (5) which, due to the action on the brake pedal, controls two independent brake circuits on the towing vehicle. Compressed air is supplied in one circuit to the brake cylinders of the front wheels (12), and in the second circuit via the brake force regulator (11) to the brake cylinders of the rear wheels (13).

The brake force regulator (11) controls the air pressure supplied to the rear wheel brake cylinders based on the change in the rear axle loading. By releasing the compressed air, the brake mechanism returns to its initial state. The parking brake is controlled by moving the control valve handle (6) to the appropriate position. The parking brake can be used as an auxiliary brake; then the control valve is designed so that it continuously changes the braking force, in order to ensure braking with a deceleration of 0.25 g until stopping. The relay valve (10) has the role of ensuring

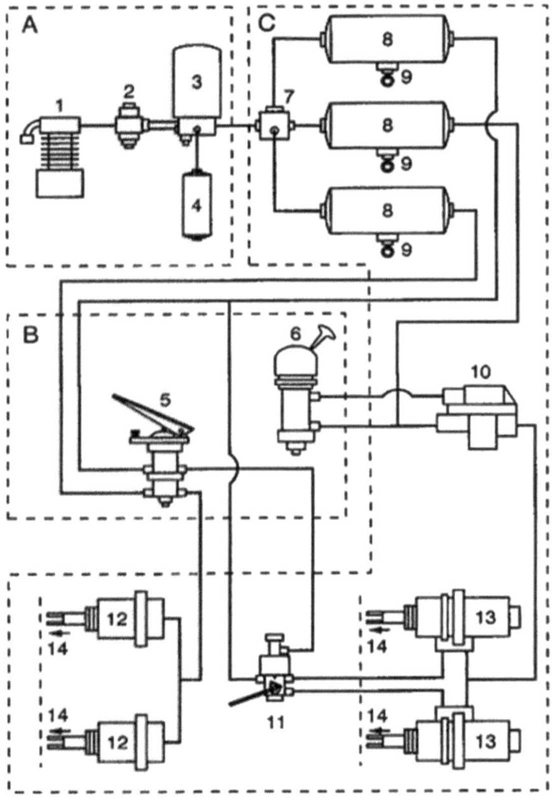

Fig. 11.14 Diagram of a two-circuit pneumatic brake [8]. **a** Power supply module, **b** Control module, c Transmission module; 1—compressor, 2—pressure regulator, 3—air dryer, 4—regeneration air tank, 5—service brake shoe and valve, 6—handbrake lever and valve, 7—four-circuit safety valve, 8—compressed air tanks, 9—valve for draining condensate, 10—relay valve for filling and emptying the parking brake cylinders, 11—brake force regulator, 12—brake cylinders of the front wheels, 13—brake cylinders of the rear wheels, 14—push rods of the brake cylinders

fast filling and emptying of the brake cylinders, thus achieving a quick response of the brakes.

EBS—Electronic Braking System

Modern trucks use, in addition to pneumatic braking systems, also additional systems such as hydrodynamic brakes, electromagnetic brakes, and motor brakes, as well as active braking systems (WABCO, KNORR, HALDEX, …). The electronically regulated braking system (EBS-*Electronic Braking System*) ensures optimal harmonization of the braking forces between the wheels, as well as between the towing vehicle and the trailer, Fig. 11.15.

Features of the EBS system compared to classic pneumatic brake systems:

- The response speed of the brake installation gives a shorter stopping distance of the vehicle,
- Compatibility of the braking of the towing vehicle and the trailer during long-term braking,
- Integration with other vehicle systems (ABS—anti-lock braking system, EBD—brake force distribution, ASR—drive wheel slip prevention system, …),

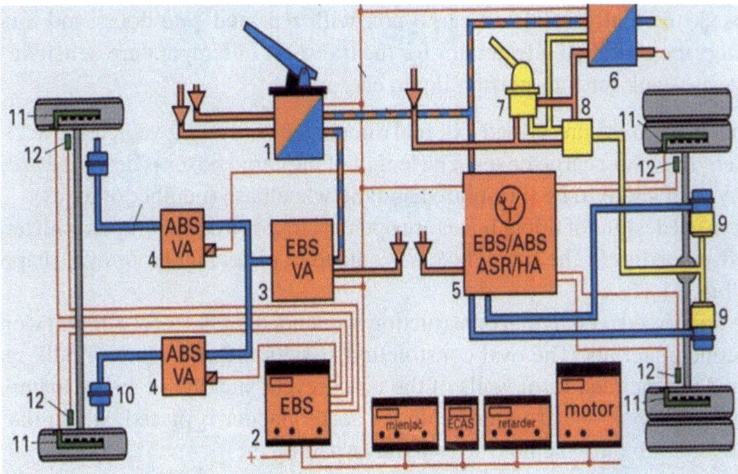

Fig. 11.15 Scheme of the electronic braking system (EBS) [9]. 1—brake pedal, 2—EBS control unit, 3—front axle modulator, 4—ABS front axle valve, 5—EBS-ABS rear axle modulator, 6—trailer control valve, 7—parking brake valve, 8—trailer relay valve, 9, 10—pneumatic brake cylinder, 11—wheel rotation sensor, 12—sensors of brake linings wear

- Equal wear of the brake linings on the axles.

Retarders (permanent brakes) play a significant role in long-term braking. The retarder must maintain constant speed of the loaded truck of 30 km/h, in the appropriate gear on a 7% downhill grade and a 6 km traveling distance. When maintaining constant speed of a tipper truck 32–41 t, for example 30 km/h on a 7% downhill, the required braking power is about 180 kW. Depending on the available braking power, this can be achieved in gear IV, with a power reserve of 60 kW. Long-term brakes are controlled automatically, which relieves the driver and increases the driving safety.

11.4 Features of the Tipper Superstructure with a Box

Vehicle superstructure software packages offer truck and trailer design, with the ability to add equipment such as cranes (front/rear), rear lift platforms, container carriers, etc.

The general designs of superstructures with a box for transporting construction cargo are:

- Open cargo box with sides,
- Open cargo box with self-unloading,
- Cargo box with tarps and tarpaulin,
- Closed cargo box with rear (and/or side) door,

- Closed, thermally insulated cargo box with rear and side doors and a cooling device, the so-called refrigerator for the transport of temperature-sensitive cargo,
- Cistern—tank for transporting liquid cargo.

In addition to the prescribed external dimensions of freight vehicles, trucks differ in the length of their cargo boxes. The length of the cargo boxes is designed according to the type of cargo to be transported and the wheelbase (number of axles).

The usual designs of tipper boxes are open or closed structures (open with tarpaulin and self-unloading). The tipper box is designed in a rectangular or oval shape (*Half Pipe*), Fig. 11.16.

The reinforced rectangular construction is intended for the most difficult work with large stone materials. The oval construction centers the unloading of bulk material well, and the inclined front walls of the box prevent sticking of sticky material. On the side of the box—next to the cabin, an access ladder is placed, which the driver can use to climb onto the box, stretch the tarpaulin, etc.

Tilting of the cargo box

There are three ways of tilting the box:

- One-sided tipper—tilting to the rear,
- Double-sided tipper—tilting to the rear and one side,
- Three-sided tipper—tilting to the rear and both sides of the vehicle.

The hydraulic cylinder for lifting or the box lifting mechanism is placed in the middle of the box or on the front of the box, Fig. 11.17. Additionally, with larger and heavier tipper boxes, *scissors are installed to stabilize* the box during lifting and unloading, which maintain the angle of the box lifting and reduce torsional stresses.

Fig. 11.16 Tipper truck with rectangular and oval box

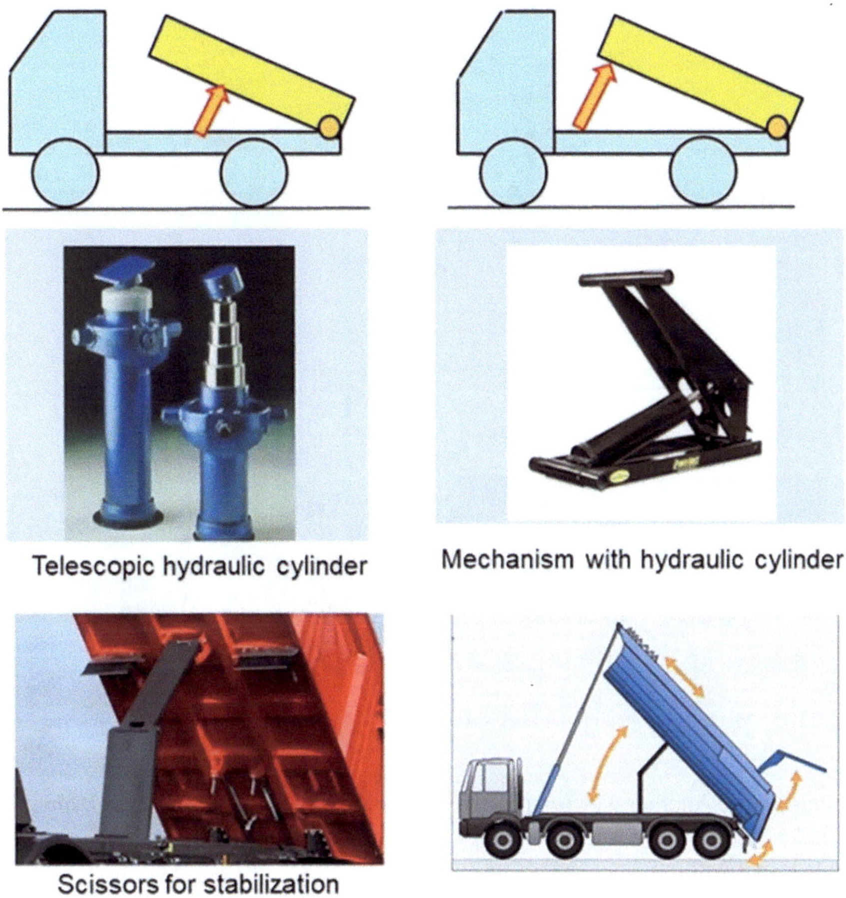

Fig. 11.17 The position of the hydraulic cylinder and the mechanism for lifting and stabilizing the box

Opening of the back side

The back side of the box can be opened in different ways. Opening can be divided into opening the side downwards or upwards and opening manually or with the help of a hydraulic cylinder. When opening the side downwards, it can hang freely and the load falls over it, the side does not allow material to be scattered under the vehicle (*a*), Fig. 11.18.

The side can also be held by a chain (*b*) or a spring, which then does not allow the side to open completely, but only up to an angle of 90°, i.e., the open side is then a continuation of the box platform. In this way, it is possible to unload the material slightly further from the rear end of the vehicle. In the case of free opening of the side upwards (*c*), the side is opened by bulk material that pushes it under the influence of gravitational force. The loose material then touches the inner part of the side when it

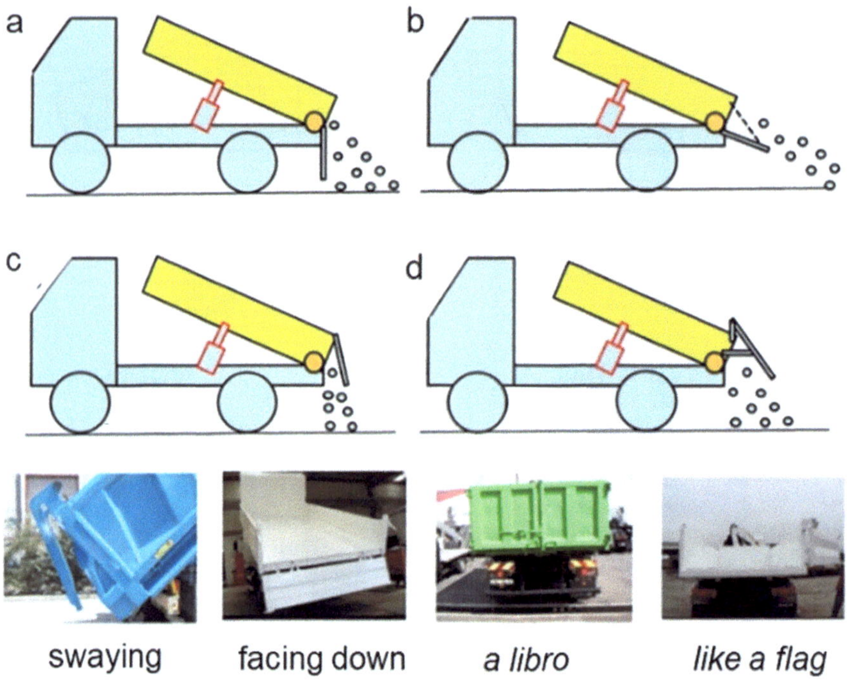

Fig. 11.18 Methods of opening the back side

is shaken out. Another way to open the side upwards is with the help of a hydraulic cylinder (*d*) and in this way it allows the loose material to fall freely without touching the side and without unnecessary friction.

Conceptualizing the tipper superstructure

In the following text, examples of two different concepts of tipper superstructures are presented. There are also different concepts depending on the manufacturer. Two concepts for the design of tipper boxes for bulk cargo are:

1. *Concept K1*—three-sided tipper superstructure of a rectangular shape

The K1 concept is a three-sided tipper, which enables the load to be shifted to the rear and the sides of the vehicle. Such tipper superstructures have a load capacity of up to 20 t. The side is opened manually, the side panels open downwards, and the back side has the option of opening both downwards and upwards (swinging the side). An advanced solution is the hydraulic opening of the side (*Bordmatic type*, or other).

Fig. 11.19 Three-sided tipper superstructure with box TK 160 (*Bordmatic*—hydraulic side opening) and TK 220 "V" sides reinforced with "jet rips" [10]

The sides of the box are made of high-strength steel, which increases the mass of the superstructure, but contributes to the durability of the interior of the cargo area. Also, the box floor is made of high-strength wear-resistant steel *Hardox 450*, and the sides are made of standard steel or the stronger option of *Weldox 700* steel.

The box supports at the ends have ball joints with a wedge or are made in the form of fork joints. Manual placement of the wedges selects the side on which the cargo is unloaded. The hydraulic cylinder is placed in the middle of the box. The advanced tipper side dumper is identified by the front cylinder and the auxiliary cylinder under the box. Three-sided tipper superstructures with a large volume box of 20 m^3, are installed on vehicles with 3 and 4 axles (6 × 4 and 8 × 4), Fig. 11.19.

Examples of three-sided tipper boxes (*TK models, PK-Palfinger Kran*)

TK 160—for vehicles 26 t
Box volume 10.2 m^3
Steel rectangular box:
– Length 4.8 m or 5.0 m, side height 0.85 m
– Floor, *Hardox HX 450* steel, 5 mm thick
– Front side with cabin protection
– Sides, steel *S355* (3 mm), side thickness 80 mm, reinforced with "jet rips"
– Two-sided opening of the side panels(180° or 160° —optional and swinging)
– Two-part side panels with a removable central column
– Swinging back side (opening of the bottom edge automatic mechanical)
Options:
– *Bordmatic*—hydraulic side opening (BM160)
– *Weldox 700*, 3.2 mm thick (increases weight by 100 kg)
– Reinforced side "V" side profiles
– Pneumatic opening of the side panels from the cabin
– Opening of the back side «a libro» (like a *book*)
– Aluminum sides, instead of steel, 60 mm thick

TK 200—for vehicles 33 t
Box volume 14.2 m^3 Steel rectangular box: - Length of 4.8 m or 5.2 m, side height 1 m - Floor *Hardox HX 450*, 6 mm thick - Front side with cabin protection (except for the crane) - "V" version sides, made of *Weldox 700*, 4 mm sheet - Two-sided opening of the side panels (180º or 160º - swinging option) - Two-part sides with a central column - Swinging back side (opening the lower edge automatically mechanically) - Middle column removable, front and rear fixed

2. Concept *K2*—rear tipper superstructure with box, rectangular and oval shape

(a) Rectangular shape of the box

The *K2 concept* is a one-sided tipper superstructure that unloads bulk cargo only at the rear of the vehicle. A rectangular box is used. A hydraulic telescopic cylinder for lifting the box is placed on the front part. The front side provides cabin protection. The side pages are fixed, reinforced with "jet rips", and the back side is oscillating (opening the lower edge upwards). For demanding work in the quarry, the box option without the classic rear side or lifting side (*Mulden tipper*) is used. The floor material is wear-resistant (*Hardox 450/500*), and the reinforced sides are made of *Weldox 700 steel*, Fig. 11.20.

Examples of rear tipper boxes (*SK models, PK-Palfinger Kran*)

Fig. 11.20 Rear tipper superstructures with boxes SK 160 and SK 200 M, swinging rear side or *Mulden tipper* rear side (*PK-Palfinger Kran*) [10]

SK 160—for vehicles 26 t
Box volume 11–17 m^3 Steel rectangular box: – Floor, *Hardox HX 450* steel, 6 mm thick – Sides made of *Weldox 700* steel, 4 mm thick, side thickness 80 mm – Fixed sides, reinforced with "jet rips" – Swinging back side (opening the lower edge automatically mechanically) – Fixed front side with cabin protection – The option of an asphalt hopper or rear backfill protection, 400 mm long (instead of the standard 200 mm)

SK 200 M—for vehicles 33 t—*Mulden*—for work in quarries
Box volume 21 m^3 Steel rectangular box: – The length of the kipper box is from 4.8 m to 5.2 m, the height of the sides is 1 m – Floor, *Hardox HX 450* steel, 8 mm thick – Sides made of Weldox 700 steel, 5 mm thick, side thickness 80 mm – Fixed sides, reinforced with "jet rips" – Without rear side or hydraulic opening

Regarding the volume of the box and the total weight of the truck, there is a distinction between the medium and heavy class of tipper trucks (*Meiller* division), Table 11.1.

(b) Oval shape of the box (*Half Pipe*)

The K2 concept is also a larger volume single-sided tipper superstructure that centers the bulk load well only at the rear of the vehicle. An oval-shaped box (*half pipe*) is used because there is no need for sides, with a load capacity of up to 20 t. For easier mounting and the box shape, a hydraulic telescopic cylinder is placed on the front part of the box. The rear side opens upwards—by means of a hydraulic cylinder. The advantage of this concept is that there is no need to manually open the rear side, so

Table 11.1 Rear tipper trucks (*H models*) [11]

Class		Medium	Heavy
3 axles (6 × 4) 	Tipper box volume[a]	Up to 18 m^3	Up to 24 m^3
	Total mass	26–33 t	33–40 t
	Models	H330	H336
4 axles (8 × 4) 	Tipper box volume[a]	Up to 20 m^3	Up to 28 m^3
	Total mass	41 t	41–50 t
	Models	H436	H450
	Side panels	1.2–2.0 m	1.6–2.0 m

[a] Rear side options

a deeper box can be installed in case of need for a larger volume of cargo space, but the downside is that the center of gravity of the cargo then also rises. The material of the box floor is also high-strength wear-resistant steel (*Hardox 450*, *500*), and the side panels and the rear side are made of *Weldox 700* steel, Fig. 11.21.

Example of a rear tipper HP box (*PK-Palfinger Kran*)

HP 200—for vehicles 33 t
HP 220—for vehicles 32–41 t
Box volume 17 or 18 m^3
Steel oval box:
– Box length from 5.3 mm or 5.8 m, height 1.5 m
– Floor, steel *Hardox HX 450*, thickness 6 mm (HP 200) or 8 mm (HP 220)
– Fixed front side with cabin protection
– "V" sides made of *Weldox 700* steel
– Rear and front side 5 mm thick, side 4 mm thick (HP 200)
– Back and front side 6 mm thick, side 5 mm thick (HP 220)
– Swinging rear side with automatic mechanical opening and closing
Options:
– Hydraulic opening of the rear side
– "roll-up" tarpaulin with winding mechanism (PVC or neoprene—for asphalt)
– Side panels made of *Weldox 700*, 6 mm thick, weight increase by 120 kg
– Rear side of *Weldox 700*, thickness 6 mm, weight increase by 60 kg

An "insulated" box (usually 18–22 m^3 in volume) is used for the transport of bitumen, with a special system for fixing the insulation and protective stainless steel sheets. The box is sandblasted and varnished before installing the insulating panels on the sheet metal, which is heated to 180 °C. The functionality of the insulation enables the optimum bitumen temperature to be maintained for a certain period. The cooling rate depends on the external conditions and the time elapsed before unloading. Owing to the special materials used to maintain the quality of the thermal insulation of the boxes, the quality of transport can be improved compared to traditional boxes. The back side can be made watertight because of using a pneumatic seal.

Fig. 11.21 Rear tipper boxes *Half Pipe* HP 200 and HP 220, rear side swinging or with hydraulic opening [10]

Table 11.2 *Half Pipe*—rear tipper trucks (*P models*) [11]

Class		Light	Medium	Heavy
3 axles (6 × 4)	Tipper box volume[a]		Up to 15 m^3	Up to 21 m^3
	Total mass		26–33 t	33–40 t
	Models		P330	P336
4 axles (8 × 4)	Tipper box volume[a]	Up to 17 m^3	Up to 20 m^3	Up to 28 m^3
	Total mass	32–41 t	41 t	41–50 t
	Models	P430	P436	P450
5 axles (10 × 4)	Tipper box volume[a]	Up to 17 m^3	Up to 20 m^3	
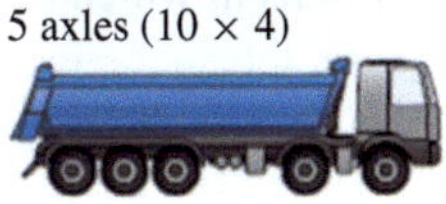	Total mass	32–41 t	41 t	
	Models	P330	P336	
	Side panels	1.2–1.5 m	1.2–2.0 m	1.6–2.0 m

[a] Rear side options

Regarding the volume of the box and the total weight of the truck, the light, medium and heavy classes of *Half Pipe* tippers (*Meiller division*) are distinguished, Table 11.2.

11.5 Construction of Tipper Superstructure with Box

The tipper superstructure with box is constructed in accordance with the rules and regulations valid in the EU. Also, the manufacturer's instructions for the specific truck must be followed.

General requirements for the development of a tipper superstructure with a box:

- As little mass as possible,
- As high load capacity of the structure as possible,
- As low the position of the center of gravity of the superstructure as possible,
- As great stability against overturning during unloading as possible,
- Ease of use.

The superstructure can be constructed using specialized programs for superstructures and configuration of trucks and trailers. The construction of tipper superstructure with box and the development of other superstructures is facilitated by the use of software packages, such as *TrailerWIN, Solidworks,* or others.

According to the data on the undercarriage, first an undercarriage model is developed or a ready-made model is downloaded, on which the tipper superstructure model with a box will be placed. First, an auxiliary frame (often *an auxiliary chassis or substructure*) is constructed, which will reinforce the basic vehicle frame (*vehicle chassis*).

Here is an example of constructing a tipper superstructure of a smaller truck with a box and some possible solutions for individual parts, assemblies, and devices There are also many other solutions used by the manufacturers. Superstructures can be performed in many different ways depending on the needs and requirements of the customer, the type of cargo that will be transported, and the conditions in which the vehicle will be used.

11.5.1 Designing the Auxiliary Frame of Trucks

The auxiliary frame takes the load from the tipper box and the load and transfers it evenly to the surface of the base frame. The auxiliary frame is part of the tipper superstructure that connects the box to the vehicle chassis. It can be made in the form of two models: *fixed*, which is firmly connected to the basic frame, or *portable* when replacing the box, Fig. 11.22.

The auxiliary frame increases the height of the vehicle center of gravity but provides adequate rigidity and reduces strain. It consists of two side longitudinal supports, transverse and diagonal stiffeners, a hydraulic cylinder support for lifting the box, and two transverse supports with balls on which the box rests. The profile of the side supports and stiffeners is usually a steel U profile with high resistance to bending and torsion. During construction, the necessary dimensions and the connecting method are defined.

The calculation of the maximum bending moment of the interacting connection of two frames (*interaction substructure*) is performed as a beam on two supports, which is acted upon by concentrated loading forces, and graphically represented by a bending moment diagram, Fig. 11.23.

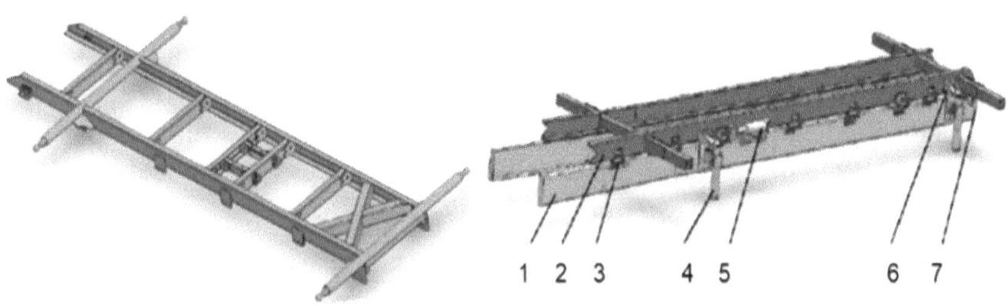

Fig. 11.22 Subframe models, fixed and portable; 1-main frame, 2-subframe, 3-locking bushings, 4-lifting/lowering cylinders, 5-moving cylinders, 6-lifting/lowering stand, 7-safety pins

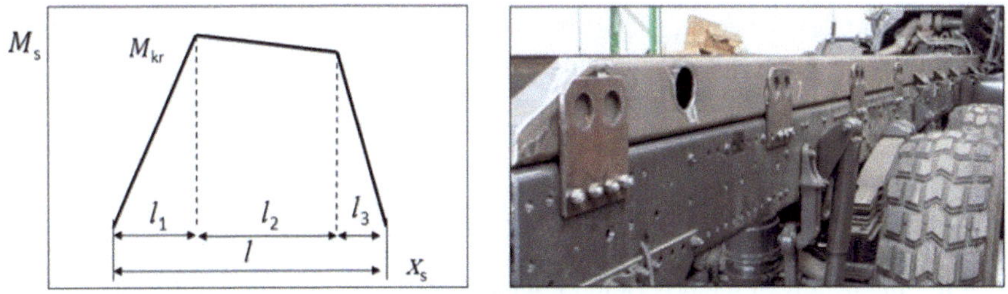

Fig. 11.23 Diagram of bending moment and connection of the main and auxiliary frame [12]

As the auxiliary frame is placed on the base frame of the vehicle, the interacting connection of both frames is considered in the calculation of the cross-sectional resistance of the frame. It is such a substructure in which interconnected elements behave as a single unit. For connection, rigid fastening elements are used, which must be placed close to each other. Cross-sectional areas and moments of inertia (data, *Solidworks or other software*) are sufficient to verify the maximum stresses at the critical section ($\sigma_i = M_{kr}/W_i$).

Method of connecting the main and auxiliary frames

The auxiliary frame is connected to the main frame with rigid and elastic connections. The location of the largest bending moment determines the critical location, so the rigid connection is placed at that location (l_1). From that place, elastic connections are placed towards the front of the vehicle, and rigid connections towards the rear. An example of the layout of connections is shown in Fig. 11.24.

By using rigid connections between the main and auxiliary chassis, an interacting substructure is obtained where the interconnected elements behave as a whole. Such a construction has higher resistance to deformations (bending, twisting) and enables taking on higher loadings, which leads to an increase in the vehicle carrying capacity.

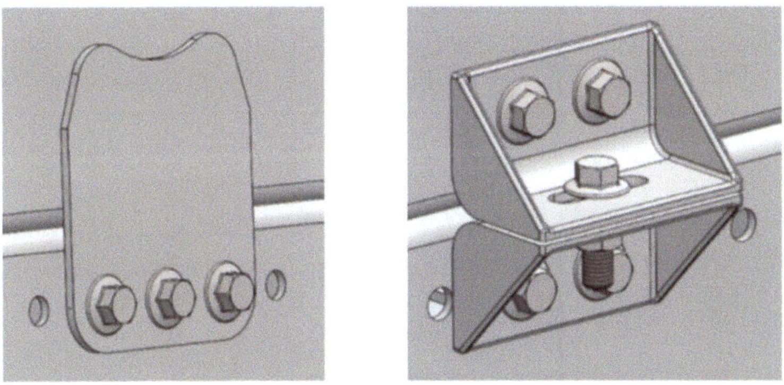

Fig. 11.24 Basic forms of connection of two frames: rigid and flexible connection

Both rigid and flexible connections are connected with high-quality bolt connection. Bolts and materials are used according to the manufacturer's recommendations.

11.5.2 Designing of the Tipper Truck Box

Overall dimensions of the box (length, width and height) are determined depending on the dimensions of the base vehicle. The tipper truck box consists of the box frame (1) and the sides of the box (2), Fig. 11.25.

The box frame (1) forms the base network of longitudinal and transverse supports, Fig. 11.26. Two longitudinal supports and two transverse U-profile supports form the main supports of the structure. The profile height of the main longitudinal and transverse supports decreases towards the front and back part of the frame. Transverse supports pass through the longitudinal supports. Two main transverse supports reinforce the frame network and determine the position of the bed of the telescopic hydraulic cylinder of the tipper box. External profiles for connection with box floor are welded to the base network of profiles.

The upper part of the tipper box consists of a floor and four sides (2) with a mechanism for opening and closing. The front side is fixed to the box frame by welding the side posts, made of rectangular profiles, and cannot be opened, while the side panels can be opened downwards, and the rear ones both downwards and upwards. All sides are of the same height, and when they are folded, you get the interior cargo space.

The floor of the box is connected to the external profiles by welding. Parts of the box are made of high-strength steel *Hardox 450/500* or *Weldox 700* (*SSAB Oxelősund*

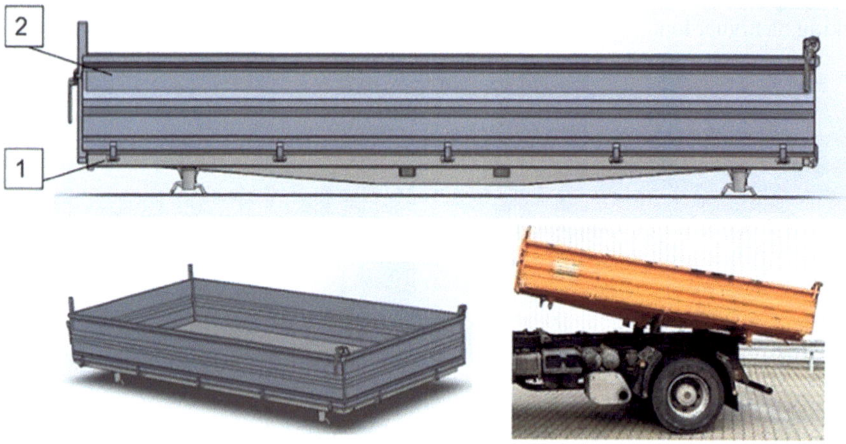

Fig. 11.25 The shape of a smaller box of a tipper of truck (LxWxH)

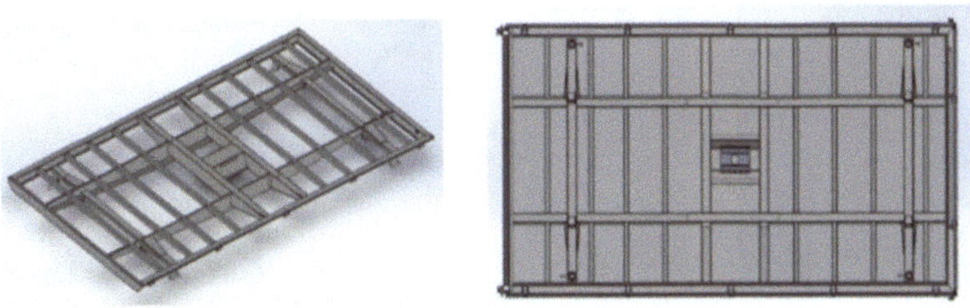

Fig. 11.26 Small box frame model [13]

AB). These materials, especially Weldox 700, have extremely good mechanical properties such as high strength and toughness, and they also have good flexibility and weldability.

Additional semicircular supports that provide the support of the box are welded to the two transverse supports, Fig. 11.27. The cup-shaped parts are fitted on the spherical heads of the auxiliary frame. The choice of the unloading side is made possible by spindle locks (lock), which ensure the connection of the box frame and the auxiliary frame:

- Unloading to the rear side—locks at the back;
- Unloading on the side—locks on the same side.

Since the box is supported on only two frame supports during unloading, the reaction in one support is half of the total reaction force. The reactions in one spherical support are determined vertically and horizontally, and the resultant force is determined. The spherical support is attached to a tube of round cross-section and supported at a certain distance from the end of the sphere on which the forces act. When unloading the cargo at the rear end of the vehicle, the box rests on the balls of the rear transverse support.

Hinges for closing and opening the sides are placed on the back and the sides of the box frame, Fig. 11.28.

The sides can be opened or closed by moving the lever. When you want to open the rear side upwards (swinging side), another mechanism is used, which is placed

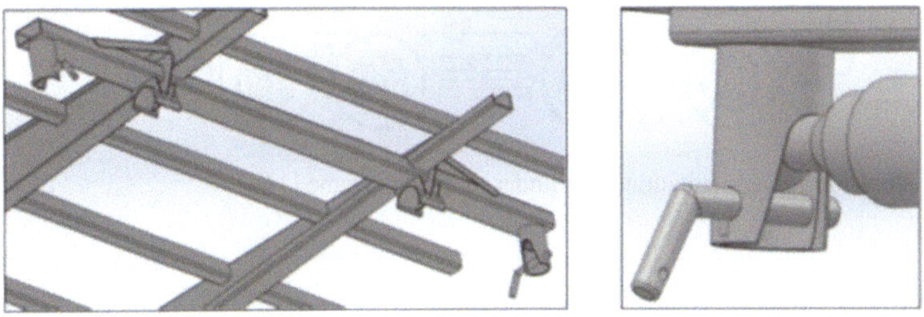

Fig. 11.27 Supports, joints and locks of the box frame

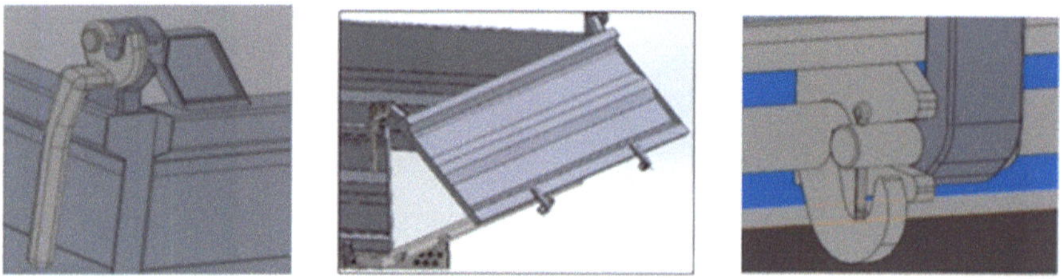

Fig. 11.28 Mechanism for opening and closing the lateral and back sides [13]

on the back part of the box, under the back side. The mechanism consists of a rod onto which the so-called hooks are welded.

Construction of tipper box and hydraulic lifting cylinder

The tipper box is connected to the hydraulic cylinder in the middle of the box frame. The hydraulic cylinder is connected as shown in Fig. 11.29. The bearing of the cylinder holder on the box into which the ball fits, is placed on the support bed and secured with a connection of a specific shape. The horseshoe-shaped receiving element secures the ball from falling out.

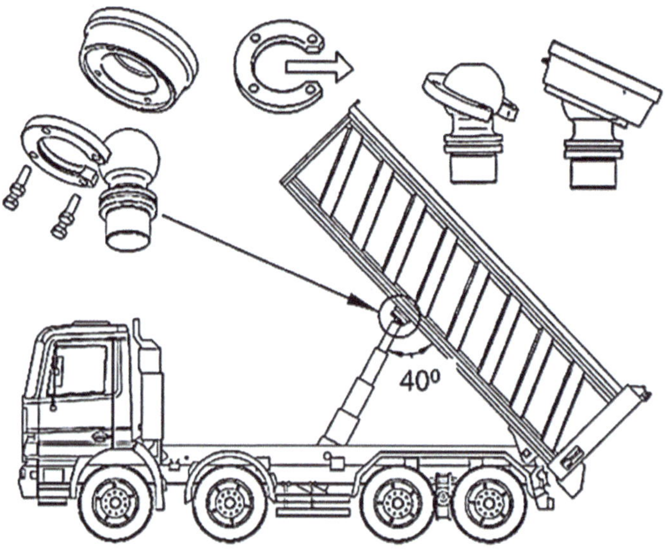

Fig. 11.29 Connecting the hydraulic cylinder to the box frame [14]

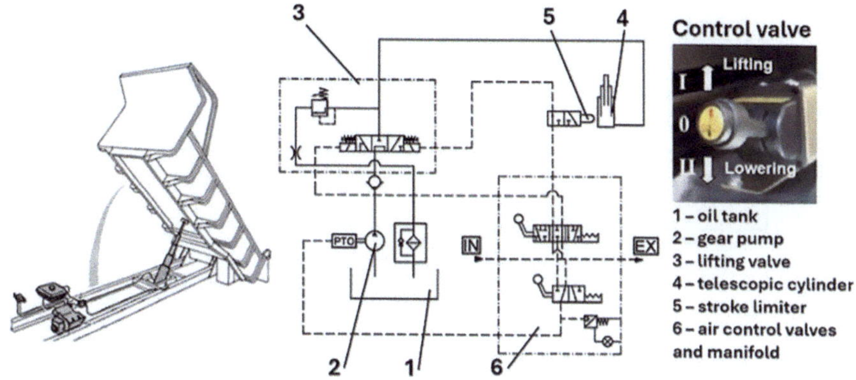

Fig. 11.30 Hydraulic installation of tipper truck [14]

11.5.3 Hydraulic Installation of the Tipper of Truck

The hydraulic installation of the tipper truck enables the lifting of the box using a telescopic hydraulic cylinder. The cylinder is placed in the middle of the box or on the front of the box. The diagram of the hydraulic installation is shown in Fig. 11.30.

The gear pump is driven by the transmission output (PTO—*Power Take-Off*). On most vehicles, it is necessary to stop the vehicle and activate the parking brake in order to engage the transmission outlet. The raising or lowering of the cylinder is provided by an oil distributor, which is operated by means of an air control valve. When the air control valve handle (6) is moved to the lifting position, the compressed air activates the lifting valve (3) and positions it in place to allow hydraulic oil to fill the cylinder which then lifts the cargo box. In the end position of the cargo box, the lift is stopped by the limit stroke valve, and the oil returns to the tank. By moving the handle to the lowering position (left position of valve 3), the lowering of the box starts and the oil returns from the cylinder to the tank.

The operating principle of the hydraulic telescopic cylinder is based on the order of raising the pistons, depending on their surface. This means that the first piston with the largest diameter (first stage) is lifted first, followed by the rest of the smaller pistons. At the same time, when extracting individual pistons, different pressure is required ($p_1 < p_2 < p_3$). The speed of cylinder extraction increases according to the order of activation of the pistons ($v_1 < v_2 < v_3$). This reduces the cycle time necessary for raising and lowering the tipper box.

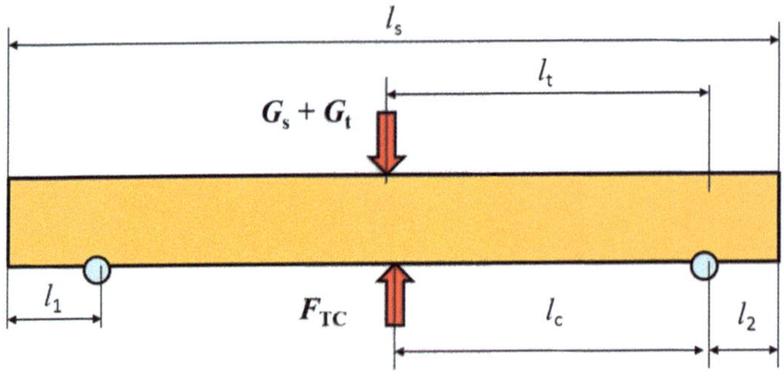

Fig. 11.31 The loading of the box at the beginning of lifting

Calculation of the hydraulic cylinder

The choice of the hydraulic cylinder is performed according to the force required to lift the tipper box with the load and according to the required length of the cylinder extraction. The calculation of the required force of the hydraulic cylinder is performed for the case of tilting the box to the rear. Lateral tilting requires less force than rearward tilting. The required cylinder extension length is sufficient only for the case of tilting to the rear side, because in this case a longer hydraulic cylinder extension length is required.

At the beginning of raising the box, the hydraulic cylinder acts with a vertical force upwards, opposite to the weight of the box and the weight of the load (cargo), i.e., in the case of the highest load capacity, Fig. 11.31. The force of the hydraulic cylinder (F_{TC}) acts along the axis of the cylinder on arm l_c, and forces G_s and G_t in the center of gravity of the mass of the empty frame box and the mass of the largest load.

Assuming that the force arms (G_s, G_t, F_{TC}) are equal (lc = lt), the required force of the hydraulic cylinder is: $F_{TC} = G_s + G_t$, Fig. 11.32.

G_s Weight of the box
G_t Max cargo weight

The maximum load weight (G_t) can be estimated as follows:

1. Permissible carrying capacity of superstructure and cargo = maximum permitted weight—weight of the base vehicle
2. Maximum cargo weight = permissible loading capacity of superstructure and cargo—weight of auxiliary frame—weight of empty box—weight of hydraulic cylinder and installation

The weight of the empty box (G_s) can be read in *Solidworks* or another similar program in which the box is designed.

When tilting the box of 45°, the hydraulic cylinder has the largest angle (β). In doing so, the required length of cylinder extraction (L_{TC}) can be read from the program model.

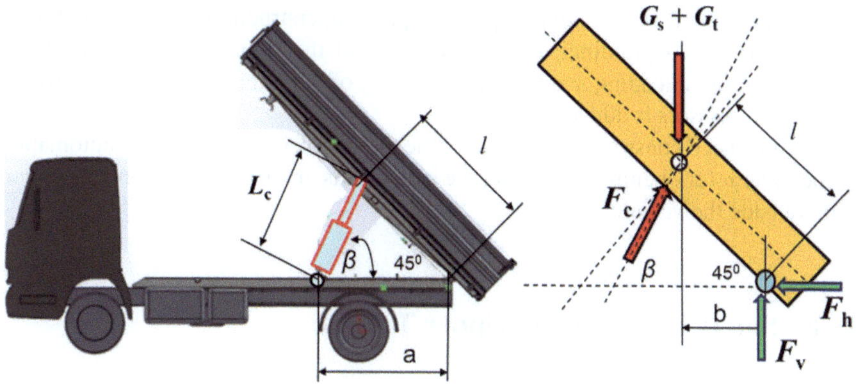

Fig. 11.32 The force of the hydraulic cylinder during rear tilting of the box and hydraulic cylinder stroke length

Selection of hydraulic cylinder

According to the manufacturer's production program and based on the required force and length (F_{TC}, L_{TC}), a hydraulic telescopic cylinder is selected. The position of the support of the hydraulic cylinder on the auxiliary frame is selected so that the upper cylinder support is in the middle of the box, Fig. 11.33.

The relevant data of the telescopic hydraulic cylinder are:

- Lifting force/carrying capacity
- Maximum length of extraction
- Number of extraction stages
- Maximum working oil pressure
- The diameter of the ball and socket
- Required amount of oil
- Weight of hydraulic cylinder and installation.

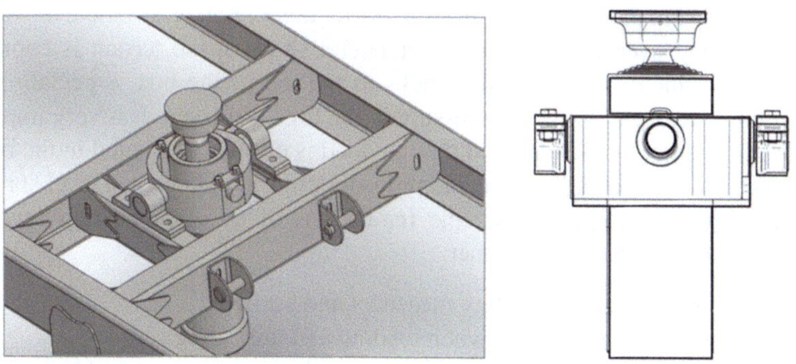

Fig. 11.33 Position and bearing of the telescopic hydraulic cylinder [13, 14]

For operational safety, some types of tipper superstructures use two hydraulic cylinders in the longitudinal or transverse axis of the vehicle. To achieve greater stability during unloading and to reduce twisting of the frame, a torsion stabilizer in the form of scissors is added.

The advanced construction concept of the tipper box includes also the automation of the system for opening and lifting the side panels and the remote control method from outside the vehicle.

11.6 Safety in the Use of Tipper Trucks

The safety of a loaded truck in traffic and safety during unloading require the driver's attention. Driving in a curve and the tilting of the superstructure at the same time causes the center of gravity of the vehicle to shift, which affects the lateral stability. The stability of the truck against lateral tilting can be evaluated by the vehicle stability parameters.

The lateral acceleration of the vehicle $a_b = (s/h_T)g$, is defined as a criterion of vehicle stability (s—wheel track, h_T—height of the vehicle's center of gravity). Trucks and trailers are subject to transverse stability control with the help of a rollover lateral acceleration threshold as a measure of rollover resistance ($a_b = 0.35\ g\ \mathrm{m/s^2}$). A larger threshold means greater lateral stability of the vehicle. For example, tank trucks with a higher lateral acceleration threshold are more resistant to overturning ($a_b = 0.4\ g\ \mathrm{m/s^2}$). Due to an improperly distributed load or an uncontrolled increase in the height of the center of gravity, the lateral acceleration threshold can fall below 0.25 g and cause the vehicle to overturn.

Unloading a fully loaded truck with a high center of gravity, especially on a side tilt can be risky because of the possibility of overturning, Fig. 11.34. Unloading is a risky operation, especially when the ground is uneven or when the load is unloaded into a pit. Greater risks can arise, for example, when the cargo is wet, frozen in the box, or gets stuck in the box. When the tipper box is lifted on a side tilt, the inertial force of the loaded box can cause the truck to sway and overturn.

With the load in the raised box, in case of irregular or reckless vehicle movement, the vehicle may overturn. Even a minor lateral slope of the terrain is enough to overturn the vehicle if the material sticks to the top of the box, especially if the suspension is damaged or the air pressure in the tires is uneven. It is very important that the truck is not overloaded and that the load is evenly distributed in the box, in accordance with the load diagram, and unloading must be done on a level surface.

When certifying tipper trucks, the regulations of the national regulations on occupational safety must also be met:

- Rulebook on occupational safety measures and norms
- Regulations on safety at work when loading and unloading cargo
- Regulations on environmental testing and machines with increased hazards
- Regulations on the list of machines and devices with increased risk.

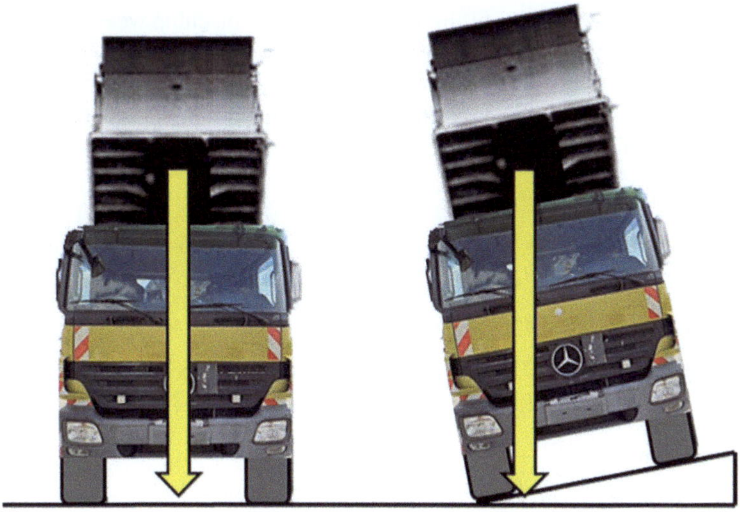

Fig. 11.34 Not allowed operation of driving on a side slope with the tipper box raised (lifting cargo and moving vehicles) [15]

Number of tipper trucks

With the aim of optimal operation of the mechanization during soil excavation, the work of backhoes and trucks is coordinated, and the required number of excavators and trucks is proposed. The number of trucks (N) is determined based on the criteria of the required time for transporting the material, and the required working time of the backhoe during loading. This means that an analysis of the compliance of the number of trucks and backhoes should be carried out, according to the following principle:

1. One type of backhoe and several types of trucks. The analysis involves coordinating the work of the selected backhoe and several types of trucks, which usually depends on the available means funds that the company already owns or will acquire certain trucks.
2. One type of truck and several types of backhoe. Analysis of the most favorable excavator based on the principle of calculating the number of trucks needed to work with excavators of different capacities, along with their performance characteristics and loading time (U_p, $T_{u,\,uk}$).

Required number of tipper trucks

Let the input data be as follows:

- Excavated soil is taken to a landfill 10 km away. The total volume of soil for transport is 1500 m^3.
- One backhoe with a deep shovel, the volume of the loading shovel (bucket) $V = 0.95$ m^3, the working cycle time of the excavator is $T_c = 24$ s. Corrective data are:

looseness factor $k_r = 0.87$; filling factor $k_p = 0.85$; machine wear and tear factor $k_d = 1.0$; effective operation of the machine $k_v = 0.81$, rotation of the machine $k_z = 0.93$.
- Tipper truck 8 × 4, loading box capacity 17 m^3, maximum speed 90 km/h, empty truck speed 80 km/h and full truck speed 50 km/h.

Backhoe efficiency

- Theoretical effect: $U_t = V/T_c$ (m^3/h); U_t=(0.95*3600)/24=142.5 m^3/h
- Practical effect: $U_p = U_t\, k_d\, k_z\, k_p\, k_r\, k_v$ (m^3/h)

  $Up = 142.5{*}1{*}0.93{*}0.85{*}0.87{*}0.81 = 79.38$ m^3/h

- Daily effect: $U_{p,\,dn} = U_p$* shift (m^3/day), $U_{p,\,dn} = 79.38{*}10 = 793.80$ m^3/day.
- Total time required for loading: $T_{u,\,uk} = 1500/79.38 = 18.90$ h.

Truck duty cycle

- Loading time: $T_u = 17/79.38 = 0.252$ h = 770.98 s.
- Driving time of a full truck: $T_{vp} = 10/50 = 0.2$ h = 720 s.
- Unloading time with maneuver: $T_{is} = 5$ min = 300 s.
- Return time: $T_p = 10/80 = 0.125$ h = 450 s.
- Total driving cycle time: $T_c = T_u + T_{vp} + T_{is} + T_p$.

  $T_c = 907 + 720 + 300 + 450 = 2377$ s = 39.6 min.

Optimal number of tipper trucks

Required number of trucks for coordinated work with a backhoe:

$$N_{dn} = T_c/T_u$$
$$N_{dn} = 2377/770.98 = 3.08 \sim 3\,\text{trucks}$$

The result can be compared with trucks of different capacity, using a larger excavator bucket, and thus evaluate the optimal choice of the number of trucks. When calculating the number of trucks, the costs were not considered here as relevant for the selection of the means.

References

1. Directive 2007/46/EC and Regulations 167/2013 and 168/2013. Technical categories of motor vehicles. EU documents
2. www.mercedes-benz-trucks.com
3. Pravilnik o tehnickim uvjetima vozila u prometu na cestama (2020) Ministarstvo pomorstva, prometa i infrastrukure, Hrvatska
4. www.trailerwin.com/product
5. Rumbak M (2020) Konstrukcija nadogradnje komunalnog vozila prilagođenog manjim mjestima, Diplomski rad, Fakultet strojarstva i brodogradnje Zagreb

6. www.lasiportal.de
7. www.vbi.truck.volvo.com/portal/perfman/index.htm#010manual/100tractiveeffort.htm
8. Mikulic D (2020) Motorna vozila, Teorija kretanja i konstrukcija. Veleuciliste Velika Gorica, Velika Gorica
9. Fachkunde Kraftfahrzeugtechnik (2001) 27. Auflage. Nourney, Volmer GmbH&Co, Hann-Gruiten Germany, Copyright 2001: Verlag Europa -Lehrmittel
10. www.pk.hr/pk_proizvodi/straznji-half-pipe-kiperi
11. www.meiller.com/en/products/rear-tippers
12. www.vacuum-truck.com
13. Abicic L (2021) Nadogradnja za rasute terete za vozila N2 kategorija. Diplomski rad, Fakultet strojarstva i brodogradnje Zagreb
14. www.rositeh.hr
15. www.hsseworld.com/dump-truck-safety-tips

Chapter 12
Maintenance of Construction Machines

12.1 Maintenance System

A modern maintenance system of construction machines ensures their reliability and availability throughout their lifetime. Machine manufacturers, authorized sales service organizations and machine users (construction companies, entrepreneurs) achieve business results when they cooperate with each other in the procurement, use, and maintenance of the machines.

The characteristics of a modern maintenance system of construction machines are the concept of maintenance, organization of maintenance and maintenance technology, and the way of managing maintenance, Fig. 12.1.

The concept is a very important feature of the maintenance system that affects its quality. It can be defined as the principle of making a decision on the time and method of carrying out maintenance activities. Maintenance of construction machines includes three concepts: preventive, corrective, and combined maintenance, which will be most dealt with in this chapter.

The organization of the maintenance service is based on specialized maintenance departments and field teams. The normal activities of the maintenance department are the organization of the maintenance process, monitoring of the state of the machines, preparation and implementation of the technological maintenance process, organization of transport and support for securing the supply of spare parts.

The maintenance technology defines the maintenance method. The technology determines how the procedure should be performed for each level of maintenance: with what tools and equipment, in what order, in what time, with which professional staff. The application of a certain technology depends mostly on the structural complexity of the machines, the equipment of the workshop and the training of the employees.

Management ensures the supervision of the implementation of the maintenance system. Organizational departments are responsible for conducting the business in

D. Mikulić and D. Mück, *Construction Machines*,
https://doi.org/10.1007/978-3-031-67845-5_12

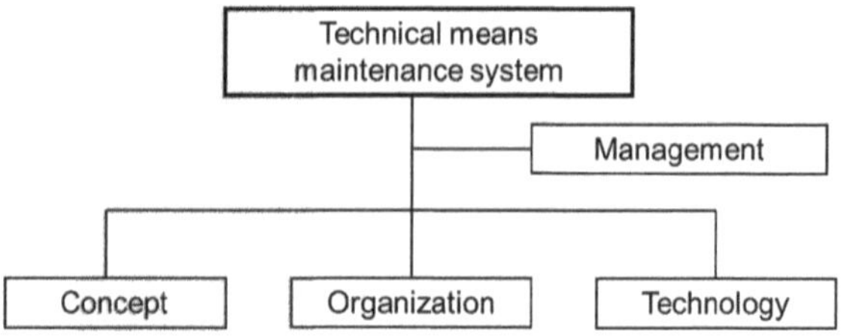

Fig. 12.1 Maintenance system of technical means

their professional areas, such as the finance and administration department, the sales and marketing department, the field and workshop service department, the customer service department, the department for the purchase and sales of spare parts and spare parts stock. A modern information system is used to apply the predictive analytics of system operations and management.

According to the above, there are many factors in the system that affect the quality maintenance of machines. Maintenance of construction machinery is therefore a complex organizational and technological system. Each system element has its own characteristics and state parameters. System elements are interconnected and depend on each other, they are defined by limitations, the most important of which are maintenance costs and costs due to unavailability or downtime.

The user of construction machinery can rely on an external maintenance service, with the help of authorized representatives and repairers. Sometimes, the entire maintenance system is developed by the user of the machines. However, practice shows a real option, i.e. a combination of these two legal entities, the user's own maintenance (maintenance of lower complexity) and external maintenance by an authorized service provider.

Sales and service representatives offer machinery and equipment as an integral part of the construction process and accept payment for rent and service based on the number of working hours or scope of work. Owned machines that are rarely used are usually expensive machines. For example, it is considered that for the use of a machine under 1000 operating hours per year it is best to rent a machine rather than buy it. That is why machines are rented because then there are no purchase costs, interests, insurance, or maintenance costs. It is expected that in the future, renting machines will become one of the better investment decisions.

12.2 Reliability and Availability

Any malfunction of the machine does not mean at the same time its failure. The machine can perform its intended function with some permissible level of malfunction. Technical means have active diagnostic systems that warn of malfunctions and predict the possibility of further use of the machine up to a certain limit or the obligation to intervene. Good knowledge of the logic diagrams of a machine is also required for accurate fault localization.

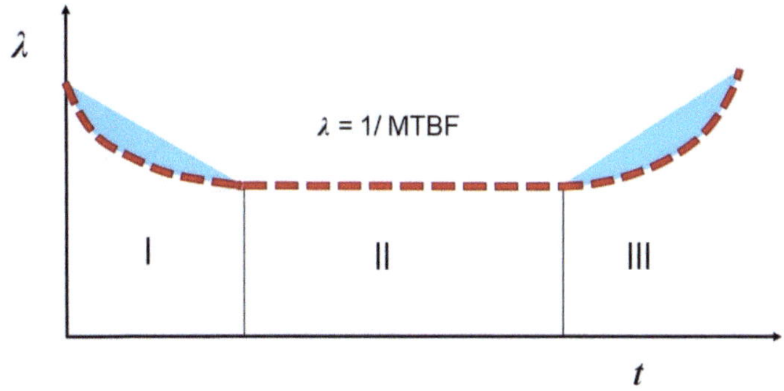

Fig. 12.2 Failure intensity curve during the lifetime of a machine

Malfunction can be technical and functional. Technical malfunction is the state of the machine which, with certain limitations or in its entirety, still provides the ability to be used. Functional malfunction is the failure of the machine, and it occurs when the asset is no longer safe for use and when further use can cause very expensive failure and long downtime.

Reliability of machines

The working life of technical equipment has three periods in which the value of failure intensity and, accordingly, the reliability differ, Fig. 12.2.

Reliability is the probability that the machine will operate without functional failure in the expected manner under the intended operating conditions and for a specified time, with minimal interruptions due to the need for maintenance. The basic reliability parameter is the failure intensity (λ), that is, the mean time between failures (MTBF).

The first period (I) of the failure intensity curve is characterized by "children's diseases", i.e. it is the period of running-in of the machine in which the machine warranty is usually valid.

In the second period (II.) of the working life, random machine failures occur (attributed to internal factors). This usually takes 2/3 of the working life of the machine. In that period, there is an approximately constant intensity of machine failure ($\lambda = \text{const.}$). This means that the frequency of such failures can be monitored, as well as the reliability of the machine.

In the third period (III) of the working life, frequent failures occur due to age, wear, corrosion, fatigue, etc., so intensive preventive maintenance is necessary.

Failure intensity λ in period II is independent of time, and usually expressed in the number of failures per unit of time; for example, per working hour (1/h), $\lambda = 1/MTBF$, and the mean (average) time between failures, $MTBF = 1/\lambda$. The reliability of the machine for a certain time can be set as an exponential function of probability: $R = e^{-\lambda t}$

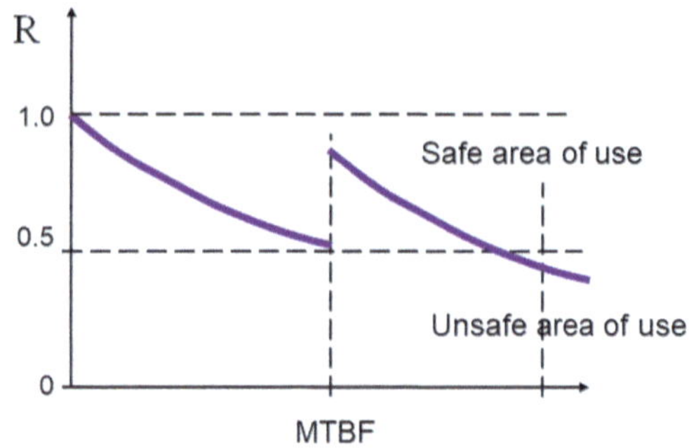

Fig. 12.3 Cyclic function of the exponential distribution of machine reliability [1]

Remark Exponential probability distribution is more suitable for electronic systems and normal distribution for mechanical systems. The Weibull distribution can be used, with the change of certain parameters, to describe the failure distributions in all three periods of life.

By undertaking maintenance in MTBF reference states, the assumption is to achieve the cyclic reliability function in the life of the machine, and therefore, its operational availability for work, Fig. 12.3.

Availability in the theory of maintenance is also expressed as the probability that the machine will work satisfactorily in the future under certain conditions, maintenance, repairs and administration. In practice, the actual availability of machines differs from the declared availability by the machine manufacturer. Therefore, the ratio of actual and declared availability can be considered.

The following parameters are used to calculate the availability of machines: technical, achieved, and operational availability.

Technical (inherent) availability:

$$A_i = \mathrm{MTBF}/(\mathrm{MTBF} + \mathrm{MTTR})(\%)$$

MTBF Mean time between failures, reliability parameter.
MTTR Mean active repair time, a parameter of the benefit of corrective maintenance.

Availability reached:

$$A_a = \mathrm{MTBM}/(\mathrm{MTBM} + \mathrm{M})(\%)$$

MTBM Mean time between maintenance.
M Medium active time of preventive and corrective maintenance.

Operational availability is the probability that the machine will be ready for work when needed:

$$A_o = \mathrm{MTBM}/(\mathrm{MTBM} + \mathrm{MDT})(\%)$$

MTBM Mean time between all maintenance actions (preventive, corrective) is a parameter of convenience for machine maintenance that most affects its availability.

MDT Mean downtime time due to maintenance (repair time, waiting for r/d, losses, etc.).

Machine working life

Working life of a technical means is limited by repair costs or a low level of reliability and availability. This prevents endless and economically unjustified overhauls. The total costs of a technical asset during its lifetime can be presented and analyzed in the form of an "iceberg".

It is an assumption that the working life of a machine should be equal to the time of use in which the costs of repairs do not exceed the price of the machine itself at the time of failure. The working life model consists of using MTBF (*Mean Time Between Failure*) of the key machine components and their prices to determine the working life until mid-overhaul using mathematical expectation [1]:

$$M(t) = \sum MTBF_i g_i = \sum (1/\lambda) g_i$$

$MTBF_i$ Mean (average) time between failure of key parts and assemblies.

g_i Repair prices of key parts and assemblies, $g_i = C_{ki}/C; \sum g_i = 1$.

In cases where the machines are defective for some reason for a long time, it is necessary to evaluate the profitability of the repair. Based on this, a decision on investment repair (overhaul) can be made. Instead of purchasing a new machine, investment repair provides a more economical option for machine utilization. In practice, construction machines are overhauled mainly with intermediate overhaul and general overhaul of the engine and individual components. It is considered that the average expected working life of a construction machine is 10,000 working hours (approximately 10 years) [1]. However, in practice, a general overhaul of the machine is often recommended at 20,000 operating hours, which depends on the machine model.

12.3 Concept of Maintenance

The concept of maintenance assumes maintenance policies: preventive maintenance, corrective, and combined maintenance, Fig. 12.4.

Features of the concept of preventive, corrective, and combined maintenance.

1. Preventive maintenance is a method of maintenance at predetermined maintenance intervals. Activities include regular service activities, diagnostics, replacement of parts according to recommended times or working hours, lubrication and cleaning of equipment. Another way of preventive maintenance is condition

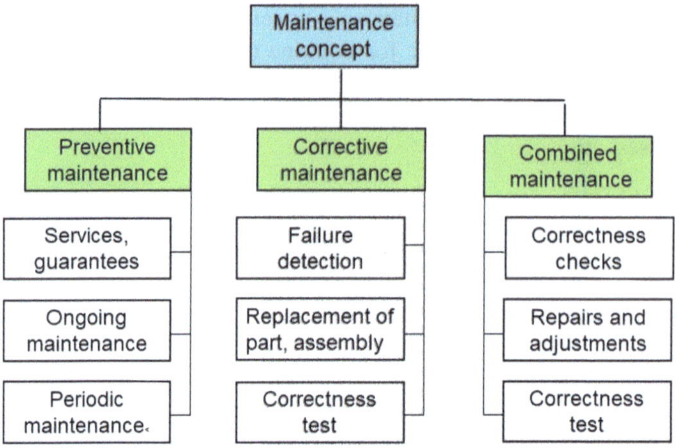

Fig. 12.4 Elements of maintenance concept [2]

maintenance where there is no predefined interval but changes until it reaches certain parameters (e.g. brake pads and discs, etc.) This is a common concept of preventive maintenance. More intensive preventive maintenance is required for older machines with a higher number of working hours, in order to avoid expensive repairs and to increase availability.

2. Corrective maintenance is a method of maintenance after failure has occurred. When a failure occurs, repairs, replacement of parts or assemblies, and overhaul of machines are carried out in order to restore the functionality of the machine. Corrective maintenance often requires immediate interventions to minimize the wasted time and reduce the possibility of additional downtime. The main procedures in corrective maintenance are: failure detection, condition diagnosis, corrective action (repair) and condition check test after repair and before handing over the machine to the user.
3. Combined maintenance is a practically an actual way of maintenance that combines the previous activities of preventive and corrective maintenance. For example, after machine failure and repair, preventive maintenance activities, such as diagnostics and adjustment, are carried out. Also, with preventive maintenance, an additional failure can be observed that requires repair or replacement of parts, which results in combined maintenance procedures.

According to the complexity of maintenance, the following levels of such maintenance can be distinguished:

I. *level of maintenance*, includes user maintenance, servicing, cleaning and lubrication and replacement of individual parts in field conditions or service inspections by users in the field (lower level of complexity of preventive maintenance).

II. *level of maintenance*, performed at the user's site or in workshops, certified mechanics, tools and spare parts, after labor resource (combined maintenance, complex repairs, combination of user and authorized service).
III. *level of maintenance*, is carried out in the organization of the user, authorized service or manufacturer (medium or general overhaul, in its scope of work).

This concept of maintaining construction machinery is planned, organized and implemented in practice through basic (weekly), ongoing (monthly) and periodic maintenance. Renowned manufacturers of construction machinery practically classify this concept in their servicing models (*Basic, Standard* and *Premium*) as an efficient way of maintaining machinery at machine users.

The users of construction machinery can have continuous remote monitoring of the state of the machines. *Condition monitoring telematics* ascertains the condition of the machine on construction sites based on the state of its key functions. This technology and prevention resulted from simple GPS location technology. In addition to the position of the machine, working time, fuel consumption and improper use and protection of the machine, maintenance telematics offers monitoring of the condition and their remote diagnosis of the machine, especially during the warranty period. For example, integrated telematics equipment, for instance on the engine, enables remote troubleshooting that can initiate diagnostic testing and determine potential problems while the machine is running. If a repair is required, the user is warned of the problem and of the arrival of the intervention team.

12.4 Organization of Maintenance

Construction machines are first of all heavy and expensive machines, poorly mobile, at a relatively long distance from roads and service centers. Therefore, machine maintenance services are performed mainly in the field, at the construction site (about 90%), Fig. 12.5. More extensive work that cannot be done in the field is carried out in the service workshop of an authorized or customer service center. The decision on the place of service is made for each individual case in agreement with the user of the machine.

Some characteristics of services that provide maintenance services:

- The service department is equipped with mobile workshops. These are usually vans that contain the necessary tools and equipment for servicing and repairs; mechanical, hydraulic, electronic components.
- There is no serious competition among maintenance services for construction machinery, since each of them is specialized in its own brand and does not compete in the field of another brand.
- In the maintenance of machines, no large amount of stock of spare parts is created in the warehouse, but there is an available supply of individual parts within 24 h.
- The services have an established ISO 9001 quality management system.

Fig. 12.5 Field maintenance of machines (*Liebherr, Caterpillar*) [3, 4]

The field organization of machine maintenance in several construction projects is shown in Fig. 12.6. Each group of machines at these locations (*Project 01* and *Project 02*) is under the supervision of a field team (TE), at a relatively acceptable distance from the camp (L). A departmental organization can be considered a sufficient organization to manage the maintenance process of a group of different machines.

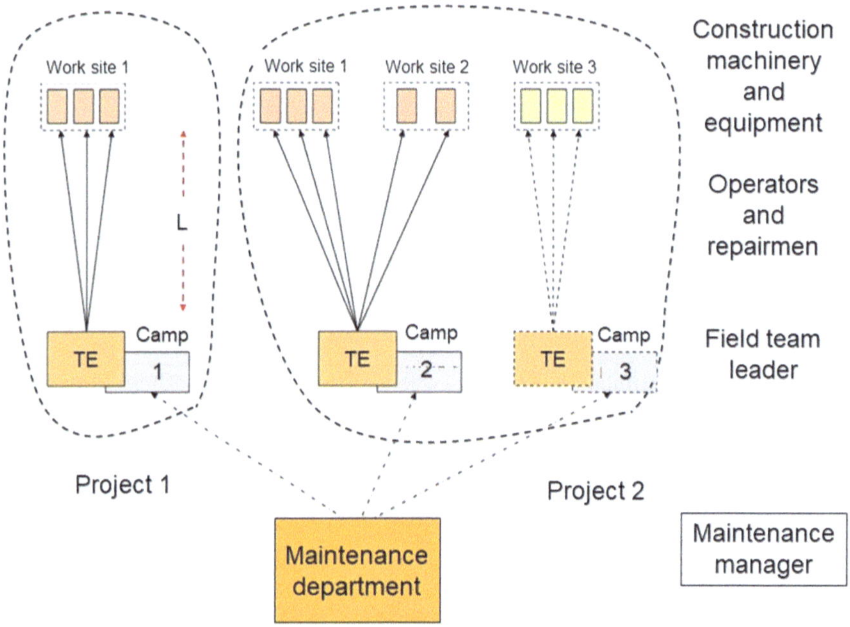

Fig. 12.6 Field organization of maintenance

Maintenance organization structure

The maintenance organization includes the application of the concept of maintenance, through basic, current, and periodic maintenance, Table 12.1.

Basic maintenance consists of inspections and cleaning, during which the correctness and adjustment of the machine is checked weekly during use, and warnings are given of possible malfunctions and failures.

Current maintenance includes preventive and corrective maintenance in field conditions. According to the maintenance plan, after a certain number of working hours, worn parts are replaced and accidental failures repaired. According to the complexity of maintenance technology, current maintenance belongs to levels I and II of maintenance by machine users. Preventive maintenance of the main hydraulic assemblies and power units is extremely important on a larger resource of working hours (which are mostly not prescribed by the manufacturer) and can cause major downtime of the machines.

Periodic maintenance is the maintenance of machines according to condition in order to prepare for the upcoming construction project. It is characterized by a greater scope of defection works, fluid quality analysis, and greater maintenance complexity. It can be classified into. maintenance level II and partly level III with the application of combined maintenance. After approximately one year, that is, every 1000 working hours and based on periodic maintenance undertaken, it is considered that the level of reliability and availability of the machines has been raised.

An example of maintenance organization

Representatives of well-known manufacturers of construction machinery offer different types of service contracts to machine users. For example, dealers of *Caterpillar* (*Cat*) machines offer several CVA service contracts (*Customer Value Agreements*) to machine users. They include the mentioned basic, current, and periodic

Table 12.1 Structure of maintenance activities

Basic maintenance	1. Preventive examinations, weekly or monthly examination at the user's	Checking the correctness and adjustment of the machine Level I
Current (ongoing) maintenance	1. Replacement of worn parts and elimination of observed malfunctions 2. Repair of sudden terminations	Preventive and corrective maintenance in the field or in the workshop, after a certain number of working hours Level II
Periodic (cyclic) maintenance	1. Maintenance according to condition and repair 2. Replacement of worn parts and assemblies, adjustment	Preventive maintenance, every 1000 working hours, with the use of original lubricants, when the machines are not in operation Level III Combined maintenance

Fig. 12.7 Service center for maintenance of construction machines

maintenance, practically classified into three models of modern service: *Basic, Standard,* and *Premium.* Service contracts are considered to represent an efficient way of organization, logistics, and ensuring the availability of machines and reducing maintenance costs. The modern sales and service center offers all the necessary logistics and maintenance management of construction machinery, Fig. 12.7. Satellite or mobile network coverage enables comprehensive and flexible remote monitoring, usage monitoring, and maintenance management.

The *Basic option* is offered to owners who maintain their machines themselves, providing the following support according to the maintenance plan:

- Set of Cat parts and lubricants for preventive maintenance that is coordinated with the regular maintenance schedule.
- Delivery of parts and fluid for regular maintenance to the user's address.
- Visit of an expert by appointment to check the key components and systems, assess correctness and solution methods.

The *Standard option* offers maintenance which provides the following support:

- Preventive maintenance programs that are coordinated with the regular maintenance schedule of each machine.
- Trained repairmen who ensure the quality of maintenance and availability of the machine.
- Consulting on management and working with machines to increase competitiveness.
- Access data, operating hours, location, fuel consumption and diagnostic trouble codes.

- Expert service once a year to check the key components and systems of the machine according to a detailed checklist.
- Determining signs of wear or malfunction, and proposing a solution.
- SOS analysis of fluid samples.

The *Premium option* is offered for the maintenance of new machines with a 3-year warranty (up to 6000 working hours). The option includes advanced machine health monitoring data, reports, and expert advice to machine users. In addition to all the benefits of the *Standard* option, the *Premium* option also provides comprehensive machine protection, which includes fluids and fluid analysis. The analysis of fluid samples provides valuable information, important for the management of machines and the reduction of operating costs (analysis of the degree of wear, analysis of the state and purity of the oil, oil contamination, etc.). This predicative maintenance is used for condition diagnosis and prognosis and repair before failure. It is important to detect problems at an early stage and avoid unnecessary downtime of the machine by eliminating minor malfunctions before they become critical. For example, it is better to replace the bearings or seals or oil at the planned time than to wait for the machine to fail.

12.5 Maintenance Technology

Maintenance technology is the process of carrying out maintenance work in a safe manner, with the application of knowledge, skills, tools, and materials. Authorized representatives of the construction machines offer the use of servicing technology to machine users. On the example of a machine for moving cargo, the procedure for applying regular maintenance technology follows.

An example of a telehandler

Basic maintenance of a telehandler is performed according to the resource of working hours. Maintenance every 50 h is maintenance once a week, ongoing maintenance after 500 working hours (monthly maintenance), and every 1000 working hours once a year. The maintenance plan is displayed most often in tabular presentation.

"Service master"

The assistance in machine maintenance is provided by the "Service master" diagnostic device, Fig. 12.8. It is used to read errors, monitor the operation of machine parts in real time and change the program settings in the central units of the engine, transmission, etc.

"Service master" is used for diagnostics and monitoring of engine operation, diagnostics and monitoring of transmission operation and adjustment of the rollover protection system. It consists of a laptop (A), and a program that comes on a CD (B), a converter (C) and connecting cables (D). Establishing a connection with the central

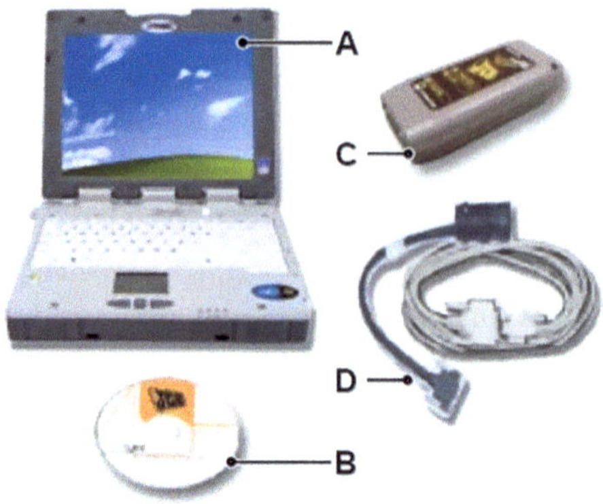

Fig. 12.8 Diagnostic device "Service master" [5]

units of the machine is achieved through the socket on the machine and connected to the CAN system.

When monitoring the engine operation in real time, the data are displayed on the screen within dark windows, Fig. 12.9. Thus, after starting the engine, the following can be monitored: engine hours (A), turbocharger pressure (B), air temperature (C), fuel pressure in the high-pressure pipe (D), engine oil pressure (E), engine oil level (F), engine speed (G), engine temperature (H), and fuel temperature (J). The display on the screen is also an electrical diagram, so when searching for malfunctions, there is no need to waste time looking for a suitable diagram. Reading of possible errors is done by starting the corresponding icon in the program. Along with the error, the number of times the error was repeated (*black box*) and the possible causes and tips for eliminating the failure are displayed.

Checking the correctness of the load indicator

The load indicator as part of the rollover protection system can fail. After its replacement, it is important to perform calibration so that the rollover protection system works properly. The system is calibrated using the "Service master" diagnostic device.

Checking hydraulic oil pressure and flow

When searching for and removing malfunctions of the hydraulic system, a flow meter is used, Fig. 12.10, a device that measures the pressure and oil flow at the place where the measuring part is installed. It is also required when adjusting the pressure and flow of optional connections.

Hydraulic oil

The hydraulic installation is the dominant structure of the function of the construction machine. The speed at which the executive components work is determined by the flow rate of the hydraulic oil, while the hydraulic pressure determines the force acting on the working device. One of the main causes of hydraulic failure is contamination of the hydraulic installation. It is believed that 75% of failures are due to contaminated

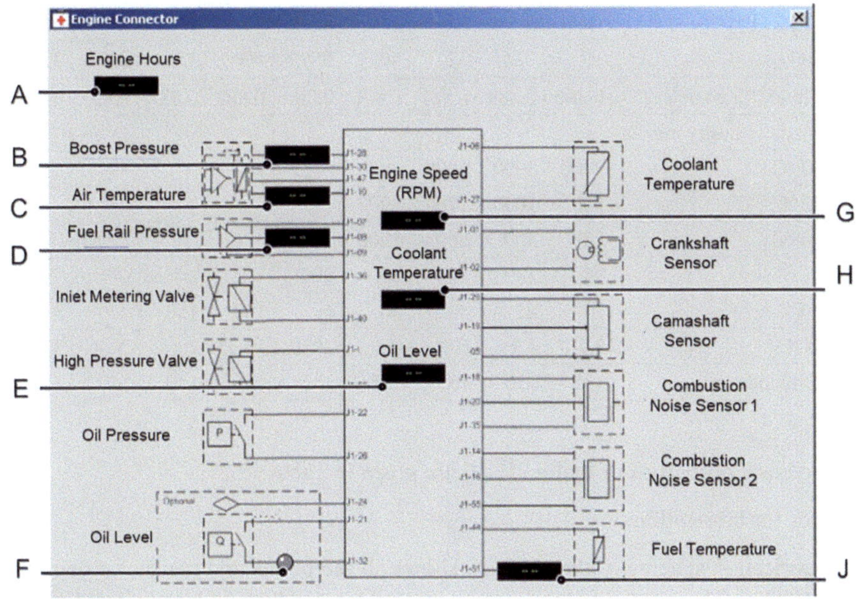

Fig. 12.9 Graphic interface of the "Service master" software [6]

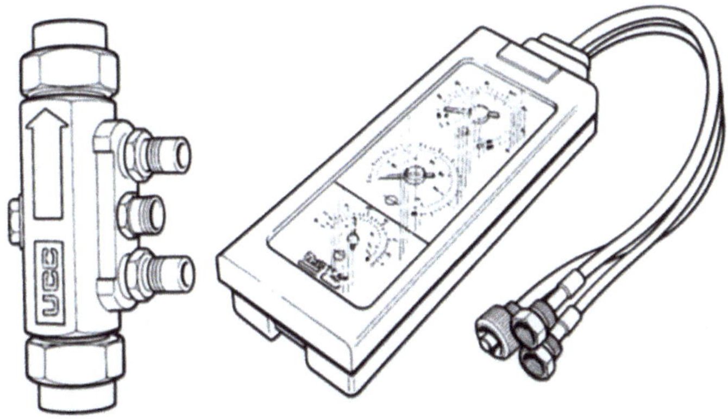

Fig. 12.10 Devices for checking hydraulic oil pressure and flow

or old hydraulic fluid. As a result, damage to one component can cause damage to other components. For example, the most common reason for overheating are the dirty coolers, the maintenance of which is the domain of the operator.

Classes of hydraulic oil

- Hydraulic oil viscosity marks: ISO VG 10-320 (ISO 3448)
- Viscosity oils often used: VG 46, VG 68

Table 12.2 Physical and chemical characteristics of hydraulic oil [7]

ISO VG	15	22	32	46	68	100	150	220	320
Density 15 °C, g/cm^3	0.886	0.889	0.881	0.876	0.880	0.882	0.887	0.895	0.901
Kinematic viscosity mm^2/s									
at −30 °C	2000	3000	6000						
at −20 °C	500	1000	1700	3900					
at −10 °C	220	400	660	1200	2000				
at 0 °C	95	180	300	500	900				
at 40 °C	15	22	32	46	68	100	150	220	320
at 100 °C	3.7	4.9	6.4	8.2	10.2	14.33	18.3	24.4	32.3
Viscosity index	135	150	155	155	135	145	140	140	140

The basic properties of hydraulic oil are given in Table 12.2.

Greases for lubrication

The lubrication scheme of the machine shows all the points and places for ongoing maintenance. The joints are lubricated using grease nipples, with universal lithium grease, resistant to moisture, water, and temperature through grease nipples. The key places for lubrication are roller and sliding bearings, both undercarriage components and the lever mechanism. In addition to manual lubrication of the bearings, there is a variant of the automatic central lubrication system, which ensures increased service life of the bearings, reduces general maintenance costs as well as energy losses due to increased friction in the bearings.

Figure 12.11 shows the position of the grease nipples or the component parts of the machine that need to be lubricated every 50 h of operation (weekly).

Lithium lubricating grease is used for bearing lubrication: LIS EP1, LIS EP2 and LIS EP3 [7]:

LIS EP 1. Recommended for central lubrication systems. Depending on the working conditions, it can be used in the range of working temperatures from -−30 to 110 °C, short-term 120 °C.

LIS EP 2. Recommended for the lubrication of all versions of roller and sliding bearings that are exposed to high and shock loads.

LIS EP 3. Recommended for the lubrication of all versions of roller and sliding bearings that are exposed to heavy and shock loads, such as construction and agricultural machinery.

Lubrication of the telescopic boom, i.e. the parts of the telescope that slide on the slides, must be lubricated every 500 working hours. The telescopic boom is extended at ground level all the way to the end and then the area is lubricated with grease, which is very thin and can be applied by spraying. The property of this grease is that dirt sticks to it less, which reduces the wear of the crane slide.

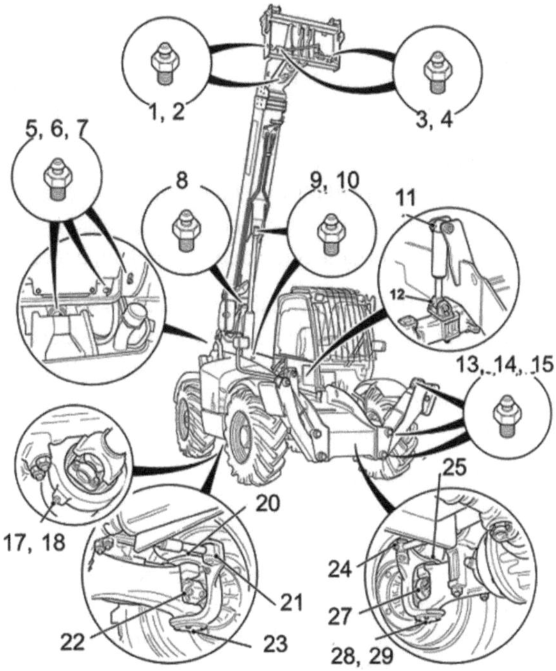

Fig. 12.11 Lubrication scheme of a telehandler [5]. 1, 2—Fork tilt cylinder bolts, 3 and 4—Quick coupler, 5, 6 and 7—Telescope handle, 8, 9 and 10—Telescope lift cylinder, 11 and 12—Machine tilt cylinder, 13, 14 and 15—Stabilizer cylinder and suspension, 17 and 18—PTO support, 20, 21, 22 and 23—Steering supports on the rear axle, 24, 25, 26, 27, 28 and 29—Steering supports on the front axle

12.6 Maintenance Management

The maintenance management sets goals and determines the business policy of the maintenance company. The tasks of the administration are business management, planning and organization of business, as well as supervision of all areas of business, from financial, technical to environmental and marketing. Organizational services or departments are responsible for conducting business in their professional areas. For example, the department for finance and administration manages tasks related to financial, accounting, and other operations, monitoring of legal norms and regulations, personnel planning, especially implementation of plans set by the administration. The sales department performs marketing activities and sales of machines and equipment. The service department implements the technology of maintaining the correctness of machines in the field and in the workshop. Customer service advises the users on maintenance, and performs tasks related to guarantees provided by the machine manufacturer or an authorized service center. The service for the purchase and sale of spare parts manages the purchase and sale of fluids, lubricants, pneumatics and the necessary materials for business, especially spare parts stock.

Business management is connected to the IT support system. Using information computer software for maintenance management helps to make efficient business decisions. That is why it is necessary to collect accurate and timely information and their qualitative analysis. The following information can be grouped:

- For management (maintenance concept),
- For planning (availability of capacity for preventive/corrective maintenance),
- For supervision (maintenance quality, costs, executors),
- For training (organization and maintenance technologies).

The evaluation of the condition of the machine park using the information system provides fundamental information for making decisions about machine maintenance, optimizing resources, reducing the risk of malfunctions, failure, and extending the life of the machines.

Elements of condition assessment:

- Machine structure, age,
- Workshop capacities,
- The structure of the professional force,
- The correctness of machines, malfunctions, and cancellations,
- Realized resources, number of working hours, km,
- Quality of machines and services, reliability, ease of maintenance,
- Customer satisfaction,
- Costs, spare parts, fuel, oil, lubricants, energy.

Maintenance requirements can be classified according to the priority and importance in order to establish a maintenance plan, which includes a service schedule, replacement of parts, repairs, and other necessary maintenance activities.

Quality management implies the establishment of quality policy and objectives, and quality management system. All procedures in the maintenance system are aimed towards the customer, i.e. meeting the needs of the user/party.

Use of the IT system

Business areas are supported using an IT system that uses high-quality software as well as appropriate IT equipment. In addition to the usual software for managing commercial and accounting affairs, specific computer programs are also used. Individual business units of the service are interconnected, Fig. 12.12.

Jobs supported by the IT system:

Administration and planning

- Creation of statistical analyses for daily, monthly, or annual operations;
- Determination of operational values;
- Implementation of control functions of the administration;
- Planning the sale of machines, parts, service operation.

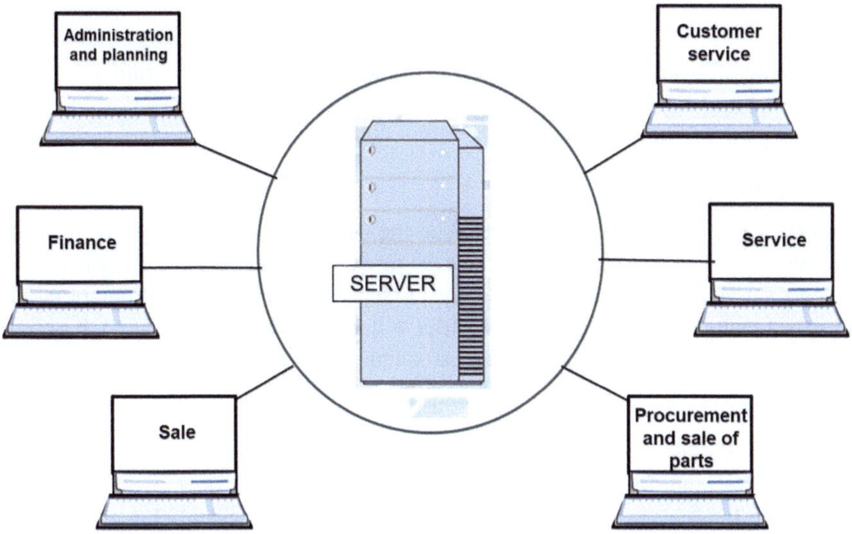

Fig. 12.12 Networked business units in the IT system [8]

Procurement and sale of parts, equipment, materials

- Management of business with spare parts, equipment, materials;
- Procurement and inventory management;
- Sales statistics;
- Management of supplier databases;
- Maintaining a database on purchase and sale prices.

Customer service

- Management of user/client databases;
- Creation of offers and work orders;
- Estimation of repair costs;
- Application of working time databases;
- Creation of the necessary documentation with the work order;
- Creation of account;
- Financial transactions used in sales, cooperation with banks;
- Sales promotion.

Finances, administration

- Bookkeeping;
- Human resources;
- Financial transactions.

Service

- Keeping of technical documentation in the service, electrical diagrams, technical information, digital service booklet;
- Processing of warranty orders.

Opening a work order

A work order, which is a contract between the user and the service, is opened with the help of the IT system, and it contains the most important data about the machine, the user's remarks, a list of works that need to be done and the agreed time for taking over the machine. Based on the work order, a list of spare parts is formed, based on which the spare parts service supplies parts and material.

Application of IT system in service operation

The IT systems that are used for the technical part of the service operation are mostly created by the manufacturer of a particular brand of machine. They are used with the help of the Internet, because a large amount of data and constant changes to it are only possible online, directly on the manufacturer's server.

Such systems provide insight into:

- Electrical schemes,
- The work instructions,
- Standard times,
- Detailed technical data of the machine,
- Technical information,
- Control sheets for individual service interventions,
- Diagnostic and measurement procedures,
- Reading and deleting memorized errors,
- Service activities or recalls,
- Computer programming,
- Current operating parameters of individual machine systems.

12.7 Supply of Spare Parts

Spare parts are usually divided into: capital parts, non-standard and standard parts [7, 8].

Capital spare parts are key parts of the machine with a high price value (with a long service life), which cause the functional failure of the machine. These parts are usually obtained from the original manufacturer just prior to installation. However, with the construction machinery, it is usually a longer delivery time, which causes machine unavailability and downtime costs.

Non-standard parts are parts that must be purchased from the original parts manufacturer but are not counted as capital parts because they are relatively low in price.

These are parts that are usually produced by the manufacturer only for a certain type of machine or engine and cannot be found at suppliers that cover several brands of machines.

Standard parts are available on the market and have a short delivery time. Standard parts do not need to be sourced from the original manufacturer. For example, engine oil filters, brake pads, etc.

The time of unavailability of construction machinery in practice mostly depends on the waiting time for the capital and non-standard spare parts, as well as on the complexity of the repair [2]. To solve the problem of waiting for spare parts, it is necessary to ensure sufficient stocks of these parts.

However, it is not good to have a too large stock of spare parts for the following reasons:

- Stocks are “dead” capital,
- Storage costs are not negligible,
- Parts and materials deteriorate over time.

The quantity of spare parts for maintenance mostly depends on the severity of the failure. Greater intensity of machine use also causes greater intensity of their failure. Therefore, it is very important to ensure a stock of capital spare parts based on the intensity of the failure. The second period of the working life of a machine is the basis for the calculation of the stock of spare parts. The basic parameter of this reliability period according to the exponential probability distribution is the constant failure intensity (λ).

The purpose of stock assurance is to provide a quick balance of fluctuating demand with the procurement of spare parts, for the following reasons:

- *Long process of procurement of spare parts (s/p),*
- *Maintaining the availability of construction machines for the execution of works,*
- *Changing working conditions,*
- *Economic and physical conditions at the time of the need for spare parts.*

Storage organization can be pyramidal—multi-level (higher-level warehouse supplies several lower-level warehouses) and decentralized—single-level (warehouse for one type of s/p or for several types of s/p). For specific field locations, a decentralized warehouse within the field station (camp) is used.

Spare parts stock management (s/p)

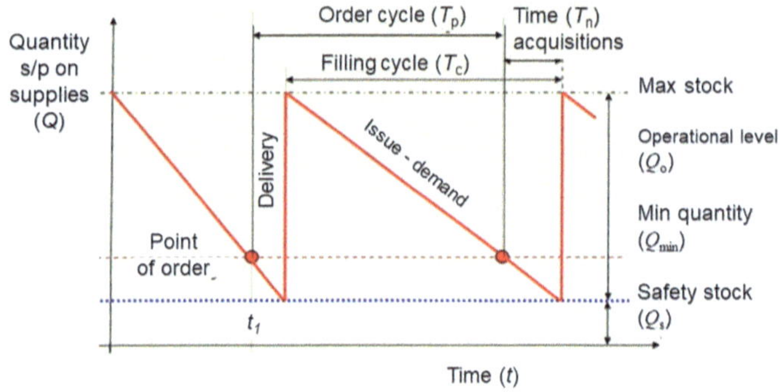

Fig. 12.13 Diagram of spare parts inventory management

The process of spare parts inventory management is graphically shown in the diagram in Fig. 12.13.

Operational stock level (Q_o)	The amount of s/p required for maintenance in the period between ordering and receiving
Minimum quantity (Q_{min})	Stock level at time of procurement
Point of order (t_1)	The moment at time of procurement
Time of procurement (T_n)	Time between order date and receipt (includes: administrative time, production time, delivery time, control, and receipt)
Order cycle (T_p)	The interval between individual checks of the current stock level / between individual orders, i.e. consumption time
Filling cycle (T_c)	The interval between individual replenishments.
Order size (Q)	Quantity of units for which procurement is initiated
Safety stock (Q_s)	Quantity of s/p to ensure the demand.

Calculation of spare parts stock

The calculation model is based on a statistical/stochastic spare parts inventory budget model.

The *Poisson* probability distribution is applied, which determines the reliability that the spare part will be in stock:

$$P(r)_i = \left[\,\Sigma(N\, n_i\, \lambda_i\, t)^r / r!\right] exp(-N\, n_i\, \lambda_i\, t) = P_z$$

$P(r)_i$	Probability that there will be "r" requests for some spare part
N	Number of certain machines (for which s/p supplies are provided)
n_i	Number of parts which are counted as s/p in the machine
λ_i	Failure intensity of the part
t	Time of machine operation for which s/p supplies are calculated
r_i	Quantity (number) of spare parts for that part

P_z Required reliability of the spare part stock (probability).

For the practical use of the equation, the term "expected demand" is defined:

$$p = N * n_i * \lambda_i * t$$

Procedure:

- *For each s/p important for the function, find the failure intensity (λ_i) and set a requirement for stock reliability (P_z);*
- *Determine the assortment of spare parts that are essential for the function ("i"";*
- *Determine the number of the same parts (n_i) and the number of machines (N);*
- *Determine the working time in the period for which stocks are calculated (t);*
- *Determine the number of parts (r) which satisfies the equation for the reliability of stock P_z.*

For simplified calculation of spare parts stocks with this model, ready-made reliability nomograms, tables, or ready-made computer programs are used.

Nomogram for determining the quantity of spare parts (λ = constant), Fig. 12.14.

Examples of spare parts inventory calculations

Example 1 In the vibrating roller, the complex assembly has a failure intensity of $\lambda = 110*10–6\ h^{-1}$. How many pieces of that assembly should be kept in reserve in the warehouse, in order to fulfill the need for its replacement with 99% reliability, for a duration of 350 days per 5 h of daily use?

Solution: Fig. 12.14.

1. *Expected demand:* $p = N*ni*\lambda_i*t = 1*1*110*10^{-6}(350*5) = 0.192$.

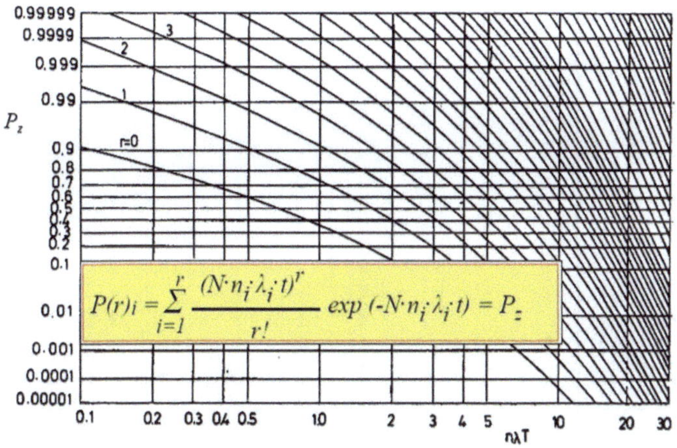

Fig. 12.14 Nomogram for determining the quantities of spare parts [7]

2. In the nomogram, a vertical line should be drawn on the calculated demand, up to the intersection with the required reliability (i.e. up to 0.99 curve).
3. Read "*r*" for which the reliability condition is satisfied (adopt the first larger integer).
4. It is therefore, necessary to have 2 pieces in reserve, $Q = 2$ s/p, which guarantee stock reliability 0.999.

Example 2 There are 5 components of the same type in the crawler excavator, with a failure rate of $\lambda = 28*10–6\ h^{-1}$. How many pieces of such parts should be kept in reserve when maintaining 10 such machines, to meet the need for its replacement with 95% reliability for a duration of 6 months?

Solution: Fig. 12.15.

1. *Expected demand*: $p = N*ni*\lambda_i*t = 10*5*28*10^{-6}\ (6*30*6) = 1.522$
2. At the intersection of the line with the first higher value of demand (1.6), a reliability of 95% (probability 0.95) is required.

Expected demand, *p*	,70	,75	,80	,85	,90	,95	,99	,995	,999
50	54	55	56	57	59	62	67	69	73
40	43	44	45	47	48	51	55	57	61
30	33	34	35	36	37	39	43	45	48
25	27	28	29	30	32	33	37	39	42
20	22	23	24	25	26	28	31	32	35
18	20	21	22	22	24	25	29	30	32
16	18	19	19	20	21	23	26	27	30
14	16	16	17	18	19	20	23	25	27
12	14	14	15	16	17	18	21	22	24
10	12	12	13	13	14	15	18	19	21
9,0	10	11	11	12	13	14	17	18	20
8,0	9	10	10	11	12	13	15	16	18
7,0	8	9	9	10	10	12	14	15	16
6,0	7	8	8	9	9	10	12	13	15
5,0	6	6	7	7	8	9	11	12	13
4,0	5	5	6	6	7	8	9	10	11
3,0	4	4	4	5	5	6	8	8	10
2,5	3	3	4	4	5	6	7	7	9
2,0	3	3	3	3	4	5	6	6	8
1,8	2	3	3	3	4	4	6	6	7
1,6	2	2	3	3	3	4	5	6	7
1,4	2	2	2	3	3	4	5	5	6
1,2	2	2	2	2	3	3	4	5	6
1,0	1	2	2	2	2	3	4	4	5
0,9	1	1	2	2	2	3	4	4	5
0,8	1	1	1	2	2	2	3	4	5
0,7	1	1	1	2	2	2	3	4	4
0,6	1	1	1	1	2	2	3	3	4
0,5	1	1	1	1	1	2	3	3	4
0,4	1	1	1	1	1	2	2	3	3
0,3	0	1	1	1	1	1	2	2	3
0,2	0	0	0	1	1	1	2	2	3
0,1	0	0	0	0	0	1	1	1	2
0,05	0	0	0	0	0	0	1	1	2

Required reliability, P_z

Fig. 12.15 Stock determination, *reliability—demand*

3. Required quantity of spare parts, $Q = 4$ pieces.

Determining the stock of spare parts (λ = constant), Fig. 12.15.

References

1. Mikulic D (1992) Doprinos modeliranju pouzdanosti gradevinskih strojeva, Disertacija. Fakultet strojarstva i brodogradnje, Zagreb
2. Matijascic Z (2012) Logisticko inzenjerstvo. Veleuciliste Velika Gorica, Velika Gorica
3. Liebherr Brochure PR 7341 (2022). www.api.aecweb.com.br/cls/catalogos/liebherr/trator-de-esteira-pr734l.pdf
4. CVA Customer Value Agreement, Caterpillar (2023). www.warrencat.com/service/field-service/ www.teknoxgroup.com/fileadmin/user_upload/CVA_Teknoxgroup_CRO.pdf
5. Technical Data Brochure JCB 540-170 (2012). www.coppard.co.uk/managed/assets/documents/JCB Loadall_540-170_Specifications.pdf
6. Linaric I (2012) Poznavanje i odrzavanje telehendlera. Završni rad, Veleuciliste Velika Gorica, Velika Gorica
7. Barković M (2021) Pricuvni dijelovi za odrzavanje. Veleuciliste Velika Gorica, Velika Gorica
8. Cerlek S, Cerlek N (2016) Organizacija servisa motornih vozila, Veleuciliste Velika Gorica, Velika Gorica

Correction to: Construction Machines

Correction to:
D. Mikulić and D. Mück, *Construction Machines*,
https://doi.org/10.1007/978-3-031-67845-5

The original version of the book was inadvertently published with incorrect author sequence, which has now been corrected. The book has been updated with the change.

The updated version of this book can be found at
https://doi.org/10.1007/978-3-031-67845-5

D. Mikulić and D. Mück, *Construction Machines*,
https://doi.org/10.1007/978-3-031-67845-5_13

MIX
Papier aus verantwortungsvollen Quellen
Paper from responsible sources
FSC® C105338

If you have any concerns about our products,
you can contact us on
ProductSafety@springernature.com

In case Publisher is established outside the EU,
the EU authorized representative is:
Springer Nature Customer Service Center GmbH
Europaplatz 3, 69115 Heidelberg, Germany

Printed by Libri Plureos GmbH
in Hamburg, Germany